BUS

ACPL ITEM
DISCARDED

D0742810

Non-Newtonian Flow in the Process Industries

Fundamentals and Engineering Applications

Non-Newtonian Flow in the Process Industries

Fundamentals and Engineering Applications

R.P. Chhabra
Department of Chemical Engineering
Indian Institute of Technology
Kanpur 208 016
India

and

J.F. Richardson
Department of Chemical and Biological Process Engineering
University of Wales, Swansea
Swansea SA2 8PP
Great Britain

OXFORD   AUCKLAND   BOSTON   JOHANNESBURG   MELBOURNE   NEW DELHI

Butterworth-Heinemann
Linacre House, Jordan Hill, Oxford OX2 8DP
225 Wildwood Avenue, Woburn, MA 01801-2041
A division of Reed Educational and Professional Publishing Ltd

A member of the Reed Elsevier plc group

First published 1999

© R.P. Chhabra and J.F. Richardson 1999

All rights reserved. No part of this publication
may be reproduced in any material form (including
photocopying or storing in any medium by electronic
means and whether or not transiently or incidentally
to some other use of this publication) without the
written permission of the copyright holder except
in accordance with the provisions of the Copyright,
Designs and Patents Act 1988 or under the terms of a
licence issued by the Copyright Licensing Agency Ltd,
90 Tottenham Court Road, London, England W1P 9HE.
Applications for the copyright holder's written permission
to reproduce any part of this publication should be addressed
to the publishers

British Library Cataloguing in Publication Data
A catalogue record for this book is available from the British Library

Library of Congress Cataloguing in Publication Data
A catalogue record for this book is available from the Library of Congress

ISBN 0 7506 3770 6

Typeset by Laser Words, Madras, India
Printed in Great Britain

FOR EVERY TITLE THAT WE PUBLISH, BUTTERWORTH-HEINEMANN
WILL PAY FOR BTCV TO PLANT AND CARE FOR A TREE.

Contents

Preface xi
Acknowledgements xv

1 Non-Newtonian fluid behaviour 1

  1.1 Introduction 1

  1.2 Classification of fluid behaviour 1
    1.2.1 Definition of a Newtonian fluid 1
    1.2.2 Non-Newtonian fluid behaviour 5

  1.3 Time-independent fluid behaviour 6
    1.3.1 Shear-thinning or pseudoplastic fluids 6
    1.3.2 Viscoplastic fluid behaviour 11
    1.3.3 Shear-thickening or dilatant fluid behaviour 14

  1.4 Time-dependent fluid behaviour 15
    1.4.1 Thixotropy 16
    1.4.2 Rheopexy or negative thixotropy 17

  1.5 Visco-elastic fluid behaviour 19

  1.6 Dimensional considerations for visco-elastic fluids 28

  1.7 Further Reading 34

  1.8 References 34

  1.9 Nomenclature 36

2 Rheometry for non-Newtonian fluids 37

  2.1 Introduction 37

  2.2 Capillary viscometers 37
    2.2.1 Analysis of data and treatment of results 38

  2.3 Rotational viscometers 42
    2.3.1 The concentric cylinder geometry 42
    2.3.2 The wide-gap rotational viscometer: determination of the flow curve for a non-Newtonian fluid 44
    2.3.3 The cone-and-plate geometry 47

2.3.4 The parallel plate geometry   48
2.3.5 Moisture loss prevention – the vapour hood   49

2.4 The controlled stress rheometer   50

2.5 Yield stress measurements   52

2.6 Normal stress measurements   56

2.7 Oscillatory shear measurements   57
2.7.1 Fourier transform mechanical spectroscopy (FTMS)   60

2.8 High frequency techniques   63
2.8.1 Resonance-based techniques   64
2.8.2 Pulse propagation techniques   64

2.9 The relaxation time spectrum   65

2.10 Extensional flow measurements   66
2.10.1 Lubricated planar-stagnation die-flows   67
2.10.2 Filament-stretching techniques   67
2.10.3 Other 'simple' methods   68
2.11 Further reading   69
2.12 References   69
2.13 Nomenclature   71

3 Flow in pipes and in conduits of non-circular cross-sections   70

3.1 Introduction   73

3.2 Laminar flow in circular tubes   74
3.2.1 Power-law fluids   74
3.2.2 Bingham plastic and yield-pseudoplastic fluids   78
3.2.3 Average kinetic energy of fluid   82
3.2.4 Generalised approach for laminar flow of time-independent fluids
        83
3.2.5 Generalised Reynolds number for the flow of time-independent fluids
        86

3.3 Criteria for transition from laminar to turbulent flow   90

3.4 Friction factors for transitional and turbulent conditions   95
3.4.1 Power-law fluids   96
3.4.2 Viscoplastic fluids   101
3.4.3 Bowen's general scale-up method   104
3.4.4 Effect of pipe roughness   111
3.4.5 Velocity profiles in turbulent flow of power-law fluids   111

3.5 Laminar flow between two infinite parallel plates   118

3.6 Laminar flow in a concentric annulus   122
3.6.1 Power-law fluids   124
3.6.2 Bingham plastic fluids   127

3.7 Laminar flow of inelastic fluids in non-circular ducts   133

3.8 Miscellaneous frictional losses   140
    3.8.1 Sudden enlargement   140
    3.8.2 Entrance effects for flow in tubes   142
    3.8.3 Minor losses in fittings   145
    3.8.4 Flow measurement   146

3.9 Selection of pumps   149
    3.9.1 Positive displacement pumps   149
    3.9.2 Centrifugal pumps   153
    3.9.3 Screw pumps   155

3.10 Further reading   157

3.11 References   157

3.12 Nomenclature   159

4 Flow of multi-phase mixtures in pipes   162

    4.1 Introduction   162

    4.2 Two-phase gas–non-Newtonian liquid flow   163
        4.2.1 Introduction   163
        4.2.2 Flow patterns   164
        4.2.3 Prediction of flow patterns   166
        4.2.4 Holdup   168
        4.2.5 Frictional pressure drop   177
        4.2.6 Practical applications and optimum gas flowrate for maximum power
            saving   193

    4.3 Two-phase liquid–solid flow (hydraulic transport)   197

    4.4 Further reading   202

    4.5 References   202

    4.6 Nomenclature   204

5 Particulate systems   206

    5.1 Introduction   206

    5.2 Drag force on a sphere   207
        5.2.1 Drag on a sphere in power-law fluids   208
        5.2.2 Drag on a sphere in viscoplastic fluids   211
        5.2.3 Drag in visco-elastic fluids   215
        5.2.4 Terminal falling velocities   216
        5.2.5 Effect of container boundaries   219
        5.2.6 Hindered settling   221

    5.3 Effect of particle shape on terminal falling velocity and drag force   223

5.4 Motion of bubbles and drops   224

5.5 Flow of a liquid through beds of particles   228

5.6 Flow through packed beds of particles (porous media) 230
     5.6.1 Porous media   230
     5.6.2 Prediction of pressure gradient for flow through packed beds   232
     5.6.3 Wall effects   240
     5.6.4 Effect of particle shape   241
     5.6.5 Dispersion in packed beds   242
     5.6.6 Mass transfer in packed beds   245
     5.6.7 Visco-elastic and surface effects in packed beds   246

5.7 Liquid–solid fluidisation   249
     5.7.1 Effect of liquid velocity on pressure gradient   249
     5.7.2 Minimum fluidising velocity   251
     5.7.3 Bed expansion characteristics   252
     5.7.4 Effect of particle shape   253
     5.7.5 Dispersion in fluidised beds   254
     5.7.6 Liquid–solid mass transfer in fluidised beds   254

5.8 Further reading   255

5.9 References   255

5.10 Nomenclature   258

6 Heat transfer characteristics of non-Newtonian fluids in pipes   260

    6.1 Introduction   260

    6.2 Thermo-physical properties   261

    6.3 Laminar flow in circular tubes   264

    6.4 Fully-developed heat transfer to power-law fluids in laminar flow   265

    6.5 Isothermal tube wall   267
        6.5.1 Theoretical analysis   267
        6.5.2 Experimental results and correlations   272

    6.6 Constant heat flux at tube wall   277
        6.6.1 Theoretical treatments   277
        6.6.2 Experimental results and correlations   277

    6.7 Effect of temperature-dependent physical properties on heat transfer   281

    6.8 Effect of viscous energy dissipation   283

    6.9 Heat transfer in transitional and turbulent flow in pipes   285

    6.10 Further reading   285

    6.11 References   286

    6.12 Nomenclature   287

7 Momentum, heat and mass transfer in boundary layers   289

   7.1 Introduction   289

   7.2 Integral momentum equation   291

   7.3 Laminar boundary layer flow of power-law liquids over a plate   293
      7.3.1 Shear stress and frictional drag on the plane immersed surface   295

   7.4 Laminar boundary layer flow of Bingham plastic fluids over a plate   297
      7.4.1 Shear stress and drag force on the immersed plate   299

   7.5 Transition criterion and turbulent boundary layer flow   302
      7.5.1 Transition criterion   302
      7.5.2 Turbulent boundary layer flow   302

   7.6 Heat transfer in boundary layers   303
      7.6.1 Heat transfer in laminar flow of a power-law fluid over an isothermal
         plane surface   306

   7.7 Mass transfer in laminar boundary layer flow of power-law fluids   311

   7.8 Boundary layers for visco-elastic fluids   313

   7.9 Practical correlations for heat and mass transfer   314
      7.9.1 Spheres   314
      7.9.2 Cylinders in cross-flow   315

  7.10 Heat and mass transfer by free convection   318
      7.10.1 Vertical plates   318
      7.10.2 Isothermal spheres   319
      7.10.3 Horizontal cylinders   319

  7.11 Further reading   321

  7.12 References   321

  7.13 Nomenclature   322

8 Liquid mixing   324

   8.1 Introduction   324
      8.1.1 Single-phase liquid mixing   324
      8.1.2 Mixing of immiscible liquids   325
      8.1.3 Gas–liquid dispersion and mixing   325
      8.1.4 Liquid–solid mixing   325
      8.1.5 Gas–liquid–solid mixing   326
      8.1.6 Solid–solid mixing   326
      8.1.7 Miscellaneous mixing applications   326

   8.2 Liquid mixing   327
      8.2.1 Mixing mechanisms   327
      8.2.2 Scale-up of stirred vessels   331
      8.2.3 Power consumption in stirred vessels   332

        8.2.4 Flow patterns in stirred tanks   346
        8.2.5 Rate and time of mixing   356

    8.3 Gas–liquid mixing   359
        8.3.1 Power consumption   362
        8.3.2 Bubble size and hold up   363
        8.3.3 Mass transfer coefficient   364

    8.4 Heat transfer   365
        8.4.1 Helical cooling coils   366
        8.4.2 Jacketed vessels   369

    8.5 Mixing equipment and its selection   374
        8.5.1 Mechanical agitation   374
        8.5.2 Portable mixers   383

    8.6 Mixing in continuous systems   384
        8.6.1 Extruders   384
        8.6.2 Static mixers   385

    8.7 Further reading   388

    8.8 References   389

    8.9 Nomenclature   391

Problems   393

Index   423

Preface

Non-Newtonian flow and rheology are subjects which are essentially inter-disciplinary in their nature and which are also wide in their areas of application. Indeed non-Newtonian fluid behaviour is encountered in almost all the chemical and allied processing industries. The factors which determine the rheological characteristics of a material are highly complex, and their full understanding necessitates a contribution from physicists, chemists and applied mathematicians, amongst others, few of whom may have regarded the subject as central to their disciplines. Furthermore, the areas of application are also extremely broad and diverse, and require an important input from engineers with a wide range of backgrounds, though chemical and process engineers, by virtue of their role in the handling and processing of complex materials (such as foams, slurries, emulsions, polymer melts and solutions, etc.), have a dominant interest. Furthermore, the subject is of interest both to highly theoretical mathematicians and scientists and to practising engineers with very different cultural backgrounds.

Owing to this inter-disciplinary nature of the subject, communication across subject boundaries has been poor and continues to pose difficulties, and therefore, much of the literature, including books, is directed to a relatively narrow readership with the result that the engineer faced with the problem of processing such rheological complex fluids, or of designing a material with rheological properties appropriate to its end use, is not well served by the available literature. Nor does he have access to information presented in a form which is readily intelligible to the non-specialist. This book is intended to bridge this gap but, at the same time, is written in such a way as to provide an entrée to the specialist literature for the benefit of scientists and engineers with a wide range of backgrounds. Non-Newtonian flow and rheology is an area with many pitfalls for the unwary, and it is hoped that this book will not only forewarn readers but will also equip them to avoid some of the hazards.

Coverage of topics is extensive and this book offers an unique selection of material. There are eight chapters in all.

The introductory material, *Chapter 1*, introduces the reader to the range of non-Newtonian characteristics displayed by materials encountered in every day life as well as in technology. A selection of simple fluid models which are used extensively in process design calculations is included here.

Chapter 2 deals with the characterisation of materials and the measurementa of their rheological properties using a range of commercially available instruments. The importance of adequate rheological characterisation of a material under conditions as close as possible to that in the envisaged application cannot be overemphasised here. Stress is laid on the dangers of extrapolation beyond the range of variables covered in the experimental characterisation. Dr. P.R. Williams (Reader, Department of Chemical and Biological Process Engineering, University of Wales, Swansea, U.K.) who has contributed this chapter is in the forefront of the development of novel instrumentations in the field.

The flow of non-Newtonian fluids in circular and non-circular ducts encompassing both laminar and turbulent regimes is presented in *Chapter 3*. Issues relating to the transition from laminar to turbulent flow, minor losses in fittings and flow in pumps, as well as metering of flow, are also discussed in this chapter.

Chapter 4 deals with the highly complex but industrially important topic of multiphase systems – gas/non-Newtonian liquid and solid/non-Newtonian liquids – in pipes.

A thorough treatment of particulate systems ranging from the behaviour of particles and drops in non-Newtonian liquids to the flow in packed and fluidised beds is presented in *Chapter 5*.

The heating or cooling of process streams is frequently required. *Chapter 6* discusses the fundamentals of convective heat transfer to non-Newtonian fluids in circular and non-circular tubes under a range of boundary and flow conditions.

The basics of the boundary layer flow are introduced in *Chapter 7*. Heat and mass transfer in boundary layers, and practical correlations for the estimation of transfer coefficients are included. Limited information on heat transfer from variously shaped objects – plates, cylinders and spheres – immersed in non-Newtonian fluids is also included here.

The final *Chapter 8* deals with the mixing of highly viscous and/or non-Newtonian substances, with particular emphasis on the estimation of power consumption and mixing time, and on equipment selection.

At each stage, considerable effort has been made to present the most reliable and generally accepted methods for calculations, as the contemporary literature is inundated with conflicting information. This applies especially in regard to the estimation of pressure gradients for turbulent flow in pipes. In addition, a list of specialist and/or advanced sources of information has been provided in each chapter as "Further Reading".

In each chapter a number of worked examples has been presented, which, we believe, are essential to a proper understanding of the methods of treatment given in the text. It is desirable for both a student and a practising engineer to understand an appropriate illustrative example before tackling fresh practical problems himself. Engineering problems require a numerical answer and it is thus essential for the reader to become familiar with the various techniques so that the most appropriate answer can be obtained by systematic methods rather than by intuition. Further exercises which the reader may wish to tackle are given at the end of the book.

Incompressibility of the fluid has generally been assumed throughout the book, albeit this is not always stated explicitly. This is a satisfactory approximation for most non-Newtonian substances, notable exceptions being the cases of foams and froths. Likewise, the assumption of isotropy is also reasonable in most cases except perhaps for liquid crystals and for fibre filled polymer matrices. Finally, although the slip effects are known to be important in some multiphase systems (suspensions, emulsions, etc.) and in narrow channels, the usual no-slip boundary condition is regarded as a good approximation in the type of engineering flow situations dealt with in this book.

In part, the writing of this book was inspired by the work of W.L. Wilkinson: *Non-Newtonian Fluids*, published by Pergamon Press in 1960 and J.M. Smith's contribution to early editions of *Chemical Engineering*, Volume 3. Both of these works are now long out-of-print, and it is hoped that readers will find this present book to be a welcome successor.

R.P. Chhabra
J.F. Richardson

Acknowledgements

The inspiration for this book originated in two works which have long been out-of-print and which have been of great value to those working and studying in the field of non-Newtonian technology. They are W.L. Wilkinson's excellent introductory book, *Non-Newtonian Flow* (Pergamon Press, 1960), and J.M. Smith's chapter in the first two editions of Coulson and Richardson's *Chemical Engineering, Volume 3* (Pergamon Press, 1970 and 1978). The original intention was that R.P. Chhabra would join with the above two authors in the preparation of a successor but, unfortunately, neither of them had the necessary time available to devote to the task, and Raj Chhabra agreed to proceed on his own with my assistance. We would like to thank Bill Wilkinson and John Smith for their encouragement and support.

The chapter on Rheological Measurements has been prepared by P.R. Williams, Reader in the Department of Chemical and Biological Process Engineering at the University of Wales, Swansea – an expert in the field. Thanks are due also to Dr D.G. Peacock, formerly of the School of Pharmacy, University of London, for work on the compilation and processing of the Index.

J.F. Richardson
June 1999

Chapter 1
Non-Newtonian fluid behaviour

1.1 Introduction

One may classify fluids in two different ways; either according to their response to the externally applied pressure or according to the effects produced under the action of a shear stress. The first scheme of classification leads to the so called 'compressible' and 'incompressible' fluids, depending upon whether or not the volume of an element of fluid is dependent on its pressure. While compressibility influences the flow characteristics of gases, liquids can normally be regarded as incompressible and it is their response to shearing which is of greater importance. In this chapter, the flow characteristics of single phase liquids, solutions and pseudo-homogeneous mixtures (such as slurries, emulsions, gas–liquid dispersions) which may be treated as a continuum if they are stable in the absence of turbulent eddies are considered depending upon their response to externally imposed shearing action.

1.2 Classification of fluid behaviour

1.2.1 Definition of a Newtonian fluid

Consider a thin layer of a fluid contained between two parallel planes a distance dy apart, as shown in Figure 1.1. Now, if under steady state conditions, the fluid is subjected to a shear by the application of a force F as shown, this will be balanced by an equal and opposite internal frictional force in the fluid. For an incompressible Newtonian fluid in laminar flow, the resulting shear stress is equal to the product of the shear rate and the viscosity of the fluid medium. In this simple case, the shear rate may be expressed as the velocity gradient in the direction perpendicular to that of the shear force, i.e.

$$\frac{F}{A} = \tau_{yx} = \mu \left(-\frac{dV_x}{dy} \right) = \mu \dot{\gamma}_{yx} \tag{1.1}$$

Note that the first subscript on both τ and $\dot{\gamma}$ indicates the direction normal to that of shearing force, while the second subscript refers to the direction of the force and the flow. By considering the equilibrium of a fluid layer, it can

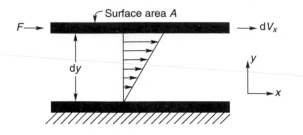

Figure 1.1   *Schematic representation of unidirectional shearing flow*

readily be seen that at any shear plane there are two equal and opposite shear stresses–a positive one on the slower moving fluid and a negative one on the faster moving fluid layer. The negative sign on the right hand side of equation (1.1) indicates that τ_{yx} is a measure of the resistance to motion. One can also view the situation from a different standpoint as: for an incompressible fluid of density ρ, equation (1.1) can be written as:

$$\tau_{yx} = -\frac{\mu}{\rho}\frac{\mathrm{d}}{\mathrm{d}y}(\rho V_x) \qquad (1.2)$$

The quantity 'ρV_x' is the momentum in the x-direction per unit volume of the fluid and hence τ_{yx} represents the momentum flux in the y-direction and the negative sign indicates that the momentum transfer occurs in the direction of decreasing velocity which is also in line with the Fourier's law of heat transfer and Fick's law of diffusive mass transfer.

The constant of proportionality, μ (or the ratio of the shear stress to the rate of shear) which is called the Newtonian viscosity is, by definition, independent of shear rate ($\dot{\gamma}_{yx}$) or shear stress (τ_{yx}) and depends only on the material and its temperature and pressure. The plot of shear stress (τ_{yx}) against shear rate ($\dot{\gamma}_{yx}$) for a Newtonian fluid, the so-called 'flow curve' or 'rheogram', is therefore a straight line of slope, μ, and passing through the origin; the single constant, μ, thus completely characterises the flow behaviour of a Newtonian fluid at a fixed temperature and pressure. Gases, simple organic liquids, solutions of low molecular weight inorganic salts, molten metals and salts are all Newtonian fluids. The shear stress–shear rate data shown in Figure 1.2 demonstrate the Newtonian fluid behaviour of a cooking oil and a corn syrup; the values of the viscosity for some substances encountered in everyday life are given in Table 1.1.

Figure 1.1 and equation (1.1) represent the simplest case wherein the velocity vector which has only one component, in the x-direction varies only in the y-direction. Such a flow configuration is known as simple shear flow. For the more complex case of three dimensional flow, it is necessary to set up the appropriate partial differential equations. For instance, the more general case of an incompressible Newtonian fluid may be expressed – for the x-plane – as

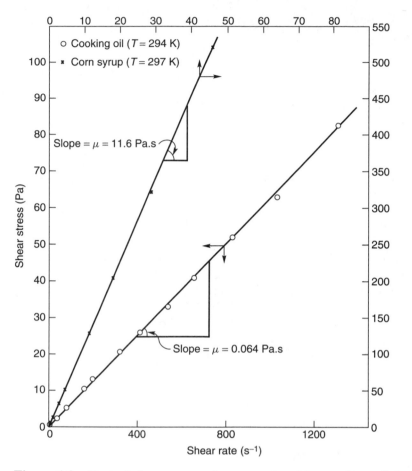

Figure 1.2 *Typical shear stress–shear rate data for a cooking oil and a corn syrup*

follows [Bird *et al.*, 1960, 1987]:

$$\tau_{xx} = -2\mu \frac{\partial V_x}{\partial x} + \frac{2}{3}\mu \left(\frac{\partial V_x}{\partial x} + \frac{\partial V_y}{\partial y} + \frac{\partial V_z}{\partial z} \right) \qquad (1.3)$$

$$\tau_{xy} = -\mu \left(\frac{\partial V_x}{\partial y} + \frac{\partial V_y}{\partial x} \right) \qquad (1.4)$$

$$\tau_{xz} = -\mu \left(\frac{\partial V_x}{\partial z} + \frac{\partial V_z}{\partial x} \right) \qquad (1.5)$$

Similar sets of equations can be drawn up for the forces acting on the y- and z-planes; in each case, there are two (in-plane) shearing components and a

Table 1.1 *Typical viscosity values at room temperature*

Substance	μ (mPa·s)
Air	10^{-2}
Benzene	0.65
Water	1
Molten sodium chloride (1173 K)	1.01
Ethyl alcohol	1.20
Mercury (293 K)	1.55
Molten lead (673 K)	2.33
Ethylene glycol	20
Olive oil	100
Castor oil	600
100% Glycerine (293 K)	1500
Honey	10^4
Corn syrup	10^5
Bitumen	10^{11}
Molten glass	10^{15}

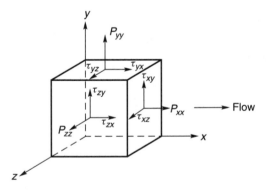

Figure 1.3 *Stress components in three dimensional flow*

normal component. Figure 1.3 shows the nine stress components schematically in an element of fluid. By considering the equilibrium of a fluid element, it can easily be shown that $\tau_{yx} = \tau_{xy}$; $\tau_{xz} = \tau_{zx}$ and $\tau_{yz} = \tau_{zy}$. The normal stresses can be visualised as being made up of two components: isotropic pressure and a contribution due to flow, i.e.

$$P_{xx} = -p + \tau_{xx} \qquad (1.6a)$$

$$P_{yy} = -p + \tau_{yy} \qquad (1.6b)$$

$$P_{zz} = -p + \tau_{zz} \qquad (1.6c)$$

where τ_{xx}, τ_{yy}, τ_{zz}, contributions arising from flow, are known as deviatoric normal stresses for Newtonian fluids and as extra stresses for non-Newtonian fluids. For an incompressible Newtonian fluid, the isotropic pressure is given by

$$p = -\tfrac{1}{3}(P_{xx} + P_{yy} + P_{zz}) \qquad (1.7)$$

From equations (1.6) and (1.7), it follows that

$$\tau_{xx} + \tau_{yy} + \tau_{zz} = 0 \qquad (1.8)$$

For a Newtonian fluid in simple shearing motion, the deviatoric normal stress components are identically zero, i.e.

$$\tau_{xx} = \tau_{yy} = \tau_{zz} = 0 \qquad (1.9)$$

Thus, the complete definition of a Newtonian fluid is that it not only possesses a constant viscosity but it also satisfies the condition of equation (1.9), or simply that it satisfies the complete Navier–Stokes equations. Thus, for instance, the so-called constant viscosity Boger fluids [Boger, 1976; Prilutski *et al.*, 1983] which display constant shear viscosity but do not conform to equation (1.9) must be classed as non-Newtonian fluids.

1.2.2 Non-Newtonian fluid behaviour

A non-Newtonian fluid is one whose flow curve (shear stress versus shear rate) is non-linear or does not pass through the origin, i.e. where the apparent viscosity, shear stress divided by shear rate, is not constant at a given temperature and pressure but is dependent on flow conditions such as flow geometry, shear rate, etc. and sometimes even on the kinematic history of the fluid element under consideration. Such materials may be conveniently grouped into three general classes:

(1) fluids for which the rate of shear at any point is determined only by the value of the shear stress at that point at that instant; these fluids are variously known as 'time independent', 'purely viscous', 'inelastic' or 'generalised Newtonian fluids', (GNF);
(2) more complex fluids for which the relation between shear stress and shear rate depends, in addition, upon the duration of shearing and their kinematic history; they are called 'time-dependent fluids', and finally,

(3) substances exhibiting characteristics of both ideal fluids and elastic solids and showing partial elastic recovery, after deformation; these are categorised as 'visco-elastic fluids'.

This classification scheme is arbitrary in that most real materials often exhibit a combination of two or even all three types of non-Newtonian features. Generally, it is, however, possible to identify the dominant non-Newtonian characteristic and to take this as the basis for the subsequent process calculations. Also, as mentioned earlier, it is convenient to define an apparent viscosity of these materials as the ratio of shear stress to shear rate, though the latter ratio is a function of the shear stress or shear rate and/or of time. Each type of non-Newtonian fluid behaviour will now be dealt with in some detail.

1.3 Time-independent fluid behaviour

In simple shear, the flow behaviour of this class of materials may be described by a constitutive relation of the form,

$$\dot{\gamma}_{yx} = f(\tau_{yx}) \tag{1.10}$$

or its inverse form,

$$\tau_{yx} = f_1(\dot{\gamma}_{yx}) \tag{1.11}$$

This equation implies that the value of $\dot{\gamma}_{yx}$ at any point within the sheared fluid is determined only by the current value of shear stress at that point or vice versa. Depending upon the form of the function in equation (1.10) or (1.11), these fluids may be further subdivided into three types:

(a) shear-thinning or pseudoplastic
(b) viscoplastic
(c) shear-thickening or dilatant

Qualitative flow curves on linear scales for these three types of fluid behaviour are shown in Figure 1.4; the linear relation typical of Newtonian fluids is also included.

1.3.1 Shear-thinning or pseudoplastic fluids

The most common type of time-independent non-Newtonian fluid behaviour observed is pseudoplasticity or shear-thinning, characterised by an apparent viscosity which decreases with increasing shear rate. Both at very low and at very high shear rates, most shear-thinning polymer solutions and melts exhibit Newtonian behaviour, i.e. shear stress–shear rate plots become straight lines,

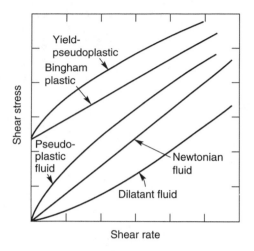

Figure 1.4 *Types of time-independent flow behaviour*

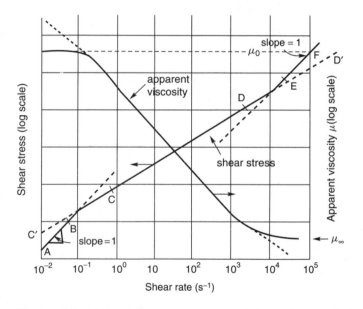

Figure 1.5 *Schematic representation of shear-thinning behaviour*

as shown schematically in Figure 1.5, and on a linear scale will pass through origin. The resulting values of the apparent viscosity at very low and high shear rates are known as the zero shear viscosity, μ_0, and the infinite shear viscosity, μ_∞, respectively. Thus, the apparent viscosity of a shear-thinning fluid decreases from μ_0 to μ_∞ with increasing shear rate. Data encompassing

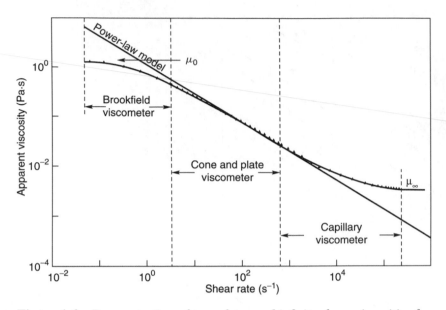

Figure 1.6 *Demonstration of zero shear and infinite shear viscosities for a shear-thinning polymer solution [Boger, 1977]*

a sufficiently wide range of shear rates to illustrate this complete spectrum of pseudoplastic behaviour are difficult to obtain, and are scarce. A single instrument will not have both the sensitivity required in the low shear rate region and the robustness at high shear rates, so that several instruments are often required to achieve this objective. Figure 1.6 shows the apparent viscosity–shear rate behaviour of an aqueous polyacrylamide solution at 293 K over almost seven decades of shear rate. The apparent viscosity of this solution drops from 1400 mPa·s to 4.2 mPa·s, and so it would hardly be justifiable to assign a single average value of viscosity for such a fluid! The values of shear rates marking the onset of the upper and lower limiting viscosities are dependent upon several factors, such as the type and concentration of polymer, its molecular weight distribution and the nature of solvent, etc. Hence, it is difficult to suggest valid generalisations but many materials exhibit their limiting viscosities at shear rates below $10^{-2}\,\mathrm{s}^{-1}$ and above $10^{5}\,\mathrm{s}^{-1}$ respectively. Generally, the range of shear rate over which the apparent viscosity is constant (in the zero-shear region) increases as molecular weight of the polymer falls, as its molecular weight distribution becomes narrower, and as polymer concentration (in solution) drops. Similarly, the rate of decrease of apparent viscosity with shear rate also varies from one material to another, as can be seen in Figure 1.7 for three aqueous solutions of chemically different polymers.

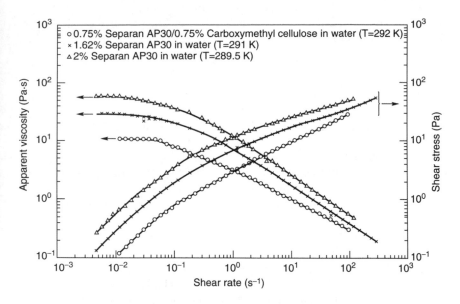

Figure 1.7 *Representative shear stress and apparent viscosity plots for three pseudoplastic polymer solutions*

Mathematical models for shear-thinning fluid behaviour

Many mathematical expressions of varying complexity and form have been proposed in the literature to model shear-thinning characteristics; some of these are straightforward attempts at curve fitting, giving empirical relationships for the shear stress (or apparent viscosity)–shear rate curves for example, while others have some theoretical basis in statistical mechanics – as an extension of the application of the kinetic theory to the liquid state or the theory of rate processes, etc. Only a selection of the more widely used viscosity models is given here; more complete descriptions of such models are available in many books [Bird *et al.*, 1987; Carreau *et al.*, 1997] and in a review paper [Bird, 1976].

(i) The power-law or Ostwald de Waele model

The relationship between shear stress and shear rate (plotted on double logarithmic coordinates) for a shear-thinning fluid can often be approximated by a straightline over a limited range of shear rate (or stress). For this part of the flow curve, an expression of the following form is applicable:

$$\tau_{yx} = m(\dot{\gamma}_{yx})^n \tag{1.12}$$

so the apparent viscosity for the so-called power-law (or Ostwald de Waele) fluid is thus given by:

$$\mu = \tau_{yx}/\dot{\gamma}_{yx} = m(\dot{\gamma}_{yx})^{n-1} \qquad (1.13)$$

For $n < 1$, the fluid exhibits shear-thinnering properties

$\quad\quad n = 1$, the fluid shows Newtonian behaviour

$\quad\quad n > 1$, the fluid shows shear-thickening behaviour

In these equations, m and n are two empirical curve-fitting parameters and are known as the fluid consistency coefficient and the flow behaviour index respectively. For a shear-thinning fluid, the index may have any value between 0 and 1. The smaller the value of n, the greater is the degree of shear-thinning. For a shear-thickening fluid, the index n will be greater than unity. When $n = 1$, equations (1.12) and (1.13) reduce to equation (1.1) which describes Newtonian fluid behaviour.

Although the power-law model offers the simplest representation of shear-thinning behaviour, it does have a number of shortcomings. Generally, it applies over only a limited range of shear rates and therefore the fitted values of m and n will depend on the range of shear rates considered. Furthermore, it does not predict the zero and infinite shear viscosities, as shown by dotted lines in Figure 1.5. Finally, it should be noted that the dimensions of the flow consistency coefficient, m, depend on the numerical value of n and therefore the m values must not be compared when the n values differ. On the other hand, the value of m can be viewed as the value of apparent viscosity at the shear rate of unity and will therefore depend on the time unit (e.g. s, minute or hour) employed. Despite these limitations, this is perhaps the most widely used model in the literature dealing with process engineering applications.

(ii) The Carreau viscosity equation

When there are significant deviations from the power-law model at very high and very low shear rates as shown in Figure 1.6, it is necessary to use a model which takes account of the limiting values of viscosities μ_0 and μ_∞.

Based on the molecular network considerations, Carreau [1972] put forward the following viscosity model which incorporates both limiting viscosities μ_0 and μ_∞:

$$\frac{\mu - \mu_\infty}{\mu_0 - \mu_\infty} = \{1 + (\lambda\dot{\gamma}_{yx})^2\}^{(n-1)/2} \qquad (1.14)$$

where $n(<1)$ and λ are two curve-fitting parameters. This model can describe shear-thinning behaviour over wide ranges of shear rates but only at the expense of the added complexity of four parameters. This model predicts Newtonian fluid behaviour $\mu = \mu_0$ when either $n = 1$ or $\lambda = 0$ or both.

(iii) The Ellis fluid model

When the deviations from the power-law model are significant only at low shear rates, it is perhaps more appropriate to use the Ellis model.

The two viscosity equations presented so far are examples of the form of equation (1.11). The three-constant Ellis model is an illustration of the inverse form, namely, equation (1.10). In simple shear, the apparent viscosity of an Ellis model fluid is given by:

$$\mu = \frac{\mu_0}{1 + (\tau_{yx}/\tau_{1/2})^{\alpha-1}} \tag{1.15}$$

In this equation, μ_0 is the zero shear viscosity and the remaining two constants $\alpha(>1)$ and $\tau_{1/2}$ are adjustable parameters. While the index α is a measure of the degree of shear-thinning behaviour (the greater the value of α, greater is the extent of shear-thinning), $\tau_{1/2}$ represents the value of shear stress at which the apparent viscosity has dropped to half its zero shear value. Equation (1.15) predicts Newtonian fluid behaviour in the limit of $\tau_{1/2} \to \infty$. This form of equation has advantages in permitting easy calculation of velocity profiles from a known stress distribution, but renders the reverse operation tedious and cumbersome.

1.3.2 Viscoplastic fluid behaviour

This type of fluid behaviour is characterised by the existence of a yield stress (τ_0) which must be exceeded before the fluid will deform or flow. Conversely, such a material will deform elastically (or flow *en masse* like a rigid body) when the externally applied stress is smaller than the yield stress. Once the magnitude of the external stress has exceeded the value of the yield stress, the flow curve may be linear or non-linear but will not pass through origin (Figure 1.4). Hence, in the absence of surface tension effects, such a material will not level out under gravity to form an absolutely flat free surface. One can, however, explain this kind of fluid behaviour by postulating that the substance at rest consists of three dimensional structures of sufficient rigidity to resist any external stress less than τ_0. For stress levels greater than τ_0, however, the structure breaks down and the substance behaves like a viscous material.

A fluid with a linear flow curve for $|\tau_{yx}| > |\tau_0|$ is called a Bingham plastic fluid and is characterised by a constant plastic viscosity (the slope of the shear stress versus shear rate curve) and a yield stress. On the other hand, a substance possessing a yield stress as well as a non-linear flow curve on linear coordinates (for $|\tau_{yx}| > |\tau_0|$), is called a 'yield-pseudoplastic' material. Figure 1.8 illustrates viscoplastic behaviour as observed in a meat extract and in a polymer solution.

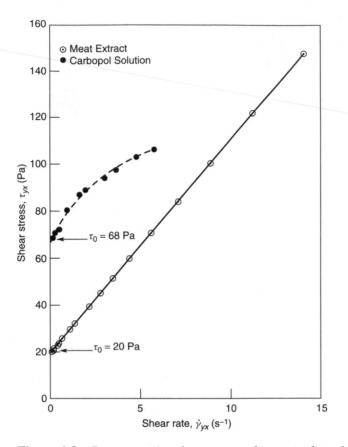

Figure 1.8 *Representative shear stress–shear rate data showing viscoplastic behaviour in a meat extract (Bingham Plastic) and in an aqueous carbopol polymer solution (yield-pseudoplastic)*

It is interesting to note that a viscoplastic material also displays an apparent viscosity which decreases with increasing shear rate. At very low shear rates, the apparent viscosity is effectively infinite at the instant immediately before the substance yields and begins to flow. It is thus possible to regard these materials as possessing a particular class of shear-thinning behaviour.

Strictly speaking, it is virtually impossible to ascertain whether any real material has a true yield stress or not, but nevertheless the concept of a yield stress has proved to be convenient in practice because some materials closely approximate to this type of flow behaviour, e.g. see [Barnes and Walters, 1985; Astarita, 1990; Schurz, 1990 and Evans, 1992]. The answer to the question whether a fluid has a yield stress or not seems to be related to the choice of a time scale of observation. Common examples of viscoplastic fluid behaviour include particulate suspensions, emulsions, foodstuffs, blood and drilling muds, etc. [Barnes, 1999]

Mathematical models for viscoplastic behaviour

Over the years, many empirical expressions have been proposed as a result of straightforward curve fitting exercises. A model based on sound theory is yet to emerge. Three commonly used models for viscoplastic fluids are briefly described here.

(i) The Bingham plastic model

This is the simplest equation describing the flow behaviour of a fluid with a yield stress and, in steady one dimensional shear, it is written as:

$$\tau_{yx} = \tau_0^B + \mu_B(\dot{\gamma}_{yx}) \quad \text{for } |\tau_{yx}| > |\tau_0^B| \tag{1.16}$$

$$\dot{\gamma}_{yx} = 0 \qquad\qquad \text{for } |\tau_{yx}| < |\tau_0^B|$$

Often, the two model parameters, τ_0^B and μ_B, are treated as curve fitting constants irrespective of whether or not the fluid possesses a true yield stress.

(ii) The Herschel–Bulkley fluid model

A simple generalisation of the Bingham plastic model to embrace the non-linear flow curve (for $|\tau_{yx}| > |\tau_0^B|$) is the three constant Herschel–Bulkley fluid model. In one dimensional steady shearing motion, it is written as:

$$\tau_{yx} = \tau_0^H + m(\dot{\gamma}_{yx})^n \quad \text{for } |\tau_{yx}| > |\tau_0^H| \tag{1.17}$$

$$\dot{\gamma}_{yx} = 0 \qquad\qquad \text{for } |\tau_{yx}| < |\tau_0^H|$$

Note that here too, the dimensions of m depend upon the value of n. With the use of the third parameter, this model provides a somewhat better fit to some experimental data.

(iii) The Casson fluid model

Many foodstuffs and biological materials, especially blood, are well described by this two constant model as:

$$(|\tau_{yx}|)^{1/2} = (|\tau_0^c|)^{1/2} + (\mu_c|\dot{\gamma}_{yx}|)^{1/2} \quad \text{for } |\tau_{yx}| > |\tau_0^c| \tag{1.18}$$

$$\dot{\gamma}_{yx} = 0 \qquad\qquad \text{for } |\tau_{yx}| < |\tau_0^c|$$

This model has often been used for describing the steady shear stress–shear rate behaviour of blood, yoghurt, tomato pureé, molten chocolate, etc. The flow behaviour of some particulate suspensions also closely approximates to this type of behaviour.

The comparative performance of these three as well as several other models for viscoplastic behaviour has been thoroughly evaluated in an extensive review paper by Bird *et al.* [1983].

1.3.3 Shear-thickening or dilatant fluid behaviour

Dilatant fluids are similar to pseudoplastic systems in that they show no yield stress but their apparent viscosity increases with increasing shear rate; thus these fluids are also called shear-thickening. This type of fluid behaviour was originally observed in concentrated suspensions and a possible explanation for their dilatant behaviour is as follows: At rest, the voidage is minimum and the liquid present is sufficient to fill the void space. At low shear rates, the liquid lubricates the motion of each particle past others and the resulting stresses are consequently small. At high shear rates, on the other hand, the material expands or dilates slightly (as also observed in the transport of sand dunes) so that there is no longer sufficient liquid to fill the increased void space and prevent direct solid–solid contacts which result in increased friction and higher shear stresses. This mechanism causes the apparent viscosity to rise rapidly with increasing rate of shear.

The term dilatant has also been used for all other fluids which exhibit increasing apparent viscosity with increasing rate of shear. Many of these, such as starch pastes, are not true suspensions and show no dilation on shearing. The above explanation therefore is not applicable but nevertheless such materials are still commonly referred to as dilatant fluids.

Of the time-independent fluids, this sub-class has received very little attention; consequently very few reliable data are available. Until recently, dilatant fluid behaviour was considered to be much less widespread in the chemical and processing industries. However, with the recent growing interest in the handling and processing of systems with high solids loadings, it is no longer so, as is evidenced by the number of recent review articles on this subject [Barnes *et al.*, 1987; Barnes, 1989; Boersma *et al.*, 1990; Goddard and Bashir, 1990]. Typical examples of materials exhibiting dilatant behaviour include concentrated suspensions of china clay, titanium dioxide [Metzner and Whitlock, 1958] and of corn flour in water. Figure 1.9 shows the dilatant behaviour of dispersions of polyvinylchloride in dioctylphthalate [Boersma *et al.*, 1990].

The limited information reported so far suggests that the apparent viscosity–shear rate data often result in linear plots on double logarithmic coordinates over a limited shear rate range and the flow behaviour may be represented by the power-law model, equation (1.13), with the flow behaviour index, n, greater than one, i.e.

$$\mu = m(\dot{\gamma}_{yx})^{n-1} \tag{1.13}$$

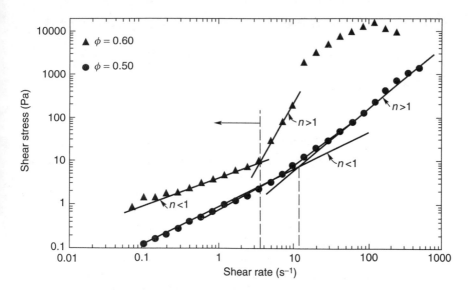

Figure 1.9   *Shear stress–shear rate behaviour of polyvinylchloride (PVC) in dioctylphthalate (DOP) dispersions at 298 K showing regions of shear-thinning and shear-thickening [Boersma et al., 1990]*

One can readily see that for $n > 1$, equation (1.13) predicts increasing viscosity with increasing shear rate. The dilatant behaviour may be observed in moderately concentrated suspensions at high shear rates, and yet, the same suspension may exhibit pseudoplastic behaviour at lower shear rates, as shown in Figure 1.9; it is not yet possible to ascertain whether these materials also display limiting apparent viscosities.

1.4 Time-dependent fluid behaviour

The flow behaviour of many industrially important materials cannot be described by a simple rheological equation like (1.12) or (1.13). In practice, apparent viscosities may depend not only on the rate of shear but also on the time for which the fluid has been subjected to shearing. For instance, when materials such as bentonite-water suspensions, red mud suspensions (waste stream from aluminium industry), crude oils and certain foodstuffs are sheared at a constant rate following a long period of rest, their apparent viscosities gradually become less as the 'internal' structure of the material is progressively broken down. As the number of structural 'linkages' capable of being broken down decreases, the rate of change of apparent viscosity with time drops progressively to zero. Conversely, as the structure breaks down, the rate at which linkages can re-form increases, so that eventually a state of dynamic equilibrium is reached when the rates of build-up and of break-down are balanced.

Time-dependent fluid behaviour may be further sub-divided into two categories: thixotropy and rheopexy or negative thixotropy.

1.4.1 Thixotropy

A material is said to exhibit thixotropy if, when it is sheared at a constant rate, its apparent viscosity (or the corresponding shear stress) decreases with the time of shearing, as can be seen in Figure 1.10 for a red mud suspension [Nguyen and Uhlherr, 1983]. If the flow curve is measured in a single experiment in which the shear rate is steadily increased at a constant rate from zero to some maximum value and then decreased at the same rate to zero again, a hysteresis loop of the form shown in Figure 1.11 is obtained; the height, shape and enclosed area of the hysteresis loop depend on the duration of shearing, the rate of increase/decrease of shear rate and the past kinematic history of the sample. No hysteresis loop is observed for time-independent fluids, that is, the enclosed area of the loop is zero.

The term 'false body' has been introduced to describe the thixotropic behaviour of viscoplastic materials. Although the thixotropy is associated with the build-up of structure at rest and breakdown of structure under shear, viscoplastic materials do not lose their solid-like properties completely and can still exhibit a yield stress, though this is usually less than the original value of the virgin sample which is regained (if at all) only after a long recovery period.

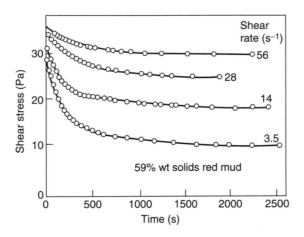

Figure 1.10   *Representative data showing thixotropy in a 59% (by weight) red mud suspension*

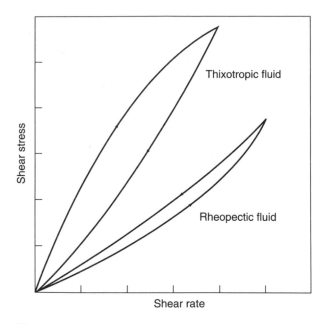

Figure 1.11   *Schematic shear stress–shear rate behaviour for time-dependent fluid behaviour*

Other examples of materials exhibiting thixotropic behaviour include concentrated suspensions, emulsions, protein solutions and food stuffs, etc. [Barnes, 1997].

1.4.2 Rheopexy or negative thixotropy

The relatively few fluids for which the apparent viscosity (or the corresponding shear stress) increases with time of shearing are said to display rheopexy or negative thixotropy. Again, hysteresis effects are observed in the flow curve, but in this case it is inverted, as compared with a thixotropic material, as can be seen in Figure 1.11.

In a rheopectic fluid the structure builds up by shear and breaks down when the material is at rest. For instance, Freundlich and Juliusberger [1935], using a 42% aqueous gypsum paste, found that, after shaking, this material re-solidified in 40 min if at rest, but in only 20 s if the container was gently rolled in the palms of hands. This indicates that gentle shearing motion (rolling) facilitates structure buildup but more intense motion destroys it. Thus, there is a critical amount of shear beyond which re-formation of structure is not induced but breakdown occurs. It is not uncommon for the same dispersion to display both thixotropy as well as rheopexy depending upon the shear rate and/or the concentration of solids. Figure 1.12 shows the gradual onset of rheopexy for a saturated polyester at 60°C [Steg and Katz, 1965]. Similar behaviour is reported to occur with suspensions of ammonium oleate, colloidal suspensions of a vanadium pentoxide at moderate shear rates [Tanner, 1988], coal-water slurries [Keller and Keller Jr, 1990] and protein solutions [Pradipasena and Rha, 1977].

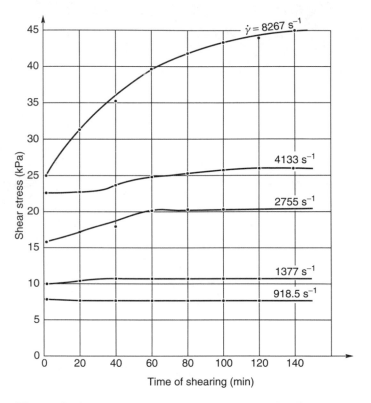

Figure 1.12   *Onset of rheopexy in a saturated polyester [Steg and Katz, 1965]*

It is not possible to put forward simple mathematical equations of general validity to describe time-dependent fluid behaviour, and it is usually necessary to make measurements over the range of conditions of interest. The conventional shear stress–shear rate curves are of limited utility unless they relate to the particular history of interest in the application. For example when the material enters a pipe slowly and with a minimum of shearing, as from a storage tank directly into the pipe, the shear stress–shear rate–time curve should be based on tests performed on samples which have been stored under identical conditions and have not been subjected to shearing by transference to another vessel for example. At the other extreme, when the material undergoes vigorous agitation and shearing, as in passage through a pump, the shear stress–shear rate curve should be obtained using highly sheared pre-mixed material. Assuming then that reliable flow property data are available, the zero shear and infinite shear flow curves can be used to form the bounds for the design of a flow system. For a fixed pressure drop, the zero shear limit (maximum apparent viscosity) will provide a lower bound and the infinite shear conditions (minimum apparent viscosity) will provide the upper bound on the flowrate. Conversely, for a fixed flowrate, the zero and infinite shear data can be used to establish the maximum and minimum pressure drops or pumping power.

For many industries (notably foodstuffs) the way in which the rheology of the materials affects their processing is much less significant than the effects that the process has on their rheology. Implicit here is the recognition of the importance of the time-dependent properties of materials which can be profoundly influenced by mechanical working on the one hand or by an aging process during a prolonged shelf life on the other.

The above brief discussion of time-dependent fluid behaviour provides an introduction to the topic, but Mewis [1979] and Barnes [1997] have given detailed accounts of recent developments in the field. Govier and Aziz [1982], moreover, have focused on the practical aspects of the flow of time-dependent fluids in pipes.

1.5 Visco-elastic fluid behaviour

In the classical theory of elasticity, the stress in a sheared body is directly proportional to the strain. For tension, Hooke's law applies and the coefficient of proportionality is known as Young's modulus, G,:

$$\tau_{yx} = -G\frac{\mathrm{d}x}{\mathrm{d}y} = G(\gamma_{yx}) \tag{1.19}$$

where $\mathrm{d}x$ is the shear displacement of two elements separated by a distance $\mathrm{d}y$. When a perfect solid is deformed elastically, it regains its original form on removal of the stress. However, if the applied stress exceeds the characteristic yield stress of the material, complete recovery will not occur and 'creep' will take place–that is, the 'solid' will have flowed.

At the other extreme, in the Newtonian fluid the shearing stress is proportional to the rate of shear, equation (1.1). Many materials show both elastic and viscous effects under appropriate circumstances. In the absence of the time-dependent behaviour mentioned in the preceding section, the material is said to be visco-elastic. Perfectly elastic deformation and perfectly viscous flow are, in effect, limiting cases of visco-elastic behaviour. For some materials, it is only these limiting conditions that are observed in practice. The elasticity of water and the viscosity of ice may generally pass unnoticed! The response of a material depends not only its structure but also on the conditions (kinematic) to which it has been subjected; thus the distinction between 'solid' and 'fluid' and between 'elastic' and 'viscous' is to some extent arbitrary and subjective.

Many materials of practical interest (such as polymer melts, polymer and soap solutions, synovial fluid) exhibit visco-elastic behaviour; they have some ability to store and recover shear energy, as shown schematically in Figure 1.13. Perhaps the most easily observed experiment is the 'soup bowl' effect. If a liquid in a dish is made to rotate by means of gentle stirring with a spoon, on removing the energy source (i.e. the spoon), the inertial circulation will die out as a result of the action of viscous forces. If the liquid is visco-elastic (as some of the proprietary soups are), the liquid will be seen to slow to a stop and then to unwind a little. This type of behaviour is closely linked to the tendency for a gel structure to form within the fluid; such an element of rigidity makes simple shear less likely to occur–the shearing forces tending to act as couples to produce rotation of the fluid elements as well as pure slip. Such incipient rotation produces a stress perpendicular to the direction of shear. Numerous other unusual phenomena often ascribed to fluid visco-elasticity include die swell, rod climbing (Weissenberg effect), tubeless siphon, and the development of secondary flows at low Reynolds numbers. Most of these have been illustrated photographically in a recent book [Boger and Walters, 1992]. A detailed treatment of visco-elastic fluid behaviour is beyond the scope of this book and interested readers are referred to several excellent books available on this subject, e.g. see [Schowalter, 1978; Bird *et al.*, 1987; Carreau *et al.*, 1997; Tanner and Walters, 1998]. Here we shall describe the 'primary' and 'secondary' normal stress differences observed in steady shearing flows which are used both to classify a material as visco-elastic or viscoinelastic as well as to quantify the importance of visco-elastic effects in an envisaged application.

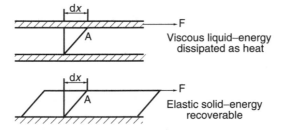

Figure 1.13   *Qualitative differences between a viscous fluid and an elastic solid*

Normal stresses in steady shear flows

Let us consider the one-dimensional shearing motion of a fluid; the stresses developed by the shearing of an infinitesimal element of fluid between two planes are shown in Figure 1.14. By nature of the steady shear flow, the components of velocity in the y- and z-directions are zero while that in the x-direction is a function of y only. Note that in addition to the shear stress, τ_{yx}, there are three normal stresses denoted by P_{xx}, P_{yy} and P_{zz} within the sheared fluid which are given by equation (1.6). Weissenberg [1947] was the first to observe that the shearing motion of a visco-elastic fluid gives rise to unequal normal stresses. Since the pressure in a non-Newtonian fluid cannot be defined by equation (1.7) the differences, $P_{xx} - P_{yy} = N_1$ and $P_{yy} - P_{zz} = N_2$, are more readily measured than the individual stresses, and it is therefore customary to express N_1 and N_2 together with τ_{yx} as functions of the shear rate $\dot{\gamma}_{yx}$ to describe the rheological behaviour of a visco-elastic material in a simple shear flow. Sometimes, the first and second normal stress differences N_1 and N_2 are expressed in terms of two coefficients, ψ_1, and ψ_2 defined as follows:

$$\psi_1 = \frac{N_1}{\dot{\gamma}_{yx}^{2}} \tag{1.20}$$

$$\text{and} \quad \psi_2 = \frac{N_2}{\dot{\gamma}_{yx}^{2}} \tag{1.21}$$

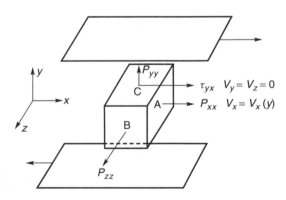

Figure 1.14   *Non-zero components of stress in one dimensional steady shearing motion of a visco-elastic fluid*

A typical dependence of the first normal stress difference on shear rate is shown in Figure 1.15 for a series of polystyrene-in-toluene solutions. Usually, the rate of decrease of ψ_1 with shear rate is greater than that of the apparent

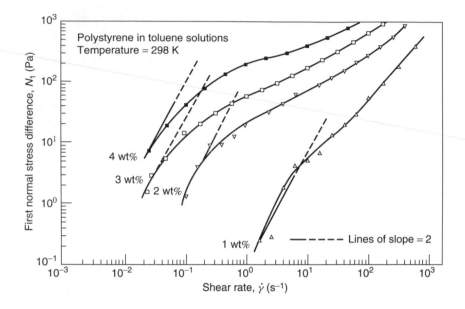

Figure 1.15   *Representative first normal stress difference data for polystyrene-in-toluene solutions at 298 K [Kulicke and Wallabaum, 1985]*

viscosity. At very low shear rates, the first normal stress difference, N_1, is expected to be proportional to the square of shear rate – that is, ψ_1 tends to a constant value ψ_0; this limiting behaviour is seen to be approached by some of the experimental data shown in Figure 1.15. It is common that the first normal stress difference N_1 is higher than the shear stress τ at the same value of shear rate. The ratio of N_1 to τ is often taken as a measure of how elastic a liquid is; specifically $(N_1/2\tau)$ is used and is called the recoverable shear. Recoverable shears greater than 0.5 are not uncommon in polymer solutions and melts. They indicate a highly elastic behaviour of the fluid. There is, however, no evidence of ψ_1 approaching a limiting value at high shear rates. It is fair to mention here that the first normal stress difference has been investigated much less extensively than the shear stress.

Even less attention has been given to the study and measurement of the second normal stress difference. The most important points to note about N_2 are that it is an order of magnitude smaller than N_1, and that it is negative. Until recently, it was thought that $N_2 = 0$; this so-called Weissenberg hypothesis is no longer believed to be correct. Some data in the literature even seem to suggest that N_2 may change sign. Typical forms of the dependence of N_2 on shear rate are shown in Figure 1.16 for the same solutions as used in Figure 1.15.

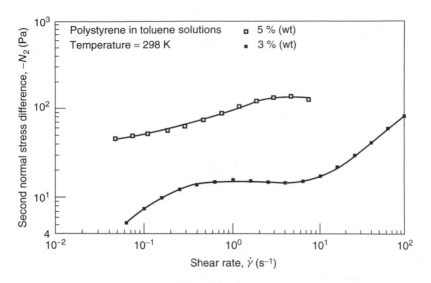

Figure 1.16 *Representative second normal stress difference data for polystyrene-in-toluene solutions at 298 K [Kulicke and Wallabaum, 1985]*

The two normal stress differences defined in this way are characteristic of a material, and as such are used to categorise a fluid either as purely viscous ($N_1 \sim 0$) or as visco-elastic, and the magnitude of N_1 in comparison with τ_{yx}, is often used as a measure of visco-elasticity.

Aside from the simple shearing motion, the response of visco-elastic materials in a variety of other well-defined flow configurations including the cessation/initiation of flow, creep, small amplitude sinusoidal shearing, etc. also lies in between that of a perfectly viscous fluid and a perfectly elastic solid. Conversely, these tests may be used to infer a variety of rheological information about a material. Detailed discussions of the subject are available in a number of books, e.g. see Walters [1975] and Makowsko [1994].

Elongational flow

Flows which result in fluids being subjected to stretching in one or more dimensions occur in many processes, fibre spinning and polymer film blowing being only two of the most common examples. Again, when two bubbles coalesce, a very similar stretching of the liquid film between them takes place until rupture occurs. Another important example of the occurrence of extensional effects is the flow of polymer solutions in porous media, as encountered in the enhanced oil recovery process, in which the fluid is stretched as the extent and shape of the flow passages change. There are three main forms of elongational flow: uniaxial, biaxial and planar, as shown schematically in Figure 1.17.

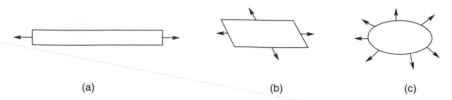

(a)   (b)   (c)

Figure 1.17   *Schematic representation of uniaxial (a), biaxial (b) and planar (c) extension*

Fibre spinning is an example of uniaxial elongation (but the stretch rate varies from point to point along the length of the fibre). Tubular film blowing which involves extruding of polymer through a slit die and pulling the emerging sheet forward and sideways is an example of biaxial extension; here, the stretch rates in the two directions can normally be specified and controlled. Another example is the manufacture of plastic tubes which may be made either by extrusion or by injection moulding, followed by heating and subjection to high pressure air for blowing to the desired size. Due to symmetry, the blowing step in an example of biaxial extension with equal rates of stretching in two directions. Irrespective of the type of extension, the sum of the volumetric rates of extension in the three directions must always be zero for an incompressible fluid.

Naturally, the mode of extension affects the way in which the fluid resists deformation and, although this resistance can be referred to loosely as being quantified in terms of an elongational or extensional viscosity (which further depends upon the type of elongational flow, i.e. uniaxial, biaxial or planar), this parameter is, in general, not necessarily constant. For the sake of simplicity, consideration may be given to the behaviour of an incompressible fluid element which is being elongated at a constant rate $\dot{\varepsilon}$ in the x-direction, as shown in Figure 1.18. For an incompressible fluid, the volume of the element must remain constant and therefore it must contract in both the y- and z-directions at the rate of $(\dot{\varepsilon}/2)$, if the system is symetrical in those directions. The normal stress P_{yy} and P_{zz} will then be equal. Under these conditions, the three components of the velocity vector V are given by:

$$V_x = \dot{\varepsilon}x, \quad V_y = -\frac{\dot{\varepsilon}}{2}y, \text{ and } V_z = -\frac{\dot{\varepsilon}}{2}z \tag{1.22}$$

and clearly, the rate of elongation in the x-direction is given by:

$$\dot{\varepsilon} = \frac{\partial V_x}{\partial x} \tag{1.23}$$

In uniaxial extension, the elongational viscosity μ_E is then defined as:

$$\mu_E = \frac{P_{xx} - P_{yy}}{\dot{\varepsilon}} = \frac{\tau_{xx} - \tau_{yy}}{\dot{\varepsilon}} \tag{1.24}$$

or P_{yy} and τ_{yy} can be replaced by P_{zz} and τ_{zz} respectively.

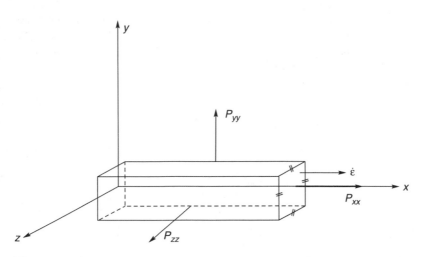

Figure 1.18   *Uniaxial extensional flow*

The earliest determinations of elongational viscosity were made for the simplest case of uniaxial extension, the stretching of a fibre or filament of liquid. Trouton [1906] and many later investigators found that, at low strain (or elongation) rates, the elongational viscosity μ_E was three times the shear viscosity μ [Barnes *et al.*, 1989]. The ratio μ_E/μ is referred to as the Trouton ratio, T_r and thus:

$$T_r = \frac{\mu_E}{\mu} \tag{1.25}$$

The value of 3 for Trouton ratio for an incompressible Newtonian fluid applies to values of shear and elongation rates. By analogy, one may define the Trouton ratio for a non-Newtonian fluid:

$$T_r = \frac{\mu_E(\dot{\varepsilon})}{\mu(\dot{\gamma})} \tag{1.26}$$

The definition of the Trouton ratio given by equation (1.26) is somewhat ambiguous, since it depends on both $\dot{\varepsilon}$ and $\dot{\gamma}$, and some convention must therefore be adopted to relate the strain rates in extension and shear. To remove this ambiguity and at the sametime to provide a convenient estimate of behaviour in extension, Jones *et al.* [1987] proposed the following definition of the Trouton ratio:

$$T_r = \frac{\mu_E(\dot{\varepsilon})}{\mu(\sqrt{3}\dot{\varepsilon})} \tag{1.27}$$

i.e., in the denominator, the shear viscosity is evaluated at $\dot{\gamma} = \sqrt{3}\dot{\varepsilon}$. They also suggested that for inelastic isotropic fluids, the Trouton ratio is equal to 3 for all values of $\dot{\varepsilon}$ and $\dot{\gamma}$, and any departure from the value of 3 can be ascribed unambiguously to visco-elasticity. In other words, equation (1.27) implies that for an inelastic shear-thinning fluid, the extensional viscosity must also decrease with increasing rate of extension (so-called "tension-thinning"). Obviously, a shear-thinning visco-elastic fluid (for which the Trouton ratio will be greater than 3) will thus have an extensional viscosity which increases with the rate of extension; this property is also called "strain-hardening". Many materials including polymer melts and solutions thus exhibit shear-thinning in simple shear and strain-hardening in uniaxial extension. Except in the limit of vanishingly small rates of deformation, there does not appear to be any simple relationship between the elongational viscosity and the other rheological properties of the fluid and, to date, its determination rests entirely on experiments which themselves are aften constrained by the difficulty of establishing and maintaining an elongational flow field for long enough for the steady state to be reached [Gupta and Sridhar, 1988; James and Walters, 1994]. Measurements made on the same fluid using different methods seldom show quantitative agreement, especially for low to medium viscosity fluids [Tirtaatmadja and Sridhar, 1993]. The Trouton ratios for biaxial and planar extensions at low strain rates have values of 6 and 9 respectively for all inelastic fluids and for Newtonian fluids under all conditions.

Mathematical models for visco-elastic behaviour

Though the results of experiments in steady and transient shear or in an elongational flow field may be used to calculate viscous and elastic properties for a fluid, in general the mathematical equations need to be quite complex in order to describe a real fluid adequately. Certainly, the most striking feature connected with the deformation of a visco-elastic substance is its simultaneous display of 'fluid-like' and 'solid-like' characteristics. It is thus not at all surprising that early attempts at the quantitative description of visco-elastic behaviour hinged on the notion of a linear combination of elastic and viscous properties by using mechanical analogues involving springs (elastic component) and dash pots (viscous action). The Maxwell model represents the corner-stone of the so-called linear visco-elastic models; though it is crude, nevertheless it does capture the salient features of visco-elastic behaviour.

A mechanical analogue of this model is obtained by series combinations of a spring and a dashpot (a vessel whose outlet contains a flow constriction over which the pressure drop is proportional to flow rate), as shown schematically in Figure 1.19. If the individual strain rates of the spring and the dashpot respectively are $\dot{\gamma}_1$ and $\dot{\gamma}_2$, then the total strain rate $\dot{\gamma}$ is given by the sum of these two components:

$$\dot{\gamma} = \dot{\gamma}_1 + \dot{\gamma}_2 = \frac{d\gamma_1}{dt} + \frac{d\gamma_2}{dt} \tag{1.28}$$

Figure 1.19 *Schematic representation of the Maxwell model*

Combining equation (1.28) with the Hooke's law of elasticity and Newton's law of viscosity, one can obtain:

$$\tau + \lambda \dot{\tau} = \mu \dot{\gamma} \tag{1.29}$$

where $\dot{\tau}$ is the time derivative of τ; μ is the viscosity of the dashpot fluid; and $\lambda (= \mu/G)$, the relaxation time, which is a characteristic of the fluid. It can easily be seen from equation (1.29) that if a Maxwell model fluid is strained to a fixed point and held there, the stress will decay as $\exp(-t/\lambda)$. An important feature of the Maxwell model is its predominantly fluid-like response. A more solid-like behaviour is obtained by considering the so-called Voigt model which is represented by the parallel arrangement of a spring and a dashpot, as shown in Figure 1.20.

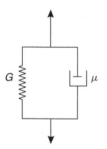

Figure 1.20 *Schematic representation of the Voigt model*

In this case, the strain in the two components is equal and the equation describing the stress–strain behaviour of this system is:

$$\tau = G\gamma + \mu \dot{\gamma} \tag{1.30}$$

If the stress is constant at τ_0 and the initial strain is zero, upon the removal of the stress, the strain decays exponentially with a time constant, $\lambda (= \mu/G)$.

The more solid-like response of this model is clear from the fact that it does not exhibit unlimited non-recoverable viscous flow and it will come to rest when the spring has taken up the load.

One of the main virtues of such linear models is that they can be conveniently superimposed by introducing a spectrum of relaxation times, as exhibited in practice, by polymeric systems or by including higher derivatives. Alternatively, using the idea of superposition, one can assume the stress to be due to the summation of a number of small partial stresses, each pertaining to a partial strain, and each stress relaxing according to some relaxation law. This approach yields the so-called 'integral' models. In addition, many other ideas have been employed to develop elementary models for visco-elastic behaviour including the dumbbell, bead-spring representations, network and kinetic theories. Invariably, all such attempts entail varying degrees of idealisation and empiricism; their most notable limitation is the restriction to small strain and strain rates.

The next generation of visco-elastic fluid models has attempted to relax the restriction of small deformation and deformation rates, thereby leading to the so-called non-linear models. Excellent critical appraisals of the developments in the field, together with the merits and de-merits of a selection of models, as well as some guidelines for selecting an appropriate equation for an envisaged process application, are available in the literature, e.g. see references [Bird *et al.*, 1987; Tanner, 1988; Larson, 1988; Barnes *et al.*, 1989; Macosko, 1994; Bird and Wiest, 1995].

1.6 Dimensional considerations for visco-elastic fluids

It has been a common practice to describe visco-elastic fluid behaviour in steady shear in terms of a shear stress (τ_{yx}) and the first normal stress difference (N_1); both of which are functions of shear rate. Generally, a fluid relaxation or characteristic time, λ_f, (or a spectrum) is defined to quantify the visco-elastic behaviour. There are several ways of defining a characteristic time by combining shear stress and the first normal stress difference, e.g. the so-called Maxwellian relaxation time is given by:

$$\lambda_f = \frac{N_1}{2\tau_{yx}\dot{\gamma}_{yx}} \tag{1.31}$$

Since, in the limit of $\dot{\gamma}_{yx} \to 0$, both $\psi_1 (= N_1/\dot{\gamma}_{yx}{}^2)$ and $\mu (= \tau_{yx}/\dot{\gamma}_{yx})$ approach constant values, λ_f also approaches a constant value. Though equation (1.31) defines a fluid characteristic time as a function of shear rate, its practical utility is severely limited by the fact that in most applications, the kinematics (or shear rate) is not known a priori. Many authors [Leider and Bird, 1972; Grimm,

1978] have obviated this difficulty by introducing the following alternative definition of λ_f:

$$\lambda_f = \left(\frac{m_1}{2m}\right)^{1/(p_1-n)} \tag{1.32}$$

This definition is based on the assumption that both N_1 and τ_{yx} can be approximated as power-law functions of shear rate in the range of conditions of interest, that is,

$$N_1 = m_1(\dot{\gamma}_{yx})^{p_1} \tag{1.33}$$

and $\tau_{yx} = m(\dot{\gamma}_{yx})^n$ \hfill (1.12)

By re-defining λ_f in this manner, it is not necessary to extend the rheological measurements to the zero-shear region. Note that in the limit of $\dot{\gamma}_{yx} \to 0$, $p_1 \to 2$, $n \to 1$ and thus equation (1.32) coincides with equation (1.31).

For Newtonian fluids, the state of flow can be described by two dimensionless groups, usually the Reynolds number, Re, (inertial forces/viscous forces) and Froude number, Fr, (inertial/gravity forces). For a visco-elastic fluid, at least one additional group involving elastic forces is required.

The Reynolds number represents the ratio of inertial to viscous forces, and it might be reasonable to expect that such a ratio would provide a useful parameter. Unfortunately, attempts to achieve meaningful correlations have not been very successful, perhaps being defeated most frequently by the complexity of natural situations and real materials. One simple parameter that may prove of value is the ratio of a characteristic time of deformation to a natural time constant for the fluid. The precise definition of these times is a matter for argument, but it is evident that for processes that involve very slow deformation of the fluid elements it is possible for the elastic forces to be released by the normal processes of relaxation as they build up. As examples of the flow of rigid (apparently infinitely viscous) materials over long periods of time, even the thickening of the lower parts of medieval glass windows is insignificant compared with the plastic flow and deformation that lead to the folded strata of geological structures. In operations that are carried out rapidly the extent of viscous flow will be minimal and the deformation will be followed by recovery when the stress is removed. To get some idea of the possible regions in which such an analysis can provide guidance, consider the flow of a 1% aqueous polyacrylamide solution. Typically, this solution might have a relaxation time of the order of 10 ms. If the fluid were flowing through a packed bed, it would be subjected to alternating acceleration and deceleration as it flowed through the interstices of the bed. With a particle size of 25 mm, say, and a superficial velocity of 0.25 m/s, the deformation or process time will be of the order of $25 \times 10^{-3}/0.25 = 0.1$ s which is greater than the fluid relaxation time. Thus, the fluid elements can adjust to the changing flow area and one would therefore not expect the elastic properties to influence the flow significantly. However, in a free jet discharge with a velocity of 30 m/s through a nozzle of 3 mm diameter, the deformation or process time of $(3 \times 10^{-3})/30 = 0.1$ ms which is 100 times smaller than the fluid response time and the fluid elements will not have sufficient time to re-adjust; it might therefore be reasonable to expect some evidence of elastic behaviour near the point of discharge from the nozzle. A parameter which might be expected to be important is the Deborah number, De, is defined as [Metzner *et al.*, 1966]:

$$De = \frac{\text{Fluid response time}}{\text{Process characteristic time}} \tag{1.34}$$

In the examples above, for the packed bed, De = 0.1 and for the free jet, De = 100. The greater the value of De, the more likely the elasticity to be of practical significance. The same material might well exhibit strongly elastic response under certain conditions of deformation and essentially an inelastic response under other conditions.

Unfortunately, this group depends on the assignment of a single characteristic time to the fluid (a relaxation time?). While this is better than no description at all, it appears to be inadequate for many visco-elastic materials which show different relaxation behaviour under differing conditions.

From the preceding discussion in this chapter, it is abundantly clear that each non-Newtonian fluid is unique in its characteristics, and the only real information about the rheology of a material comes from the experimental points or flow curves that are obtained using some form of rheometers, as discussed in Chapter 2. Provided that there are sufficient experimental points, interpolation can usually be satisfactory; extrapolation should, however, be avoided as it can frequently lead to erroneous results. Certainly, the fitting of an empirical viscosity model to limited data should not be used as a justification for extrapolating the results beyond the experimental range of shear rates or shear stresses. Similarly, it is usually possible to fit a number of different equations (e.g. the power-law and the Bingham plastic model) to a given set of data equally well, and the choice is largely based on convenience or individual preference. Frequently, it is not possible to decide whether a true yield stress exists or not. Therefore, some workers prefer to refer to an 'apparent yield stress' which is an operational parameter and its evaluation involves extrapolation of data to zero shear rate, often the value depending upon the range of data being used to evaluate it. For instance, it is therefore likely that the values of the apparent yield stress fitted in the Bingham, Herschel–Bulkley and Casson models may be quite different for the same fluid. Thus, extreme caution must be exercised in analysing, interpreting and using rheological data.

Example 1.1

The following shear stress–shear rate data were obtained for an aqueous polymer solution at 291 K.

$\dot{\gamma}_{yx}$ (s^{-1})	τ_{yx} (Pa)	$\dot{\gamma}_{yx}$ (s^{-1})	τ_{yx} (Pa)
0.14	0.12	4.43	3.08
0.176	0.14	5.57	3.79
0.222	0.17	7.02	4.68
0.28	0.21	8.83	5.41
0.352	0.28	11.12	6.53
0.443	0.35	14	8.11
0.557	0.446	17.62	9.46
0.702	0.563	22.2	11.50
0.883	0.69	27.9	13.5
1.11	0.85	35.2	16.22
1.4	1.08	44.3	18.92
1.76	1.31	55.7	22.10
2.22	1.63	70.2	26.13
2.8	2.01	88.3	30
3.52	2.53	111.2	34.8

(a) Plot the flow curve on log-log coordinates
(b) Can the power-law model fit this data over the entire range? What are the values of m and n?
(c) Can the Ellis fluid model (eq. 1.15) fit this data better than the power law model? Evaluate the values of μ_0, $\tau_{1/2}$ and α?

Solution

Figure 1.21 shows the flow curve for this polymer solution.

The plot is not linear on log-log coordinates and therefore the power-law model cannot fit the data over the whole range; however, it is possible to divide the plot in two parts, each of which can be represented by power-law model as:

$$\tau = 0.75\ \dot{\gamma}^{0.96} \quad \text{for } \dot{\gamma} < {\sim}5\,\text{s}^{-1}$$

$$\text{and} \quad \tau = 1.08\ \dot{\gamma}^{0.76} \quad 5 \le \dot{\gamma} \le 100\,\text{s}^{-1}$$

Note that at $\dot{\gamma} = 5\,\text{s}^{-1}$, both equations yield nearly equal values of the shear stress and the boundary between the two zones has been taken simply as the inter-section of the two equations. Note that if the extrapolation is based on the second equation, it will over-estimate the value of shear stress at shear rates below $5\,\text{s}^{-1}$ whereas the first equation will also over-predict shear stress at shear rates above $5\,\text{s}^{-1}$.

The fitting of the data to the Ellis model, equation (1.15), however, needs a non-linear regression approach to minimise the sum of squares. The best values are

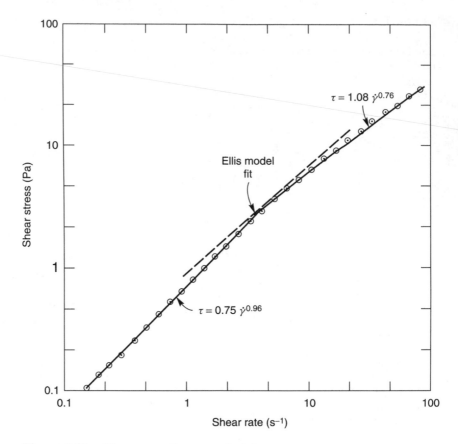

Figure 1.21   *Flow curve for example 1.1*

found to be

$$\mu_0 = 0.79\,\text{Pa·s} \quad \tau_{1/2} = 21.55\,\text{Pa} \quad \text{and} \quad \alpha = 2.03.$$

The predications of this model are also plotted in Figure 1.21 where a satisfactory fit can be seen to exist.□

This chapter is concluded by providing a list of materials displaying a spectrum of non-Newtonian flow characteristics in diverse applications to give an idea of the ubiquitous nature of such flow behaviour (Table 1.2).

Similarly, since much has been written about the importance of the measurement of rheological data in the same range of shear or deformation rates as those likely to be encountered in the envisaged application, Table 1.3 gives typical orders of magnitudes for various processing operations in which non-Newtonian fluid behaviour is likely to be significant.

Table 1.2   *Some common non-Newtonian characteristics*

Practical fluid	*Characteristics*	*Consequence of non-Newtonian behaviour*
Toothpaste	Bingham Plastic	Stays on brush and behaves more liquid like while brushing
Drilling muds	Bingham Plastic	Good lubrication properties and ability to convey derbris
Non-drip paints	Thixotropic	Thick in the tin, thin on the brush
Wallpaper paste	Pseudoplastic and Visco-elastic	Good spreadability and adhesive properties
Egg White	Visco-elastic	Easy air dispersion (whipping)
Molten polymers	Visco-elastic	Thread-forming properties
'Bouncing Putty'	Visco-elastic	Will flow if stretched slowly, but will bounce (or shatter) if hit sharply
Wet cement aggregates	Dilatant	Permit tamping operations in which small impulses produce almost complete settlement
Printing inks	Pseudoplastic	Spread easily in high speed machines yet do not run excessively at low speeds.

Table 1.3   *Shear rates typical of some familiar materials and processes [Barnes et al., 1989]*

Situation	*Typical range of shear rates (s^{-1})*	*Application*
Sedimentation of fine powders in a suspending liquid	$10^{-6} - 10^{-4}$	Medicines, paints

continued overleaf

Table 1.3   *(continued)*

Situation	Typical range of shear rates (s^{-1})	Application
Levelling due to surface tension	$10^{-2} - 10^{-1}$	Paints, printing inks
Draining under gravity	$10^{-1} - 10^{1}$	Painting and coating, Toilet bleaches
Extruders	$10^{0} - 10^{2}$	Polymers
Chewing and swallowing	$10^{1} - 10^{2}$	Foods
Dip coating	$10^{1} - 10^{2}$	Paints, confectionary
Pouring	$10^{1} - 10^{2}$	Pharmaceutical formulations
Mixing and stirring	$10^{1} - 10^{3}$	Manufacturing liquids
Pipe flow	$10^{0} - 10^{3}$	Pumping, Blood flow
Spraying and brushing	$10^{3} - 10^{4}$	Spray drying, painting, fuel atomisation, spraying of aerosols
Spreading and coating	$10^{3} - 10^{4}$	Application of nail polish and lipsticks
Rubbing	$10^{4} - 10^{5}$	Application of creams and lotions to the skin
Milling pigments in fluid bases	$10^{3} - 10^{5}$	Paints, printing inks
High speed coating	$10^{5} - 10^{6}$	Paper coating
Lubrication	$10^{3} - 10^{7}$	Gasoline engines

1.7 Further Reading

Barnes, H.A., Hutton, J.F. and Walters, K., *An Introduction to Rheology*. Elsevier, Amsterdam (1989).

Bird, R.B., Armstrong, R.C. and Hassager, O., *Dynamics of Polymeric Liquids*, Vol. 1, 2nd edn. Wiley, New York (1987).

Carreau, P.J., Dekee, D. and Chhabra, R.P., *Rheology of Polymeric Systems: Principles and Applications*. Hanser, Munich (1997).

Laba, D. (ed.) *Rheological Properties of Cosmetics and Toiletemes*, Marcel-Dekker, New York (1993).

Larson, R.G., *The structure and Rheology of Complex Fluid*, Oxford University Press, New York (1998).

Macosko, C.W., *Rheology: Principles, Measurements and Applications*, VCH, Munich (1994).

Steffe, J.F., *Rheological Methods in Food Process Engineering*, Freeman Press, East Lansing, Mi (1992).

Tanner, R.I. and Walters, K., *Rheology: An Historical Perspective*, Elsevier, Amsterdam (1998).

1.8 References

Astarita, G., *J. Rheol.* **34** (1990) 275; *ibid* **36** (1992) 1317.
Barnes, H.A., *J. Rheol.* **33** (1989) 329.
Barnes, H.A., *J. Non-Newt. Fluid Mech.* **70** (1997) 1.
Barnes, H.A., *J. Non-Newt. Fluid Mech.*, **81** (1999) 133.
Barnes, H.A., Edwards, M.F. and Woodcock, L.V., *Chem. Eng. Sci.*, **42** (1987) 591.
Barnes, H.A., Hutton, J.F. and Walters, K., *An Introduction to Rheology.* Elsevier, Amsterdam (1989).
Barnes, H.A. and Walters, K., *Rheol. Acta* **24** (1985) 323.
Bird, R.B., *Annu. Rev. Fluid Mech.* **8** (1976) 13.
Bird, R.B., Armstrong, R.C. and Hassager, O., *Dynamics of Polymeric Liquids, Vol. 1, Fluid Dynamics*, 2nd edn. Wiley, New York (1987).
Bird, R.B., Dai, G.C. and Yarusso, B.J., *Rev. Chem. Eng.* **1** (1983) 1.
Bird, R.B., Stewart, W.E. and Lightfoot, E.N., *Transport Phenomena.* Wiley, New York (1960).
Bird, R.B. and Wiest, J.M., *Annu. Rev. Fluid Mech.* **27** (1995) 169.
Boersma, W.H., Levan, J. and Stein, H.N. *AIChEJ.* **36** (1990) 321.
Boger, D.V., *J. Non. Newt. Fluid Mech.* **3** (1976) 87.
Boger, D.V., *Nature* **265** (1977) 126.
Boger, D.V. and Walters, K., *Rheological Phenomena in Focus.* Elsevier, Amsterdam (1992).
Carreau, P.J., *Trans. Soc. Rheol.* **16** (1972) 99.
Carreau, P.J., Dekee, D. and Chhabra, R.P., *Rheology of Polymeric Systems: Principles and Applications.* Hanser, Munich (1997).
Evans, I.D., *J. Rheol.* **36** (1992) 1313.
Freundlich, H. and Julisberger, F., *Trans. Faraday Soc.* **31** (1935) 920.
Goddard, J.D. and Bashir, Y., *in Recent Developments in Structured Continua II.* Longman, London (1990) Chapter 2.
Govier, G.W. and Aziz, K., *The Flow of Complex Mixtures in Pipes.* Krieger, Malabar, FL (1982).
Grimm, R.J., *AIChEJ.* **24** (1978) 427.
Gupta, R.K. and Sridhar, T., *in Rheological Measurements* (edited by Collyer A.A. and Clegg, D.W., Elsevier, Amsterdam) (1988) 211.
James, D.F. and Walters, K., *in Techniques in Rheological Measurements* (edited by Collyer, A.A., Elsevier, Amsterdam) (1994).
Jones, D.M., Walters, K. and Williams, P.R., *Rheol. Acta* **26** (1987) 20.
Keller, D.S. and Keller, Jr., D.V., *J. Rheol.* **34** (1990) 1267.
Kulicke, W.-M. and Wallabaum U., *Chem. Eng. Sci.* **40** (1985) 961.
Larson, R.G., *Constitutive Equations for Polymer Melts and Solutions* Butterworth-Heinemann, Stoneham, MA (1988).
Leider, P.J. and Bird, R.B., *Ind. Eng. Chem. Fundam.* **13** (1972) 342.
Macosko, C.W., *Rheology: Principles, Measurements and Applications.* VCH, Munich (1994).
Metzner, A.B., White J.L. and Denn, M.M., *AIChEJ.* **12** (1966) 863.
Metzner, A.B. and Whitock, M., *Trans. Soc. Rheol.* **2** (1958) 239.
Mewis, J., *J. Non-Newt. Fluid Mech.* **6** (1979) 1.
Nguyen, Q.D. and Uhlherr, P.H. T., *Proc. 3rd Nat. Conf. on Rheol.* Melbourne, Australia (1983) 63.
Pradipasena, P. and Rha, C., *J. Texture Studies.* **8** (1977) 311.
Prilutski, G., Gupta, R.K., Sridhar, T. and Ryan, M.E., *J. Non-Newt. Fluid Mech.* **12** (1983) 233.
Schowalter, W.R., *Mechanics of non-Newtonian Fluids.* Pergamon, Oxford (1978).
Schurz, J., *Rheol. Acta.* **29** (1990) 170.
Steg, I. and Katz, D., *J. Appl. Polym. Sci.* **9** (1965) 3177.
Tanner, R.I., *Engineering Rheology.* Oxford University Press, Oxford (1988).
Tanner, R.I. and Walters, K., Rheology: An Historical Perspective, Elsevier, Amsterdam (1998).
Tirtaatmadja, V. and Sridhar, T., *J. Rheol.* **37** (1993) 1081.

Trouton, F.T., *Proc. Roy. Soc.* **A77** (1906) 426.
Walters, K., *Rheometry*. Chapman and Hall, London (1975).
Weissenberg, K., *Nature* **159** (1947) 310.

1.9 Nomenclature

		Dimensions in **M, L, T**
A	area (m^2)	$\mathbf{L}^2$
De	Deborah number ($-$)	$\mathbf{M}^0\mathbf{L}^0\mathbf{T}^0$
F	force (N)	$\mathbf{MLT}^{-2}$
G	Young's modulus (Pa)	$\mathbf{ML}^{-1}\mathbf{T}^{-2}$
m	power-law consistency coefficient (Pa·s^n)	$\mathbf{ML}^{-1}\mathbf{T}^{n-2}$
m_1	power-law consistency coefficient for first normal stress difference (Pa·s^{p_1})	$\mathbf{ML}^{-1}\mathbf{T}^{p_1-2}$
N_1	first normal stress difference (Pa)	$\mathbf{ML}^{-1}\mathbf{T}^{-2}$
N_2	second normal stress difference (Pa)	$\mathbf{ML}^{-1}\mathbf{T}^{-2}$
n	power-law index ($-$)	$\mathbf{M}^0\mathbf{L}^0\mathbf{T}^0$
p	pressure (Pa)	$\mathbf{ML}^{-1}\mathbf{T}^{-2}$
P	total normal stress (Pa)	$\mathbf{ML}^{-1}\mathbf{T}^{-2}$
p_1	power law index for first normal stress difference ($-$)	$\mathbf{M}^0\mathbf{L}^0\mathbf{T}^0$
T_R	Trouton ratio ($-$)	$\mathbf{M}^0\mathbf{L}^0\mathbf{T}^0$
V	velocity (m/s)	$\mathbf{LT}^{-1}$
x, y, z	coordinate system (m)	$\mathbf{L}$

Greek letters

α	fluid parameter in Ellis fluid model ($-$)	$\mathbf{M}^0\mathbf{L}^0\mathbf{T}^0$
γ	strain ($-$)	$\mathbf{M}^0\mathbf{L}^0\mathbf{T}^0$
$\dot{\gamma}$	shear rate (1/s)	$\mathbf{T}^{-1}$
$\dot{\varepsilon}$	rate of extension (s^{-1})	$\mathbf{M}^0\mathbf{L}^0\mathbf{T}^{-1}$
λ	fluid parameter in Carreau viscosity equation or characteristic time in Maxwell model (s)	$\mathbf{T}$
μ	Newtonian or apparent viscosity (Pa·s)	$\mathbf{ML}^{-1}\mathbf{T}^{-1}$
ρ	fluid density (kg/m^3)	$\mathbf{ML}^3$
τ	component of stress tensor (Pa)	$\mathbf{ML}^{-1}\mathbf{T}^{-2}$
ϕ	volume fraction ($-$)	$\mathbf{M}^0\mathbf{L}^0\mathbf{T}^0$
ψ_1	first normal stress difference coefficient (Pa·s^2)	$\mathbf{ML}^{-1}$
ψ_2	second normal stress difference coefficient (Pa·s^2)	$\mathbf{ML}^{-1}$

Subscripts/superscripts

B	pertaining to Bingham fluid model
C	relating to Casson fluid model
E	extensional
H	relating to Herschel−Bulkley fluid model
0	zero shear
xx, yy, zz	normal stress components
$\left.\begin{array}{l} xy, yz, zx, \\ yx, zy, xz \end{array}\right\}$	shear stress components
∞	infinite shear

Chapter 2

Rheometry for non-Newtonian fluids[*]

2.1 Introduction

The rheological characterisation of non-Newtonian fluids is widely acknowledged to be far from straightforward. In some non-Newtonian systems, such
as concentrated suspensions, rheological measurements may be complicated
by non-linear, dispersive, dissipative and thixotropic mechanical properties;
and the rheometrical challenges posed by these features may be compounded
by an apparent yield stress.

For non-Newtonian fluids, even the apparently simple determination of a
shear rate versus shear stress relationship is problematical as the shear rate
can only be determined directly if it is constant (or nearly so) throughout the
measuring system employed. While very narrow shearing gap coaxial cylinder
and cone-and-plate measuring geometries provide good approximations to this
requirement, such systems are often of limited utility in the characterisation
of non-Newtonian products such as suspensions, whose particulate/aggregate
constitutents preclude the use of narrow gaps. As most measuring geometries
do not approximate to constant shear rate, various measurement strategies have
been devised to overcome this limitation. The basic features of these rheometrical approaches, and of the main instrument types for their implementation,
are considered below.

2.2 Capillary viscometers

Capillary viscometers are the most commonly used instruments for the measurement of viscosity due, in part, to their relative simplicity, low cost and (in
the case of long capillaries) accuracy. However, when pressure drives a fluid
through a pipe, the velocity is a maximum at the centre: the velocity gradient or
shear rate $\dot{\gamma}$ are a maximum at the wall and zero in the centre of the flow. The
flow is therefore non-homogeneous and capillary viscometers are restricted to
measuring *steady* shear functions, i.e. steady shear stress–shear rate behaviour
for time independent fluids [Macosko 1994]. Due to their inherent similarity to

[*] This chapter has been written by Dr. P.R. Williams, Reader, Department of Chemical and
Biological Process Engineering, University of Wales Swansea.

many process flows, which typically involve pipes, capillary viscometers are widely employed in process engineering applications and are often converted or adapted (with relative ease) to produce slit or annular flows.

2.2.1 Analysis of data and treatment of results

It is convenient to consider the case of an ideal capillary viscometer involving a fluid flowing slowly (laminar) and steadily through a long tube of radius R, at a constant temperature, and with a pressure drop $(-\Delta p)_t$ between its ends [Whorlow, 1992]. Then, for fully developed flow, the following relationship may be derived relating the shear stress at the wall of the tube, τ_w, to the volume of liquid flowing per second through any cross-section, Q, and the shear stress τ (see equation (2.1)):

$$\frac{Q}{\pi R^3} = \frac{1}{\tau_w^3} \int_0^{\tau_w} \tau^2 f(\tau) \, d\tau \tag{2.1}$$

Here $\tau_w = (R/2)(-\Delta p/L)$ where $(-\Delta p/L)$ is the magnitude of the pressure drop per unit length of tube (the pressure gradient) and the shear stress τ at any radius r is $(r/2)(-\Delta p/L)$. A graph of $Q/\pi R^3$ vs. τ_w gives a unique line, for a given material, for all values of R and $(-\Delta p/L)$.

For a Newtonian fluid, with $\dot{\gamma} = f(\tau) = \tau/\mu_N$, equation (2.1) yields the Poiseuille equation,

$$Q = \frac{\pi R^4 \left(\dfrac{-\Delta p}{L} \right)}{8\mu_N} \tag{2.2}$$

from which the viscosity μ_N can be calculated using a value of Q obtained for a single value of $(-\Delta p/L)$, and the shear rate at the tube wall is $4Q/\pi R^3$ or $(8V/D)$.

Turning to the most commonly-used model approximations to non-Newtonian flow behaviour, the following relationships are obtained for the power-law model (in the form $\mu = m\dot{\gamma}^{n-1}$), written at $r = R$:

$$\tau_w = \frac{R}{2} \left(\frac{-\Delta p}{L} \right) = m \left\{ \left(\frac{3n+1}{4n} \right) \frac{8V}{D} \right\}^n \tag{2.3}$$

and, for the Bingham model (in the form $\dot{\gamma} = f(\tau) = (\tau - \tau_0^B)/\mu_B$) of we obtain,

$$\left(\frac{8V}{D} \right) = \frac{4}{\mu_B} \left[\frac{\tau_w}{4} - \frac{\tau_0^B}{3} \left\{ 1 - \frac{1}{4} \left(\frac{\tau_0^B}{\tau_w} \right)^3 \right\} \right] \tag{2.4}$$

where μ_B is the 'plastic viscosity'.

For flow curves of *unknown* form, equation (2.1) yields (after some manipulation, see Section 3.2.5):

$$f(\tau_w) = \left(\frac{8V}{D}\right)\left\{\left(\frac{3}{4}\right) + \frac{\mathrm{d}(\log 8V/D)}{4\mathrm{d}(\log \tau_w)}\right\} \tag{2.5}$$

Various forms of this equation are used, a common form (often termed the Weissenberg–Rabinowitsch or Rabinowitsch–Mooney equation) being,

$$\dot{\gamma}_w = \dot{\gamma}_{wN}\left(\frac{3n'+1}{4n'}\right) \tag{2.6}$$

where $n' = \mathrm{d}(\log \tau_w)/\mathrm{d}(\log \dot{\gamma}_{wN})$, $\dot{\gamma}_w$ is the shear rate at the wall and $\dot{\gamma}_{wN} = (8V/D)$ is a nominal shear rate obtained by using the formula appropriate to the Newtonian fluid.

For shear-thinning fluids, the apparent shear rate at the wall is less than the true shear rate, with the converse applying near the centre of the tube [Laun, 1983]. Thus at some radius, x^*R, the true shear rate of a fluid of apparent viscosity μ equals that of a *Newtonian* fluid of the same viscosity. The stress at this radius, $x^*\tau_w$, is independent of fluid properties and thus the true viscosity at this radius equals the apparent viscosity at the wall and the viscosity calculated from the Poiseuille equation (equation (2.2)) is the true viscosity at a stress $x^*\tau_w$. Laun [1983] reports that this 'single point' method for correcting viscosity is as accurate as the Weissenberg–Rabinowitsch method.

Serious errors may be incurred due to wall slip, e.g. in the case of concentrated dispersions where the layer of particles may be more dilute near the wall than in the bulk flow: the thin, dilute layer near the wall has a much lower viscosity, resulting in an apparent slippage of the bulk fluid along the wall.

The occurrence of this phenomenon may be tested by comparing the viscosity functions obtained using capillaries of similar length-to-radius ratios, L/R, but of different radii. Any apparent wall slip may then be corrected for and the true viscosity of the fluid determined by extrpolating the results obtained to infinite pipe diameter. In the relation developed by Mooney [1931], apparent wall shear rates obtained for constant length-to-radius ratio are plotted against (L/R).

Departures from ideal flow near either end of the pipe (end effects) may be eliminated by considering the total pressure drop $(-\Delta p)_t$, between two points beyond opposite ends of a uniform bore tube in terms of $(-\Delta p/L)$, the pressure gradient in the central part of the tube of length L and an end correction, ε. The latter may be very large for non-Newtonian fluids. Using tubes of different length, a plot of $(-\Delta p)_t$ versus L yields the value of $(-\Delta p/L)$ if the upper part of the plot is linear, and ε may be found by extrapolation to zero L. Note that in applying this procedure to polymer systems it is conventional to plot pressure against L/R to construct the so-called Bagley plot [Bagley 1957].

As the construction of the Bagley plot requires considerable experimental effort, in practice a single long ($L/R \geq 60$) capillary is usually deemed to provide accurate results on the assumption that all corrections may be safely ignored. This assumption is not always warranted and the reader is referred to other texts (see 'Further Reading') for details concerning additional sources of error, which may require consideration of kinetic energy corrections, pressure dependence of viscosity, thixotropy, viscous heating, compressibility and, in the case of melts, sample fracture. A typical arrangement of a pressurised capillary viscometer is shown in Figure 2.1.

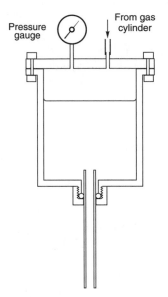

Figure 2.1 *A pressurised capillary viscometer*

Example 2.1

The following capillary viscometer data on a high pressure polyethylene melt at 190°C have been reported in the literature [A.P. Metzger and R.S. Brodkey, *J. Appl. Polymer Sci.*, 7 (1963) 399]. Obtain the true shear stress–shear rate data for this polymer. Assume the end effects to be negligible.

$\left(\dfrac{8V}{D}\right)$ (s^{-1})	τ_w (kPa)
10	22.4
20	31
50	43.5
100	57.7
200	75
400	97.3
600	111
1000	135.2
2000	164

Solution

From equation (2.5):

$$f(\tau_w) = \dot{\gamma}_w = \left\{ \frac{3n'+1}{4n'} \cdot \frac{8V}{D} \right\}$$

$$\text{and } \tau_w = \frac{R}{2} \left(\frac{-\Delta p}{L} \right)$$

where τ_w is the true shear stress at the wall irrespective of the type of fluid behaviour, whereas $(8V/D)$ is the corresponding shear rate at the wall only for Newtonian fluids. The factor $((3n'+1)/4n')$ corrects the nominal wall shear rate $(8V/D)$ for the non-Newtonian fluid behaviour.

$$\text{where} \quad n' = \frac{d \log \tau_w}{d \log(8V/D)}$$

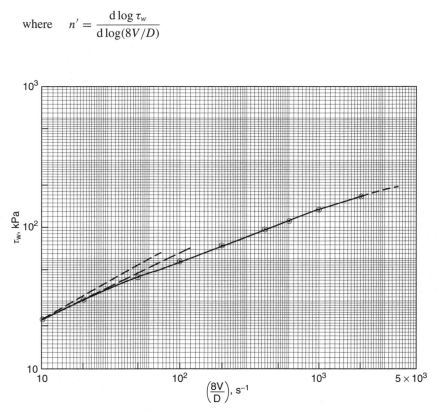

Figure 2.2 *Rheological data of Example 2.1*

Thus, the given data is first plotted on log-log coordinates as shown in Figure 2.2 and the value of n' is evaluated at each point (value of $8V/D$), as summarised in Table 2.1.

Table 2.1   *Summary of Calculations*

τ_w (kPa)	$\left(\dfrac{8V}{D}\right)$ (s^{-1})	$n' = \dfrac{d\log\tau_w}{d\log(8V/D)}$	$\dot{\gamma}_w = \left(\dfrac{3n'+1}{4n'}\right)\left(\dfrac{8V}{D}\right)$ (s^{-1})
22.4	10	0.50	12.5
31	20	0.47	25.6
43.5	50	0.43	66.6
57.7	100	0.42	135
75	200	0.40	275
97.3	400	0.36	578
111	600	0.34	891
135.2	1000	0.31	1556
164	2000	0.31	3113

As can be seen, the value of the correction factor $(3n' + 1)/(4n')$ varies from 25% to 55.6%. Thus, the values of $(\tau_w, \dot{\gamma}_w)$ represent the true shear stress–shear rate data for this polymer melt which displays shear-thinning behaviour as can be seen from the values of $n' < 1$. □

2.3 Rotational viscometers

Due to their relative importance as tools for the rheological characterisation of non-Newtonian fluid behaviour, we concentrate on this class of rheometers by considering the two main types, namely; the controlled shear rate instruments (also known as controlled *rate* devices) and controlled *stress* instruments. Both types are usually supplied with the same range of measuring geometries, principally the concentric cylinder, cone-and-plate and parallel plate systems. The relative merits, potential drawbacks, working equations and other formulae associated with these designs have been described in great detail elsewhere (e.g., see Walters, 1975; Whorlow, 1992; Macosko, 1994) and so only their most basic aspects are covered here.

2.3.1 The concentric cylinder geometry

It is appropriate to begin by considering this geometry as it was the basis of the first practical rotational rheometer. Ideally the sample is contained in a *narrow* gap between two concentric cylinders (as shown in Figure 2.3). Typically the outer 'cup' rotates and the torque T on the inner cylinder, which is usually suspended from a torsion wire or bar, is measured.

Working equations relating the measured torque to the requisite shear stress, and angular velocity (of the cup) to the requisite shear rate, are widely available

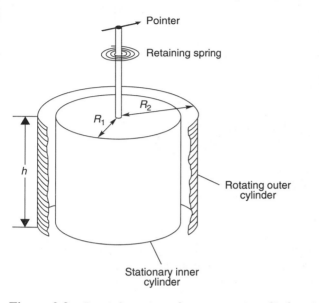

Figure 2.3   *Partial section of a concentric-cylinder viscometer*

along with their derivations (see the list of references above). It is noteworthy that the working formulae quoted in many instances ignore the curvature of the surfaces of the measuring geometry. The determination of the shear stress and shear rate within the shearing gap is thus valid only for *very* narrow gaps wherein κ, the ratio of inner to outer cylinder radii, is >0.99.

Several designs have been described which overcome end-effects due to the shear flow at the bottom of the concentric cylinder geometry. These include the recessed bottom system which usually entails trapping a bubble of air (or a low viscosity liquid such as mercury) beneath the inner cylinder of the geometry. Alternatively the 'Mooney-Ewart' design, which features a conical bottom may, with suitable choice of cone angle, cause the shear rate in the bottom to match that in the narrow gap between the sides of the cylinders, see Figure 2.4.

In this example the sample temperature is controlled by circulation of liquid through the outer cylinder housing (flow marked in and out in Figure 2.4) and h denotes the sample height within the shearing gap. The shear rate may be calculated from

$$\dot{\gamma} = \frac{R_2 \Omega}{R_2 - R_1} \tag{2.7a}$$

where R_2 and R_1 are the outer and inner cylinder radii respectively, and Ω is the angular velocity. For $k > 0.99$, the shear stress is given by:

$$\tau = T/2\pi R_1^2 h \tag{2.7b}$$

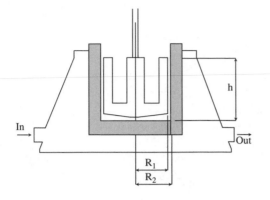

Figure 2.4   *The Mooney-Ewart geometry*

To minimise end-effects the lower end of the inner cylinder is a truncated cone. The shear rate in this region is equal to that between the cylinders if the cone angle, α, is related to the cylinder radii by:

$$\alpha = \tan^{-1} \frac{R_2 - R_1}{R_2} \qquad (2.8)$$

The main sources of error in the concentric cylinder type measuring geometry arise from end effects (see above), wall slip, inertia and secondary flows, viscous heating effects and eccentricities due to misalignment of the geometry [Macosko, 1994].

Secondary flows are of particular concern in the controlled stress instruments which usually employ a rotating *inner* cylinder, in which case inertial forces cause a small axisymmetric cellular secondary motion ('Taylor' vortices). The dissipation of energy by these vortices leads to overestimation of the torque. The stability criterion for a Newtonian fluid in a narrow gap is

$$Ta = \rho^2 \Omega^2 (R_2 - R_1)^3 R_1 / \mu^2 < 3400 \qquad (2.9)$$

where Ta is the 'Taylor' number.

In the case of non-Newtonian polymer solutions (and narrow gaps) the stability limit increases. In situations where the outer cylinder is rotating, stable Couette flow may be maintained until the onset of turbulence at a Reynolds number, Re, of *ca.* 50 000 where $Re = \rho \Omega R_2 (R_2 - R_1) / \mu$ [Van Wazer *et al.*, 1963].

2.3.2 The wide-gap rotational viscometer: determination of the flow curve for a non-Newtonian fluid

An important restriction on the use of the concentric cylinder measuring geometry for the determination of the shear rate versus shear stress relationship for a non-Newtonian fluid is the requirement, noted above, for a narrow shearing gap between the cylinders. As indicated in the introduction to this chapter, direct measurements of shear rates can only be made if the shear rate is constant (or very nearly so) throughout the shearing gap but many coaxial measuring systems do not fulfill this requirement. In addition, many (if not most) non-Newtonian fluid systems, particularly those of industrial or commercial interest such as pastes, suspensions or foods, may contain relatively large particles, or aggregates of particles. Thus the requisite shearing gap size to ensure that adequate bulk measurements are made, i.e., a gap size approximately 10–100 times the size of the largest 'particle' size [Van Wazer *et al.*, 1963], may conflict with the gap size required to ensure near constant shear rate, within the gap.

Procedures for extracting valid shear stress versus shear rate data from measurements involving wide gap coaxial cylinder systems (the Brookfield viscometer being an extreme example of wide gap devices) are therefore of considerable interest in making quantitative measurements of the flow properties of non-Newtonian process products. Most of these data-treatment procedures necessarily involve some assumption regarding the functional form of the flow curve of the material. One example is that made in the derivation of data from the Brookfield-type instrument, which assumes that the speed of rotation of the cylinder or spindle is proportional to the shear rate experienced by the fluid. This assumption implies that the flow curve is adequately described by a simple power-law (which for many shear-thinning non-Newtonian fluids may be acceptable), but this assumption is widely taken to *exclude* all fluids which display an apparent yield stress and/or non-power law type behaviour.

The starting point lies in considering the basic equation for the coaxial rotational viscometer, which has been solved by Krieger and co-workers for various sets of boundary conditions [Krieger and Maron, 1952]:

$$\Omega = \frac{1}{2} \int_{R_b}^{R_c} \frac{f(\tau)}{\tau} \, d\tau \qquad (2.10)$$

where Ω is the angular velocity of the spindle with respect to the cup, τ is the shear stress in the fluid at any point in the system, $f(\tau) = \dot{\gamma}$ is the rate of shear at the same point and the subscripts b and c refer to the bob and the cup, respectively.

The particular solution to equation (2.10) for a finite cylindrical bob rotating in an infinite cup can provide valuable quantitative rheological data for systems whose particulate constituents, and practical limitations on the size of the measuring geometry in terms of cylinder radius, preclude the use of conventional narrow gap geometries. The infinite cup boundary condition may be closely approximated by using a narrow cylindrical *spindle* (such as are supplied with instruments of the Brookfield type) in place of the more commonly used bob.

Assuming the infinite-cup boundary condition, τ_c (shear stress on the cup) in equation (2.10) becomes equal to zero and the expression may be differentiated with respect to τ_b giving:

$$\dot\gamma_b = f(\tau_b) = -2\,d\Omega/d\ln\tau_b = -2\tau_b\,d\Omega/d\tau_b \tag{2.11}$$

and thus the rate of shear may be obtained by evaluating (graphically) either of the derivatives on the right-hand side of equation (2.11).

The derivation of equation (2.11) assumes that a cup of infinite radius is filled with *fluid*. Implicitly this would exclude all systems which display a yield stress, as such systems would not behave as a fluid for values of stress below the yield value.

As many non-Newtonian systems are sufficiently 'structured' to display an apparent yield stress, this requirement would appear to severely restrict the application of what would otherwise appear to be a very useful technique. However, on closer inspection, it has been shown that for a fluid which displays a yield stress, a more general derivation than that reported by Krieger and Maron [1952] may be obtained, and that the restriction of infinite outer boundary (i.e. cup) radius may in fact be eliminated [Jacobsen, 1974].

In a system which displays yield stress behaviour, the integral in the general expression for the rate of shear need not be evaluated from the bob all the way to the cup. This is due to the fact that, for such a system, no shearing takes place where τ is less than the yield value, τ_0. Thus the integral need only be evaluated from the bob to the critical radius, R_{crit}, the radius at which $\tau = \tau_0$. This gives

$$\Omega = 1/2 \int_{R_b}^{R_{\text{crit}}} f(\tau)\,d\tau/\tau \tag{2.12}$$

where the 'critical' radius, R_{crit}, is given as:

$$R_{\text{crit}} = R_b(\tau_b/\tau_0)^{1/2} \tag{2.13}$$

This derivation relies on the fact that the condition of differentiability is not that one limit of the integral be zero (as is the case in the infinite cup solution) but that one limit be *constant*. Thus, for systems which may be described in terms of a constant value of yield stress, equation (2.12) may be differentiated, giving:

$$\dot\gamma_b = f(\tau_b) = -2\,d\Omega/d\ln\tau_b = -2\tau_b\,d\Omega/d\tau_b \tag{2.14}$$

i.e. exactly the same result is obtained as that derived for the case of the infinite cup, equation (2.11).

In practice, shear stress data are plotted against Ω and the slopes ($d\Omega/d\tau_b$) are taken at each point. Given that the graphical solution may be somewhat tedious, and that a rapid evaluation of the general form of the flow curve is often all that is required (e.g. in a product 'quality control' context), the form of the Ω versus τ_b plots is sometimes taken as giving the general form of the corresponding $\dot{\gamma}$ versus τ_b curve (although, of course, the curves will differ quantitatively). In the absence of an apparent yield stress (over the experimental time-scale) the general character of the $\dot{\gamma}$ versus τ_b curves may sometimes be correctly inferred by this procedure: the situation is quite different when the system exhibits an apparent yield stress and this situation poses a trap for the unwary.

An examination of equation (2.14) shows that for any fluid with a finite yield point, the Ω versus τ_b curve approaches the τ_b axis at zero slope, due to the requirement for such a system that the shear rate must become zero at finite τ_b. This may lead to apparent shear-thinning characteristics being ascribed to systems, *irrespective of the actual form of their flow curves above the yield point*, i.e., whether Bingham plastic, shear-thickening (with a yield stress), or shear-thinning (with a yield stress).

An instrument called the 'rotating disk indexer' (*sic.*) is also widely used in quality control applications and involves a rotating disc in a 'sea' of fluid. Williams [1979] has described a numerical method for obtaining true $\mu - \dot{\gamma}$ data with this device.

2.3.3 The cone-and-plate geometry

In the cone-and-plate geometry, the test sample is contained between an upper rotating cone and a stationary flat plate (see Figure 2.5, upper). In the example shown, the cone is 40 mm in diameter, with a cone angle of $1°\ 59'$ relative to the plate, and a truncation of 51 μm.

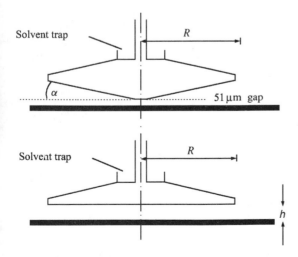

Figure 2.5   *Cone-and-plate (upper) and parallel plate (lower) geometries*

The small cone angle ($<4°$) ensures that the shear rate is constant throughout the shearing gap, this being of particular advantage when investigating time-dependent systems because all elements of the sample experience the same shear history, but the small angle can lead to serious errors arising from eccentricities and misalignment.

The small gap size dictates the practical constraints for the geometry: a gap-to-maximum particle (or aggregate) size ratio of >100 is desirable to ensure the adequate measurement of bulk material properties. This geometry is, therefore, limited to systems containing small particles or aggregates, and the strain sensitivity is fixed. Normal stress differences may be determined from pressure and thrust measurements on the plate.

The form factors for the cone-and-plate geometry are as follows:
Shear stress:

$$\tau = \frac{3T}{2\pi R^3} \qquad (2.15)$$

Shear rate:

$$\dot{\gamma} = \frac{\Omega}{\tan \alpha} \qquad (2.16)$$

where R is the radius of the cone (m), T, the torque (Nm), Ω, the angular velocity (rad/s) and α, the cone angle (rad).

The influence of geometry misalignment and other factors, such as flow instabilities arising from fluid elasticity, have been extensively studied in the case of this geometry [Macosko, 1994]. Unlike the concentric cylinder geometry, where fluid inertia causes a depression around the inner cylinder rather than the well-known 'rod-climbing' effect due to visco-elastic normal stresses, in the cone-and-plate geometry the effect of inertia is to draw the plates together, rather than push them apart [Walters, 1975].

Many experimentalists employ a 'sea' of liquid around the cone (often referred to as a 'drowned edge'), partly in an attempt to satisfy the requirement that the velocity field be maintained to the edge of the geometry.

2.3.4 The parallel plate geometry

In this measuring geometry the sample is contained between an upper rotating or oscillating flat stainless steel plate and a lower stationary plate (see Figure 2.5, lower). The upper plate in the example shown is 40 mm in diameter. In contrast to the cone-and-plate geometry, the shear strain is proportional to the gap height, h, and may be varied to adjust the sensitivity of shear rate, a feature which readily facilitates testing for wall (slip) effects [Yoshimura and Prud'homme, 1988].

The large gap sizes available can be used to overcome the limitations encountered using the cone-and-plate geometry, such as its sensitivity to eccentricities and misalignment. However, it should be borne in mind that, as in the case of the wide-gap Couette devices, shear rate is not constant. Usually the strain reported is that measured at the outer rim, which provides a maximum value of the spatially varying strain within the gap.

Loading and unloading of samples can often prove easier than in the cone-and-plate or concentric cylinder geometries, particularly in the case of highly viscous liquids or 'soft solids' such as foods, gels etc. The parallel plate geometry is particularly useful for obtaining apparent viscosity and normal stress data at high shear rates, the latter being increased either by increasing Ω or by decreasing the shearing gapsize. An additional benefit of the latter approach is that errors due to secondary flows, edge effects and shear heating may all be reduced.

Form factors for the parallel plate geometry, in terms of the apparent or Newtonian shear stress and the shear rate at $r = R$ are given below:

Shear stress:

$$\tau = \frac{2T}{\pi R^3} \qquad (2.17)$$

Shear rate:

$$\dot{\gamma} = \frac{\Omega R}{h} \qquad (2.18)$$

where h is the plate separation (m), Ω, is the angular velocity (rad/s), and R, is the plate radius (m). A full derivation of the working equations may be found elsewhere [e.g. Macosko, 1994].

2.3.5 Moisture loss prevention – the vapour hood

When dealing with high concentration samples of low volume, even low moisture loss can have a critical effect on measured rheological properties [Barnes *et al.*, 1989]. During prolonged experiments, moisture loss may be minimised by employing a vapour hood incorporating a solvent trap, as shown in Figure 2.6.

As noted above, edge effects can be encountered with each of the geometries considered here. They become of particular importance when dealing with samples which form a surface 'skin' in contact with the atmosphere, due mainly to evaporation. Conditions at the outer edge of the parallel plate and cone-and-plate geometries strongly influence the measured torque value. Stresses in this region act on a larger area and are operating at the greatest radius. To ensure homogeneous bulk sample conditions, the evaporation process at the sample surface may be minimised by employing a vapour hood, as shown in Figure 2.6 for the parallel plate system.

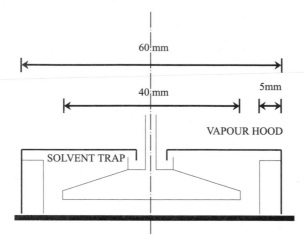

Figure 2.6   *Vapour hood employed with a parallel plate geometry*

2.4 The controlled stress rheometer

Since the mid 1980s and the advent of reliable 'second generation' controlled-stress rheometers, the controlled-stress technique has become widely established. The facility which most of this type of instrument offers, i.e. of performing three different types of test (steady shear, oscillation and creep), makes them particularly cost effective.

The instrument referred to here for illustration is a TA Instruments CSL 100 controlled-stress rheometer (TA Instruments, UK). The rheometer (typically operated under the control of a microcomputer) and ancillary equipment required for its operation, consist of the following main components (see Figure 2.7).

An electronically-controlled induction motor incorporates an air bearing, which supports and centres a rotating hollow spindle. The spindle incorporates a threaded draw rod, onto which the components of the required measuring geometry is secured and the air bearing prevents any contact between fixed and moving parts. A digital encoder consisting of a light source and a photocell is arranged either side of a transparent disc attached to the spindle. Fine lines (similar to diffraction grating lines) are photographically etched around the disc edge and, through the use of a stationary diffraction grating between the light source and the disc, diffraction patterns are set up as the disc moves under an applied torque. These are directly related to the angular displacement of the measuring system.

The non-rotating lower platen of the measuring assembly is fixed to a height-adjustable pneumatic ram which may be raised to provide the desired gap setting, with micrometer-fine adjustment. A temperature control unit is

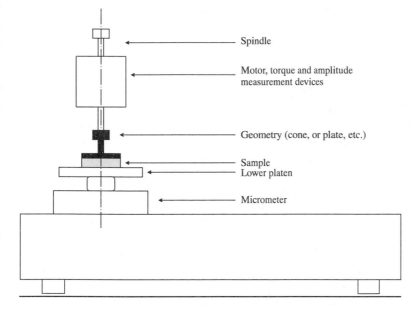

Figure 2.7 *Schematic of a TA Instruments CSL100/CSL² controlled-stress rheometer*

incorporated within the lower plate. This is usually of the peltier type, using a thermoelectric effect enabling it to function as a heat pump with no moving parts. Control of the magnitude and direction of the electrical current allows the desired temperature adjustment within the lower platen (control to 0.1°C), and thus within the sample, for cone-and-plate and parallel plate geometries. For the concentric cylinder geometry a temperature-controlled recirculating water bath is generally used.

Due in part to its ability to produce extremely low shear rates, the controlled stress technique has been found to be highly suited to the determination of apparent yield stress, and in this respect the controlled-stress instrument is widely claimed to be more successful than its controlled-shear rate-counterparts. This is usually attributed to the fact that, for suitably low stresses, the structure of the material may be preserved under the conditions of test. Indeed, the introduction of the 'second generation' of controlled-stress instruments can be said to have provoked considerable interest in, and debate surrounding, the field of yield stress determination, with some early advocates of the controlled-stress technique advancing the controversial notion of the 'yield stress myth' [Barnes and Walters, 1985].

Apart from the range of instruments described in the preceding sections, several of the inexpensive viscometers used for quality control in industry give rise to complicated flow and stress fields (which may be neither known nor uniform), but they have the great advantage that their operation is simple. In the case of Newtonian fluids, the use of such methods does not pose any problem, since the instruments can be calibrated against a standard Newtonian liquid of known viscosity. However, for non-Newtonian fluids, the analysis and interpretation of results obtained by using such devices is not simple and straightforward. Such devices can be broadly classified into two types. The first have what might loosely be called "flow constrictions", as exemplified by the Ford cup arrangement, in which the time taken for a fixed volume of liquid to drain through the constriction is measured. Such a device can cope with different ranges of viscosities by changing the size of the constriction. This robust and convenient instrument is used widely in the petroleum and oil industries. The second class of instruments involves the flow around an obstruction as in the falling ball and rolling ball methods [van Wazer *et al.*, 1963] where the time taken for the sphere to settle or roll through a known distance is measured. Although such "shop-floor" viscometers can perhaps be used for qualitative comparative purposes for purely inelastic fluids, great care needs to be exercised when attempting to characterise visco-elastic and time-dependent non-Newtonian materials even qualitatively [Barnes *et al.*, 1989; Chhabra, 1993].

2.5 Yield stress measurements

Notwithstanding the continuing debate over the very existence of a 'true' yield stress, the concept of an *apparent* yield stress has been found to be an extremely useful empiricism in many areas of science and engineering [Hartnett and Hu, 1989] (see also Chapter 1). A recent comprehensive review [Barnes, 1999] has critically assessed the various issues raised in the definition, measurement and application of apparent yield stress behaviour.

Any operational definition of apparent yield stress should take into account both the inevitable rheometrical limitations in its determination, and the characteristic time of the process to which it pertains. Such an *operational* definition has been proposed for a true yield stress in the context of the classical stress relaxation experiment [Spaans and Williams, 1995].

Notwithstanding the inherent advantages of the controlled-stress technique in yield studies, it should be borne in mind that an interpretation of the results of creep-compliance measurements in terms of a 'real' yield stress (i.e. a stress below which the sample exhibits Hookean elastic behaviour) is subject to the usual experimental limitations of machine resolution (i.e. of angular displacement) and the role of time-scale in the sample's response to applied stress.

Numerous workers have described the role of wall-slip effects on measurements made with conventional smooth-walled geometries [Barnes, 1995]. Slip can occur in suspensions at high (*ca.* 60%) solids volume fraction, and can involve fluctuating torque in a rotational viscometer under steady rotation [Cheng and Richmond, 1978; Cheng, 1986]. Given that conventional smooth walled rotational devices tend to slip when in contact with many systems which display an apparent yield stress, several workers have adopted the vane measuring geometry [Nguyen and Boger, 1985].

Typically, the vane geometry consists of a small number (usually 4) of thin blades arranged around a cylindrical shaft of small diameter. When the material in which the vane is immersed undergoes yielding, it does so within the body of the material, not at a solid boundary, thereby overcoming wall slip. An additional advantage of the vane is that on its insertion into a material there is minimal disturbance of the sample structure compared with that experienced within the narrow shearing gaps of conventional measuring geometries. This feature is important for mechanically weak structured systems, such as gels and colloidal dispersions.

Vane rheometry provides a direct measurement of the shear stress at which flow is evident under the conditions of test (i.e. within the time scale of the measurement). In a constant rate (CR) experiment the material is sheared at a low but constant rate according to the speed of rotation of the viscometer's spindle, and the corresponding torque response with time is recorded. With constant stress (CS) experiments it is the inferred *deformation* that is recorded as a function of time, under the application of a series of controlled and constant shear stresses.

The vane technique has its origins as a method for *in situ* measurements of the shear strength of soils and an important assumption in the method is that the yielding surface that results from the vane's rotation is cylindrical, and of the same diameter as the vane (corrections can be made at a later stage if this is proven otherwise but they are difficult to assess for opaque materials). This assumption dictates that the material between the blades acts as a solid cylinder of dimensions equal to those of the vane, and the issue of the yield surface of visco-elastic and plastic fluids in a vane-viscometer has been addressed in several studies (e.g. see Yan and James, 1997; Keentok *et al.*, 1985).

The maximum torsional moment (T_m) coincides with the material yielding along this cylindrical surface; thus, a torque balance at the point of yielding provides the following equation:-

Total torque = torque from the vane cylindrical shearing surface
              + torque from both vane end shearing surfaces

If τ_c is the shear stress on the cylindrical shearing surface, τ_e for each of the vane end surfaces, the torque balance can be written as:

$$T_m = (\pi D H) \left(\frac{D}{2}\right) \cdot \tau_c + 2 \int_0^{D/2} 2\pi \tau_e r^2 \, \mathrm{d}r \qquad (2.19)$$

where D is the diameter of the cylinder prescribed by the rotating vane (m); H is the vane height (m); and r is the radial distance at the vane ends (m).

To estimate the stress at the vane end surfaces it can be assumed that the stress distribution across the vane ends may either increase uniformly towards the edges or may follow a power-law expression:

(a) for a uniformly increasing stress distribution,

$$\tau_e = \left(\frac{2r}{D}\right) \cdot \tau_0 \tag{2.20}$$

(b) for a power-law stress distribution,

$$\tau_e = \left(\frac{2r}{D}\right)^s \tau_0 \tag{2.21}$$

The constant, s, is referred to as the power-law exponent and τ_0 is the yield stress.

At the point of yielding, the value for stress at the cylindrical shearing surface is τ_0. Therefore, equation (2.19) can be solved for the two cases given by equations (2.20) and (2.21):

(a) for a uniformly increasing stress distribution,

$$\frac{2T_m}{\pi \tau_0} = D^2 H + \frac{D^3}{4} \tag{2.22}$$

(b) for a power-law stress distribution,

$$\frac{2T_m}{\pi \tau_0} = D^2 H + \frac{D^3}{(s+3)} \tag{2.23}$$

For $H \gg D$ the contributions of stress from the vane ends becomes negligible. A height-to-diameter ratio, H/D, in excess of 3 is desirable, and it should be as large as is practical.

Construction of the vane with a very small shearing diameter would seem the ideal way of achieving a high $H : D$ ratio, but caution needs to be taken with small diameter vanes as it is likely that torque contributions from the resistance to rotation at the exposed shaft surfaces, above the vane, may then become important.

The vane has been used in conjunction with controlled-stress rheometers to determine apparent yield stresses in cohesive clay suspensions [James *et al.*, 1987]; a similar technique has been reported for time-independent materials [Yoshimura *et al.*, 1987]. It is important in this type of test that the material attains an equilibrium microstructural state prior to test. The time, t_e, required to achieve this may be assessed using information obtained from separate, small strain-experiments [James *et al.*, 1987].

In a typical controlled-stress vane measurement, the sample is allowed to equilibrate mechanically for a period $t > t_e$ following insertion of the vane. Subsequently a (constant) stress is applied and maintained for a period, then suddenly removed (see Figure 2.8).

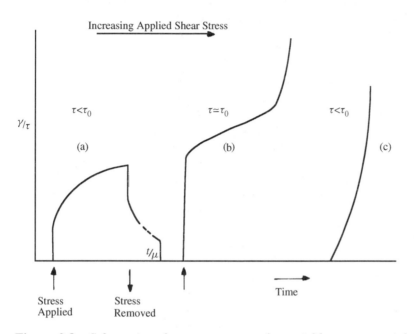

Figure 2.8 *Schematics of a creep-type test for a yield stress material*

The resulting time-dependent deformation response is recorded (in terms of the resulting angular displacement of the vane) both under applied stress, and following its removal. The material is then allowed to recover for a period $t > t_e$ and the process is repeated, at a slightly higher value of applied stress. By repeating this sequence, using gradually increased levels of stress, the elastic and viscous components of the deformation of a material may be studied from a state of virtually undisturbed structure, to the onset of an apparent yield behaviour (see Figure 2.8c).

The latter procedure is an adaptation of that usually employed to generate creep compliance data: a 'creep' yield stress is determined by extrapolating creep-rate data to zero rate, with the corresponding value of applied stress being taken as the yield stress [Lohnes *et al.*, 1972]. In addition to its use in yield stress measurement, the vane has found application as the basis of a viscometer for systems where severe cases of wall slip may be anticipated [Barnes and Carnali, 1990].

2.6 Normal stress measurements

Whorlow [1992] notes that, of the many methods which have been proposed for the measurement of various combinations of the first and second normal stress differences, N_1 and N_2 respectively, few can give reliable estimates of N_2. Combined pressure gradient and total force measurements in the cone-and-plate geometry, or combined cone-and-plate and plate-plate force measurements, appear to give reliable values [Walters, 1975]; and satisfactory results may also be obtained from techniques based on the measurement of the elevation of the surface of a liquid as it flows down an inclined open duct [Kuo and Tanner, 1974].

During rotational flow of liquids which display normal stresses, the tension along the circular streamlines is always greater than that in other directions so that streamlines tend to contract unless prevented from doing so by an appropriate pressure distribution. Determination of the pressure distribution (say, over the area of the plate in a cone-and-plate system) therefore provides a means of determining the total force exerted on the plate. Alternatively, and in practice more generally, the total force is measured directly. As the pressure distribution and total force measured in a (parallel) plate-plate system depend in a different manner on the normal stress differences, both cone-and-plate systems and plate-plate systems have been used to obtain values for both N_1 and N_2. Numerous modifications to instruments (such as the Weissenberg Rheogoniometer) to permit more accurate normal force measurements have been described, some involving the use of piezoelectric crystals as very stiff load cells [Higman, 1973].

In the work reported by Jackson and Kaye [1966] the spacing, h, between a cone and a plate was varied and the normal force measured as a function of gap size. The same method was used by Marsh and Pearson [1968] who showed that,

$$N_2(\dot{\gamma}_R) = \frac{h + R\theta}{h}\left(N_1(\dot{\gamma}_R) - \frac{2F}{\pi R^2} + \frac{h}{\pi R^2}\frac{dF}{dh}\right) \tag{2.24}$$

where the shear rate at the rim, $\dot{\gamma}_R$, is given by $\Omega R/(h + R\theta)$; θ is the cone angle. As the value of zero gap spacing is used to find N_1, equation (2.24) can only be used for non-zero values of h, and the geometry (being intermediate between that of the conventional cone-and-plate and parallel-plate systems) may not give as good estimates of N_2 as those obtained by other methods [Walters, 1975].

The direct determination of N_1 by measuring the total force on the cone in a cone-and-plate system is limited to low shear rates. Binding and Walters [1976] have described a *torsion balance rheometer* which can provide normal force and viscosity data on low-viscosity fluids at very high shear rates. This instrument is particularly useful on systems which display an apparent yield stress [Binding *et al.*, 1976]. Encouraging agreement has been reported by between the results of measurements made in a rheogoniometer, a torsion balance rheometer and a 'Stressmeter' [Lodge *et al.*, 1987], the latter instrument (which can provide data at extremely high shear rates) exploiting the 'hole pressure effect' identified by Broadbent *et al.* [1968].

For a detailed and systematic treatment of the various approaches to normal stress measurements, the reader is referred to the text by Walters [1975].

2.7 Oscillatory shear measurements

Of the techniques used to characterise the linear visco-elastic behaviour displayed by many non-Newtonian fluids, the oscillatory shear technique which involves either an applied stress or shear rate which varies harmonically with time, is perhaps the most convenient and widely used.

The definition of linear visco-elasticity may be expressed in the following form: the ratio of the applied stress to strain for any shear history is a function of time alone, and independent of stress magnitude: each stress applied to a material produces a strain which is independent of that produced by any other stresses. However, the total strain experienced by a material is equal to the sum of all the changes induced in the material by the applied stress throughout its history.

The foregoing is an expression of the Boltzmann *superposition principle* [Bird *et al.*, 1977] which may also be expressed in the following terms:

$$\tau(t) = \int_{-\infty}^{t} G(t - t')\dot{\gamma}(t') \, dt' \tag{2.25}$$

where $G(t)$ is the stress relaxation modulus and the integration is performed over all past times t' up to the current time. In visco-elastic liquids, the function $G(s)$, where $s = t - t'$, approaches zero as s approaches infinity, giving rise to the following alternative expression, in terms of strain history rather than rate of strain:

$$\tau(t) = \int_{-\infty}^{t} m(t - t')\gamma(t, t') \, dt' \tag{2.26}$$

where the 'memory function', $m(t)$, is given by $-dG(t)/dt$.

These 'constitutive' equations, involving $G(t)$, describe the response of linear visco-elastic materials to various time-dependent patterns of stress and strain in simple shear [Ferry, 1980]. In many cases it is more convenient, in terms of experimental technique, to consider the complex shear modulus, G^*, which is measured using *oscillatory* shear.

Within the region of linear visco-elastic behaviour, an imposed stress of angular frequency, ω, results in a harmonic strain of amplitude proportional to the stress amplitude, and with phase lag δ relative to the stress, which is independent of the applied stress amplitude [Ferry, 1980].

Figure 2.9 represents a controlled harmonic shear stress applied to a linear visco-elastic system, which results in a harmonic strain response waveform involving a phase lag, δ on the applied stress. This harmonic shear stress, and the resultant strain, may be conveniently expressed as complex quantities, where $\gamma^* = \gamma_m e^{i\omega t}$, $\tau^* = \tau_m e^{i(\omega t + \delta)}$, and γ_m and τ_m are the peak strain and peak stress, amplitudes respectively.

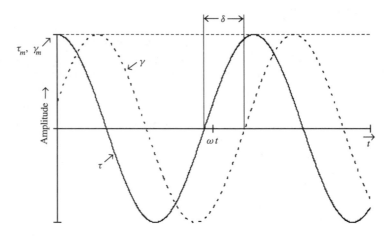

Figure 2.9   *Oscillatory shear strain (- - -) out of phase with stress (–) by a phase angle δ*

The complex shear modulus, G^*, is then defined as:

$$G^* = \frac{\tau^*}{\gamma^*} = \frac{\tau_m}{\gamma_m} \cdot \frac{e^{i(\omega t + \delta)}}{e^{i\omega t}} = \frac{\tau_m}{\gamma_m} \cdot e^{i\delta} \tag{2.27}$$

where $G^* = G' + iG''$.
Hence

$$G' = \frac{\tau_m}{\gamma_m} \cos \delta \tag{2.28}$$

and $$G'' = \frac{\tau_m}{\gamma_m} \sin \delta \tag{2.29}$$

where G' is the dynamic rigidity, defined as the stress in phase with the strain divided by the strain in a sinusoidal deformation: it is a measure of energy stored and recovered per cycle of deformation. G'' is the loss modulus, defined as the stress in quadrature (90° out of phase) with the strain, divided by the strain: it is a measure of energy dissipated per cycle. By definition:

$$\delta = \tan^{-1} \frac{G''}{G'} \quad \text{or} \quad \tan \delta = \frac{G''}{G'}$$

Phase relationships may also be expressed in terms of the complex viscosity, μ^*:

$$\mu^* = \mu' + i\mu'' \tag{2.30}$$

where the dynamic viscosity, μ', is defined as the stress in phase with the rate of strain divided by the rate of strain in a sinusoidal deformation, and μ'' as the stress in quadrature to the rate of strain in a sinusoidal deformation. These are related to G^* as follows:

$$G^* = \omega\mu'' + i\omega\mu' \tag{2.31}$$

hence

$$\mu'' = \frac{G'}{\omega} \quad \text{and} \quad \mu' = \frac{G''}{\omega} \tag{2.32}$$

Controlled strain (or, more properly, controlled displacement) oscillatory shear instruments, exemplified by the Weissenberg Rheogoniometer [Macsporran and Spiers, 1982] readily facilitate tests in which independent measurements of both changing length scales and time scales of applied deformation can be performed. In this way it is, in principle, possible to separate effects due to strain and strain *rates* as the frequency of oscillation may be held constant while the maximum (cyclic) shear strain amplitude is varied. Alternatively, the frequency of deformation can be varied at constant maximum shear strain amplitude.

Rheometers capable of performing oscillatory shear are widely available as commercial instruments, in addition to more specialised devices [Te Nijen-huis and van Donselaar, 1985]. An example of the latter is an oscillating plate rheometer [Eggers and Richmann, 1995] which requires a very small liquid sample volume (<0.3 ml) and has a (potentially) great frequency range (2 Hz to 1 kHz). This latter feature is unusual as it spans the gap between specialised devices and commercially available oscillatory shear instruments, whose frequency range is typically 10^{-3} Hz to *ca.* 10^2 Hz.

Usually, the deformation of a sample undergoing oscillatory shear is monitored by measuring the sinusoidally-varying motion of a transducer-controlled driving surface in contact with the sample. However, in turning to the subsequent calculation of shear strain amplitude in dynamic measurements, it must be recognized that conversion of experimentally determined forces and displacements to the corresponding stresses and strains experienced by a sample can involve consideration of the role of sample inertia.

Sample inertial forces will be small if the sample density, ρ, is small compared with $G'/h^2 f^2$ or $(G''/h^2 f^2)$ where h is the shearing gap thickness and f is the frequency of oscillation (in Hz). Under these conditions, the states of stress or strain may be considered uniform throughout the shearing gap and to experience no periodic spatial variation. This represents the 'gap loading' condition [Ferry, 1980].

Under 'surface loading' conditions, shear waves propagate within the medium, decaying in amplitude from the driving surface to the opposite side of the shearing gap. Thus, under 'surface loading' condition, sample inertia effects are dominant. However, it should be borne in mind that the gap loading condition is not one in which wave propagation is unimportant, and for high precision measurements the shear wavelength, λ should be at least 40-times as large as the shearing gap size, h (a ratio of λ/h of *ca.* 10 may be tolerable [Schrag, 1977].

Fluid inertia effects have been found to be very small for the cone-and-plate geometries typically supplied with these instruments. While inertial corrections are found to be unimportant for the parallel plate geometries, for shearing gaps of the order of 2 mm or less (except possibly for very 'thin' fluids), they must be taken into account in the concentric cylinder geometry (especially for high-density, mobile fluids). Evaluation methods are available for μ^* in the case of cylindrical and plane Couette flow, taking into account fluid inertia [Aschoff and Schummer, 1993].

2.7.1 *Fourier transform mechanical spectroscopy (FTMS)*

The evolution of visco-elastic properties in non-Newtonian fluids exhibiting time-dependent rheological changes is a matter of wide scientific interest, particularly so in systems undergoing gelation. The gel-point, where a three-dimensional network structure is established, may be identified rheologically by the establishment of a characteristic frequency dependence of the dynamic moduli, and an associated frequency independent loss tangent [Winter and Chambon, 1986].

This criterion for gel-point detection, and the non-equilibrium nature of systems undergoing gelation, requires that data be obtained rapidly over a wide range of frequency, prompting the development of a frequency multiplexing technique known as Fourier Transform Mechanical Spectroscopy, FTMS, which allows the measurement of G^* at several frequencies *simultaneously*, rather than consecutively, as in a conventional test [Holly *et al.*, 1988].

The technique, initially developed to measure visco-elastic properties in the curing of polymers [Malkin *et al.*, 1984], has been applied to gels (In and Prud'homme, 1993) and model visco-elastic fluids [Davies and Jones, 1994]. In a variation of the technique, dynamic mechanical properties are determined using the Fourier transform of pulsed deformations [Vratsanos and Farris, 1988].

Experimental times for determining dynamic properties depend on a material's inherent time-dependent behaviour and single-point measurements must span a time period equal to that over which the sample can respond to the imposed stress or strain. This defines a *minimum* measurement time, over which a sample must be 'quasi-stable' for meaningful rheological measurement and a 'mutation number' has been proposed which expresses the relative evolution of the measured property during the experimental time [Mours and Winter, 1994].

In FTMS experiments a sample is subjected to an imposed oscillatory stress or strain. In controlled-stress FTMS, the applied stress is conveniently expressed as an applied torque (the raw experimental parameter from which stress is obtained) as:

$$C(t) = C_o \sum_{k=1,3,5...}^{n_h} \cos k\omega t \qquad (2.33)$$

[Davies and Jones, 1994] where $\omega = 2\pi f$ (f in Hz) is the fundamental frequency of oscillation, C_o is the fundamental torque amplitude, and n_h is the highest harmonic in the series.

The individual components of an applied oscillatory torque of constant amplitude are illustrated in Figure 2.10, and the resulting complex torque signal, obtained from the superposition of these, is shown in Figure 2.11. This non-sinusoidal waveform for the applied torque results in an angular

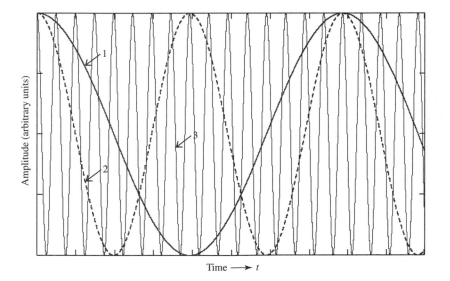

Figure 2.10 *Computer generated sinusoidal torque (or stress) signals of frequencies, 4 radian/s(curve 1), 8 radian/s(curve 2), 64 radian/s(curve 3), corresponding to integer multipliers of 1, 2 and 16, on the basis of a fundamental frequency of 4 radian/s*

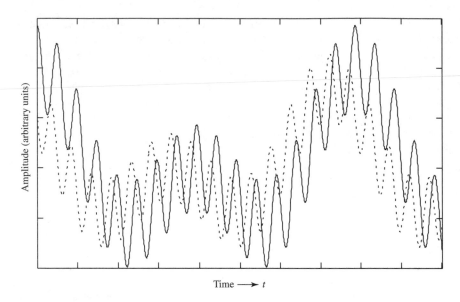

Figure 2.11   *Complex torque (stress) signal(- - -) obtained by superposition of the three waveforms shown in Figure 2.10 and a computer generated 'typical' strain response(−)*

displacement waveform to which a Fourier analysis can be applied, thereby allowing the determination of the complex visco-elastic parameters, μ^* or G^* and their components for each harmonic frequency. Basic aspects of Fourier analysis and the Fast Fourier Transform (FFT) can be found in various texts [Brigham, 1988].

The procedure for performing FTMS measurements using a controlled-stress instrument differs only slightly from conventional test procedures. The test is configured in the same manner as in a 'time sweep' but in this case the selected frequency acts as the fundamental from which further harmonics are selected, with each harmonic frequency being an integer multiple of the fundamental frequency.

An important factor in the construction of the composite waveform is the setting of the maximum strain and torque ratio for each harmonic of the fundamental. As the strain applied to the sample is the sum of the individual strains associated with each harmonic, care must be taken not to exceed the linear visco-elastic limit. For strain sensitive systems this can severely restrict the number of frequencies used.

As the value of the harmonic frequency increases, the corresponding stress amplitude decreases, resulting, eventually, in an inadequately resolved waveform. To overcome this, a torque multiplying factor is introduced which scales the fundamental torque amplitude to an adequately resolvable level. Usually an instrument's control software allows a *maximum* strain level to be designated, which should not be exceeded by any of the harmonic strain amplitudes.

2.8 High frequency techniques

In many cases, a comprehensive characterization of the rheological properties of systems, such as concentrated colloidal dispersions, can require measurements of dynamic mechanical behaviour at frequencies outside the range of conventional, commercially available, rheometers (typically 10^{-3} Hz to 10^2 Hz). In particular, consideration of the relative time scales of particle–fluid displacement and interfacial polarization mechanisms in such systems reveals the need for enhanced high frequency ranges (above *ca.* 10^2 Hz).

High frequency rheometry, which usually involves wave propagation, offers some advantages over conventional techniques due to its inherent rapidity, and the (generally) small strains which are invoked. These are particularly useful features in the context of attempts to characterize the rheology of systems undergoing gelation whose non-equilibrium properties may involve pronounced mechanical weakness and strain sensitivity.

Notwithstanding the rheometrical advantages associated with these features, attempts to exploit wave propagation in monitoring processes, such as polymer curing, have achieved only partial success, due principally to the very high frequencies employed (typically 1 MHz to 10 MHz). In studies of end-linking in polydimethylsiloxane (PDMS) curing using 10 MHz shear waves, no drastic variation in G^* has been observed in the vicinity of the gel-point [Gandelsman *et al.*, 1992]. However, a study of the same PDMS curing system using 10 MHz *longitudinal* waves has shown that the wave velocity increases during crosslinking, with a 'step-like' increase being recorded in the vicinity of the gel-point [Shefer *et al.*, 1990].

These findings illustrate an important principle in relation to the application of high-frequency techniques: the greater success of longitudinal waves over shear waves in the studies mentioned above derives from the relative length scales of the structures (e.g. particle size or dimension of polymer molecule) and the wavelengths involved in measurements. At any frequency, the wavelength of longitudinal waves considerably exceeds that of shear waves, and the former may therefore be more appropriate for probing the development of long range structural details. Alternatively, lower frequency shear waves may be used [Hodgson and Amis, 1990]. In addition to techniques which exploit bulk longitudinal and transverse waves, *surface* waves have been used to study the sol–gel transition in gelatin [Takahashi and Choi, 1996].

Inevitably, the conjunction of frequency-dependent visco-elastic properties and wave propagation leads to consideration of visco-elastic wave dispersion and its influence on conventional wave-based measurements, such as those involving resonance phenomena and pulse propagation techniques.

2.8.1 Resonance-based techniques

Resonance phenomena provide a simple method of characterizing visco-elastic properties which does not require absolute determination of force, or precise setting of shearing gaps. Many high frequency devices based on resonance have been reported [Waterman, *et al.*, 1979; Hausler *et al.*, 1996; Stoimenova *et al.*, 1996].

Basic aspects of the resonance technique may be illustrated by considering a linear visco-elastic medium between two parallel plates, one undergoing forced harmonic displacement, amplitude a $(= a_o \cos \omega t)$, the other being fixed. As ω is varied, resonances occur and a resonance bandwidth analysis yields the loss tangent, $\tan \delta$ [Whorlow, 1992; Ingard 1988].

The extent to which the resonance bandwidth analysis is susceptible to dispersion-induced errors has been considered and it has been established that serious errors may be incurred ($>1\%$ in δ) under conditions where waves are damped exponentially in one wavelength [Williams and Williams, 1994].

2.8.2 Pulse propagation techniques

Recourse to pulse propagation measurements is often prompted by their apparent simplicity, involving measurements of the 'time-of-flight' of a disturbance propagating through a visco-elastic material [Joseph *et al.*, 1986].

In plane-harmonic shear-wave propagation in a linear medium of density ρ, G' and G'' are given by,

$$G' = \frac{\rho v^2 (1 - r^2)}{(1 + r^2)^2}; \tag{2.34a}$$

$$G'' = \frac{\rho v^2 \cdot 2r}{(1 + r^2)^2} \tag{2.34b}$$

[Ferry, 1980] where ω is the angular frequency in rad/s, $r = \lambda/(2\pi x_o)$ where λ is the shear wave length, x_o is the exponential damping length and v is the shear wave phase velocity. For known ω and ρ, G' and G'' may be obtained from equation 2.34 by measurement of v and x_o.

Such simple measurements belie the complicating effects of visco-elastic wave dispersion, which may render their analysis unreliable. The tendency of pulse frequency components to travel at different velocities in dispersive media distorts the pulse, thereby influencing measurements of damping to a degree dependent on the medium and the spectral content of the pulse. The latter, in turn, depend on pulse shape. This visco-elastic wave dispersion, associated with dissipative stresses, can severely restrict the application of pulse propagation techniques in which the measured velocity v_w may correspond to a *group* velocity, U, not the requisite *phase* velocity, v. As U and v may differ significantly in visco-elastic media (in which $U > v$), serious over-estimates of elasticity may result.

In some instruments the phase velocity, v, is measured *directly* (using continuous shear waves), as in the 'virtual gap' rheometer, VGR, a multiple path shear wave interferometer which operates in the frequency range 100 Hz to *ca.* 2 kHz [Williams and Williams, 1992].

2.9 The relaxation time spectrum

The determination of the relaxation spectrum of a visco-elastic fluid from various dynamic shear measurements has been discussed by many workers (e.g. see, Orbey and Dealy, 1991; Baumgaertel and Winter, 1989; Sullivan *et al.*, 1994) and, in the case of a visco-elastic fluid, the problem of determining the relaxation spectrum from oscillatory shear measurements involves the inversion of the following pair of integral equations:

$$G' = G_c + \int_{-\infty}^{\infty} \left[\frac{H\omega^2\lambda^2}{1 + \omega^2\lambda^2} \right] \, \mathrm{d}\ln\lambda \quad \text{and} \tag{2.35a}$$

$$G'' = \int_{-\infty}^{\infty} \left[\frac{\omega\lambda}{1 + \omega^2\lambda^2} \right] \, \mathrm{d}\ln\lambda \tag{2.35b}$$

where $H(\lambda)$ denotes the *continuous* relaxation spectrum.

This is an 'ill-posed' problem and small perturbations in (measured) $G'(\omega)$ or $G''(\omega)$ can produce large perturbations in $H(\lambda)$. In addition to $H(\lambda)$, various techniques have been described to determine the *discrete* relaxation spectrum, in terms of a set of modulus-relaxation time-pairs, using the generalised Maxwell model [Ferry, 1980]. However, infinitely many parameter sets may be derived, all of which are adequate for the purpose of representing experimental data.

Methods of overcoming this problem to obtain a physically meaningful relaxation spectrum have been discussed at length in the literature: these include linear regression [Honerkamp and Weese, 1989] and non-linear regression techniques [Baumgaertel and Winter, 1989]. A commercial version of the latter is available as the software program "*IRIS*" which has been used to model the relaxation behaviour of high molecular weight polydimethylsiloxanes. The relaxation modulus, $G(t)$, can also be obtained by direct conversion from the frequency domain (to the time domain) using the Fourier transform [Kamath and Mackley, 1989].

Recently, the issue of sampling localisation in determining the relaxation spectrum has been considered [Davies and Anderssen, 1997]. It is usually assumed that G' and G'' measured over the frequency range $\omega_{min} < \omega < \omega_{max}$ yield information about the relaxation spectrum over the range of relaxation times $(\omega_{\max})^{-1} < \lambda < (\omega_{\min})^{-1}$ i.e. over the *reciprocal* frequency range but G' and G'' are (necessarily) measured over a limited range of discrete frequencies, and these measurements unavoidably involve errors. The oft-used assumption that measurements of G' and G'' over a frequency range $\omega_{\min} < \omega < \omega_{\max}$ yield information (in terms of the relaxation spectrum) over the reciprocal frequency range is misleading. In fact, the relaxation spectrum is determined on a shorter interval of relaxation times than $(\omega_{\max})^{-1} < \lambda < (\omega_{\min})^{-1}$. The correct frequency interval on which the relaxation frequency is determined is $e^{\pi/2}\omega_{\max}^{-1} < \lambda < e^{-\pi/2}\omega_{\min}^{-1}$; and thus the determination of the spectrum at a single relaxation time, λ requires measurements of G' and G'' in the frequency range $e^{-\pi/2}\lambda^{-1} < \omega < e^{\pi/2}\lambda^{-1}$ [Davies and Anderssen, 1997].

Simple moving-average formulae which can be applied to oscillatory shear data to recover estimates of the relaxation spectrum have been reported [Davies and Anderssen, 1998]. These formulae represent an improvement over previous commercial software in that they take into account the limits imposed by sampling localization and yield accurate spectra very rapidly on a PC.

2.10 Extensional flow measurements

Extensional flows may be generated within a flow field that experiences a sudden change in geometry, such as contractions, or orifice plates. The need for appropriate rheological data pertaining to such flows represents a major obstacle to the development of improved process simulation, monitoring and control for non-Newtonian fluids, and requires, not only a knowledge of fluid shear viscosity, but also an *appropriate* measure of its *extensional* viscosity, μ_E. The problematical effects of extensional viscosity in process engineering become evident when one considers that μ_E may be several *orders of magnitude* higher than the corresponding shear viscosity in some non-Newtonian liquids.

In marked contrast to measurements of shear rheological properties, such as apparent viscosity in steady shear, or of complex viscosity in small amplitude oscillatory shear, extensional viscosity measurements are far from straightforward. This is particularly so in the case of mobile elastic liquids whose rheology can mitigate against the generation of well-defined extensional flow fields.

Techniques for measuring the extensional properties of fluids can be divided (broadly) into those of the 'flow-through' and 'stagnation-point' types [Hermansky and Boger, 1995]. The former usually involve 'spinnable' fluids, a feature exploited in instruments such as the Carri-Med (now TA Instruments) EV rheometer [Ferguson and Hudson, 1990].

In spin-line experiments, fluid is delivered through a nozzle and subsequently stretched by an applied force. Procedures are available for obtaining useful semi-quantitative estimates of μ_E from spin-line experiments, and the technique is able to distinguish between fluids that are tension-thinning and tension-thickening [Jones *et al.*, 1987]. Despite being limited to low rates of strain, and generally suitable for highly viscous or elastic fluids, perhaps the most successful device of this type, is the filament-stretching technique [Sridhar *et al.*, 1991] which is described below in Section 2.10.2.

Where a fluid is not 'spinnable' the various orifice flow techniques, which involve pressure drop measurements across a contraction [Binding, 1988, 1993], can provide a means of estimating the extensional-viscosity behaviour of *shear-thinning* polymer solutions.

For low viscosity fluids such as dilute polymer solutions, the various stagnation-point devices can prove useful. In this category is the commercially available Rheometrics RFX opposing jet device, a development of earlier instruments [Cathey and Fuller, 1988], which has the potential to produce a wide range of strain rates, and which has been used to study fluids with a viscosity approaching that of water [Hermansky and Boger, 1995].

2.10.1 Lubricated planar stagnation die-flows

Lubricated dies have been proposed as a means of generating extensional flows in which the lubricant protects the sample of interest from the shear effects of the die walls, and serves to transmit forces to the walls where they can be measured. A theoretical analysis of steady, two-dimensional flow through a lubricated planar-stagnation die has shown that pure planar extension through the die is prevented by the interaction of the die shape with the normal viscous stresses at the free interface between the lubricant and the test fluid [Secor *et al.*, 1987]. Interestingly, two regions of the flow (near the inlet and oulet of the die) are nearly extensional, and are capable of being represented by an approximate expression which, in addition to being independent of the constitutive relation of the fluid, provides a simple relationship between wall pressure measurements and extensional viscosity.

In practice the technique is beset by experimental difficulties, such as the need to measure extremely small pressures. In addition, the difficulty of maintaining adequately lubricated flows requires the provision of large sample volumes. The application of converging flows for determining the extensional flow behaviour of polymer melts has been reviewed [Rides and Chakravorty, 1997].

2.10.2 Filament-stretching techniques

In this method, the sample is held between two discs, the lower of which is attached to a shaft whose movement is controlled by a computer capable of generating an exponentially varying voltage, the shaft velocity being proportional to the applied voltage. The upper disc is attached to a load measuring device and an optical system is used to measure filament diameter [Sridhar *et al.*, 1991].

A reverse-flow near the plates causes a delay in the development of the uniform cylindrical column, and a difference between the local and imposed extension rates at early times [Shipman *et al.*, 1991]. Problems may also be encountered due to adhesion of the fluid to the plates [Spiegelberg and McKinley, 1996] but this may be compensated for [Tirtaatmadja and Sridhar, 1993]. Notwithstanding these difficulties, the technique can provide meaningful extensional viscosity data for polymer solutions, as the deformation is uniform and independently imposed, and the total strain is measurable.

2.10.3 Other 'simple' methods

When a pendant drop forms slowly at the lower end of a capillary tube it ultimately falls and stretches the filament (which remains attached to the drop). For a Newtonian fluid the filament quickly thins and breaks but long filaments can be formed from visco-elastic liquids [Jones *et al.*, 1990]. The forces acting on the falling drop are determined using a force balance, and the extensional stress determined as a function of time [Jones and Rees, 1982]. The falling pendent drop technique is simple to set up and analyse, and provides consistent values of an apparent elongational viscosity.

Another simple technique involves a liquid which is slowly extruded vertically downward through a capillary into another immiscible fluid of lower density (the *submerged* pendant drop technique). The heavier ejected fluid forms a sphere at the nozzle tip, and grows until the drop is no longer supported by surface tension. The drop then falls and stretches the ligament which connects the drop to the nozzle [Matta, 1984].

It is important to note that the Trouton ratio, T_R, defined as the ratio of the extensional viscosity to the shear viscosity, involves the shear viscosity evaluated at the same magnitude of the second invariant of the rate of deformation tensor where $\dot{\varepsilon}$ is the rate of extension, i.e.:

$$T_R = \mu_E / \mu \tag{2.36}$$

where μ_E is a function of $\dot{\varepsilon}$ and μ is a function of $\dot{\gamma}$, and these values are evaluated using the $\dot{\gamma} = \dot{\varepsilon}\sqrt{3}$ equivalence.

The 'fibre-windup' technique [Padmanabhan *et al.*, 1996] can provide transient extensional-viscosity data using modified rotational shear rheometers. One end of the sample is clamped while the other end is wound around a drum at a constant rotational speed to achieve a given extension rate, with the rheometer's torque transducers being used to obtain the extensional viscosity. The technique is claimed to provide valuable extensional viscosity data for high viscosity liquids.

Another simple filament-stretching technique which has the advantage that it does not require measurements of forces acting on the filament, involves the elongation of radial filaments of a fluid on a rotating drum [Jones *et al.*, 1986]. Using a high-speed camera, photographs are taken of the fluid in its initial undeformed state, and as it stretches into a filament as the drum rotates.

A technique which involves a so-called 'open-siphon' is attractive in so far as the flow history of the fluid is apparently simple and readily calculated. Fluid is drawn up out of a beaker through an orifice into a low pressure chamber, and forces in the open siphon column are estimated from the orifice pressure drop, with corrections being made for inertia and gravity contributions [Binding *et al.*, 1990]. Although the technique is very simple, the motion in the siphon may not be entirely shear free [Matthys, 1988].

Acknowledgements

The author is grateful to G. S. Tucker, P. M. Williams and Dr. I. D. Roberts for assistance with the figures.

2.11 Further reading

James, D.F., Walters, K., A critical appraisal of available methods for the measurement of extensional properties of mobile systems. A.A. Collyer (Ed.), *Techniques in Rheological Measurements*, Elsevier, London (1993).

Macosko, C.W. and Souza Mendes P.R. *Fluid mechanics measurements in non-Newtonian fluids*, in: Goldstern, R.J. (ed.) *Fluid Mechanics Measurements*, 2nd Edition, Taylor and Francis, New York (1996).

Walters, K., *Rheometry*, Chapman and Hall, London (1975).

Whorlow, R.W., *Rheological Techniques*, 2nd ed., Ellis Horwood, London (1992).

2.12 References

Aschoff, D. and Schummer, P. *Rheol. Acta.* **37** (1993) 1237.

Bagley, E.B. *J. Appl. Phys.* **28** (1957) 624.

Barnes, H.A. *J. Non-Newt. Fluid Mech.*, **56** (1995) 221.

Barnes, H.A., Hutton, J.F. and Walters, K. *An Introduction to Rheology*. Elsevier, Amsterdam (1989).

Barnes, H.A. *J. Non-Newt. Fluid Mech* **81** (1999) 133.

Barnes, H.A. and Carnali, J.O. *J. Rheol.* **34(6)** (1990) 841.

Barnes, H.A. and Walters, K. *Rheol. Acta.* **24** (1985) 323.

Baumgaertel, M. and Winter, H.H. *Rheol. Acta* **28** (1989) 511.

Binding, D.M. *J. Non-Newt. Fluid Mech.* **27** (1988) 173.

Binding, D.M. in: *"Techniques in Rheological Measurement"* ed. A.A. Collyer, Chapman & Hall, London (1993).

Binding D.M., Hutton, J.F. and Walters, K. *Rheol. Acta* **15** (1976) 540.

Binding D.M., Jones D.M. and Walters, K. *J. Non-Newt. Fluid Mech.* **35** (1990) 121.

Binding D.M. and Walters, K. *J. Non-Newt. Fluid Mech.* **1** (1976) 259; and 277.

Bird, R.B., Hassager, O., Armstrong, R.C. and Curtiss, C.F. *Dynamics of Polymeric Liquids*; Volume 2.; Kinetic theory. John Wiley & Sons, New York (1977).

Brigham, O.E., *The FFT and its Applications*, Prentice Hall, Englewood Cliffs, NJ (1988).

Broadbent, J.M., Kaye, A., Lodge, A.S. and Vale, D.G. *Nature*, **217** (1968) 55.

Cathey, C.A. and Fuller, G.G. *J. Non-Newt. Fluid Mech.* **30** (1988) 303.

Cheng, D.C-H. *Rheol. Acta.* **25** (1986) 542.

Cheng, D.C-H. and Richmond, R.A. *Rheol. Acta.* **17** (1978) 446.

Chhabra, R.P., Bubbles, *Drops and Particles in non-Newtonian Fluids*, CRC Press, Boca Raton, Florida (1993).

Davies, A.R. and Anderssen, R.S. *J. Non-Newt. Fluid Mech.* **73** (1997) 163.

Davies, A.R. and Anderssen, R.S. *J. Non-Newt. Fluid Mech.* **79** (1998) 235.

Davies, J.M. and Jones, T.E.R. *J. Non-Newt. Fluid Mech.* **52** (1994) 177.

Eggers, F. and Richmann, K.-H. *Rheol. Acta.* **34** (1995) 483.

Ferguson, J. and Hudson, N. *J. Non-Newt. Fluid Mech.* **35** (1990) 197.

Ferry, J.D. *Visco-elastic Properties of Polymers.* 3rd Edition, Wiley, New York (1980).

Gandelsman, M. Gorodetsky, G. and Gottlieb, M. Theoretical and Applied rheology, edited by Moldenaers, P. and Keunings, R. *Proc. XII th Int. Congr. on Rheology*, Brussels (1992).

Hartnett, J.P. and Hu, R.Y.Z. *J. Rheol.* **33(4)** (1989) 671.

Hausler, K., Reinhart, W.H., Schaller, P., Dual, J., Goodbread, J. and Sayir, M. *Biorheol.* **33** (1996) 397.

Hermansky, C.G. and Boger, D.V. *J. Non-Newt. Fluid Mech.* **56** (1995) 1.

Higman, R.W. *Rheol. Acta* **12** (1973) 533.

Hodgson, D.F. and Amis, E.J., *Macromol.* **23** (1990) 2512.

Holly, E.E., Venkataraman, S.K., F. Chambon and H.H. Winter *J. Non-Newt. Fluid Mech.* **27** (1988) 17.

Honerkamp, J. and Weese, J. *Macromolecules* **22** (1989) 4372.

In, M. and Prud'homme, R.K. *Rheol. Acta* **32** (1993) 556.

Ingard, K.U. *Fundamentals of Waves and Oscillations.* Cambridge University Press, Cambridge (1988).

Jacobsen, R.T. *J. Coll. and Interf. Sci.* **48** (1974) 437.

Jackson, R. and Kaye, A. Brit. *J. Appl. Phys.* **17** (1966) 1355.

James, A.E., Williams, D.J.A. and Williams, P.R. *Rheol. Acta.* **26** (1987) 437.

Jones, T.E.R., Davies, J.M. and Thomas, A. *Rheol. Acta.* **26** (1987) 1419.

Jones, W.M., Williams, P.R. and Virdi, T.S. *J. Non-Newt. Fluid Mech.* **21** (1986) 51.

Jones, W.M., Hudson, N.E. and Ferguson, J. *J. Non-Newt. Fluid Mech.* **35** (1990) 263.

Jones, W.M. and Rees, I.J. *J. Non-Newt. Fluid Mech.* **11** (1982) 257.

Joseph, D.D., Narain, A. and Riccius, O. *J. Fluid. Mech.* **171** (1986) 289.

Kamath, V.M. and Mackley, M.R. *J. Non-Newt. Fluid Mech.* **32** (1989) 119.

Keentok, M., Milthorpe, J.F. and O'Donovan, E. *J. Non-Newt. Fluid Mech.* **17** (1985) 23.

Krieger, I.M. and Maron, S. *J. Appl. Phys.* **23** (1952) 147.

Kuo, Y. and Tanner, R.I. *Rheol. Acta* **13** (1974) 443.

Laun, H.M. *Rheol. Acta* **22** (1983) 171.

Lodge, A.S., Al-Hadithi, T.S.R. and Walters, K. *Rheol. Acta* **26** (1987) 516.

Lohnes R.A., Millan A. and Hanby R.L. *J. Soil Mech. and Found. Divn.* **98** (1972) 143.

Macosko, C.W. *Rheology: Principles, Measurements and Applications.*, VCH Publishers, New York (1994).

Macsporran, W.C. and Spiers, R.P. *Rheol. Acta.* **21** (1982) 184.

Malkin, A.Y., Begishev, V.P. and Mansurov, V.A. *Vysokomol Soedin* **26A** (1984) 869.

Marsh, B.D. and Pearson, J.R.A. *Rheol. Acta* **7** (1968) 326.

Matta, J.E. in: Mena, B., Garcia-Rejon, A., Rangel-Nafaile, C. (eds.) *Advances in Rheology* (Proc. 8th Int. Congr. on Rheology) **4** (1984) 53–60. Universidad Nacional Autonoma de Mexico.

Matthys, E.G. *J. Rheol.* **32** (1988) 773.

Mooney, M. *Trans. Soc. Rheol.* **2** (1931) 210.

Mours, M. and Winter, H.H. *Rheol. Acta.* **33** (1994) 385.

Nguyen Q.D. and Boger, D.V. *J. Rheol.* **29** (1985) 335.

Orbey, N. and Dealy, J.M. *J. Rheol.* **35** (1991) 1035.

Padmanabhan, M., Kasehagen, L.J. and Macosko, C.W. *J. Rheol.* **40** (1996) 473.

Rides, M. and Chakravorty, S. National Physical Laboratory Report CMMT (A) **80**, Aug 1997. ISSN 1361–4061.

Schrag, J.L. *Trans. Soc. Rheol.*, **21** (1977) 399.

Secor R.B., Macosko C.W. and Scriven L.E. *J. Non-Newt. Fluid Mech.* **23** (1987) 355.

Shefer, A., Gorodetsky, G. and Gottlieb, M., *Macromolecular Liquids*, Editors Safinya, C.R., Safran, S.A. and Pincus, P.A. *MRS Symp. Proc.* **177** (1990), 31.

Shipman, R.W.G., Denn, M.M. and Keunings, R. *J. Non-Newt. Fluid Mech.* **40** (1991) 281.

Spaans, R.D. and Williams, M.C. *J. Rheol.* **39** (1995) 241.

Spiegelberg, S.H. and McKinley, G.H. *J. Non-Newt. Fluid Mech.* **67** (1996) 49.

Sridhar T, Tirtaatmadja, V., Nguyen D.A. and Gupta R.K. *J. Non-Newt. Fluid Mech.* **40** (1991) 271.

Stoimenova, M., Dimitrov, V. and Okubo, T. *J. Coll. and Interface Sci.* **184** (1996) 106.

Sullivan, J.L., Gibson, R. and Wen, Y.F. *J. Macromol. Sci. -Phys.* **B33** (1994) 229.

Takahashi, H. and Choi, P.K. *Jap. J. Appl. Phys. I.* **35** (1996) 2939.

Te Nijenhuis, K. and van Donselaar, R. *Rheol. Acta.* **24** (1985) 47.

Tirtaatmadja, V. and Sridhar, T. *J. Rheol.* **37** (1993) 1081.

Van Wazer, J.R., Lyons, J.W., Lim, K.Y. and Colwell, R.E., *Viscosity and Flow Measurement*. Wiley, New York (1963).

Vratsanos, M.S. and Farris, R.G. *J. Appl. Polymer Sci.* **36** (1988) 403.

Walters, K. *Rheometry*, Chapman and Hall, London (1975).

Waterman, H.A., Oosterbroek, M. Beukema, G.J. and Altena, E.G. *Rheol Acta.* **18** (1979) 585.

Whorlow, R.W. *Rheological Techniques*, 2nd ed.; Ellis Horwood, London (1992).

Williams, P.R. and Williams, D.J.A. *J. Non-Newt. Fluid Mech.* **42** (1992) 267.

Williams, P.R. and Williams, D.J.A. *J. Rheol.* 38:4 (1994) 1211.

Williams, R.W. *Rheol. Acta.* **18** (1979) 345.

Winter, H.H. and Chambon, F. *J. Rheol.* **30** (1986) 367.

Yan, J. and James, A.E. *J. Non-Newt. Fluid Mech.* **70** (1997) 237.

Yoshimura, A.S., Princes, H.M. and Kiss, A.D., *J. Rheol.* **31** (1987) 699.

Yoshimura, A.S. and Prud'homme, R.K. *J. Rheol.* **32** (1988) 53.

2.13 Nomenclature

		Dimensions in M, L, T
D	capillary or tube diameter (m)	L
F	axial force (N)	MLT^{-2}
G'	dynamic rigidity or storage modulus (Pa)	$ML^{-1}T^{-2}$
G''	loss modulus (Pa)	$ML^{-1}T^{-2}$
G^*	complex shear modulus (Pa)	$ML^{-1}T^{-2}$
h	gap in parallel plate system (m)	L
L	length of capillary (m)	L
m	power law consistency coefficient (Pa·s^n)	$ML^{-1}T^{2-n}$
N_1	first normal stress difference (Pa)	$ML^{-1}T^{-2}$
N_2	second normal stress difference (Pa)	$ML^{-1}T^{-2}$
n	power law flow behaviour index (–)	$M^0L^0T^0$

$-\Delta p$	pressure drop (Pa)	$\mathbf{ML}^{-1}\mathbf{T}^{-2}$
$(-\Delta/v)_t$	overall pressure drop (P$_\text{a}$)	$\mathbf{ML}^{-1}\mathbf{T}^{-2}$
		Dimensions
		in $\mathbf{M, L, T}$

$\dfrac{-\Delta p}{L}$	pressure gradient (Pa/m)	$\mathbf{ML}^{-2}\mathbf{T}^{-2}$
Q	volumetric flow rate (m^3/s)	$\mathbf{L}^3\mathbf{T}^{-1}$
R	radius of capillary or of plate (m)	$\mathbf{L}$
R_1	inner radius in co-axial cylinder viscometer (m)	$\mathbf{L}$
R_2	outer radius in co-axial cylinder viscometer (m)	$\mathbf{L}$
r	radial coordinate (m)	$\mathbf{L}$
Re	Reynolds number (−)	$\mathbf{M}^0\mathbf{L}^0\mathbf{T}^0$
T	torque (N.m)	$\mathbf{ML}^2\mathbf{T}^{-2}$
Ta	Taylor number (−)	$\mathbf{M}^0\mathbf{L}^0\mathbf{T}^0$
T_R	Trouton radio (−)	$\mathbf{M}^0\mathbf{L}^0\mathbf{T}^0$
V	average velocity in capillary (m/s)	$\mathbf{LT}^{-1}$
v	shear wave phase velocity (m/s)	$\mathbf{LT}^{-1}$

Greek Symbols

γ	Shear strain	$\mathbf{M}^0\mathbf{L}^0\mathbf{T}^0$
$\dot\gamma$	shear rate (s^{-1})	$\mathbf{T}^{-1}$
δ	phase shift in oscillatory test (−)	$\mathbf{M}^0\mathbf{L}^0\mathbf{T}^0$
$\dot\varepsilon$	rate of extension (s^{-1})	$\mathbf{T}^{-1}$
θ	Cone angle of cone and plate viscosities	$\mathbf{M}^0\mathbf{L}^0\mathbf{T}^0$
λ	relaxation time (s)	$\mathbf{T}$
μ	viscosity (Pa·s)	$\mathbf{ML}^{-1}\mathbf{T}^{-1}$
ρ	fluid density (kg/m^3)	$\mathbf{ML}^{-3}$
τ	shear stress (Pa)	$\mathbf{ML}^{-1}\mathbf{T}^{-2}$
ω	frequency (Hz)	$\mathbf{T}^{-1}$
Ω	angular velocity (rad/s)	$\mathbf{T}^{-1}$

Subscripts

crit	location at which $\tau = \tau_y$
E	extensional
m	maximum value
N	Newtonian
p	plastic
R	at $r = R$
w	at wall
0	yield point

Superscripts

*	complex
′	in phase component
″	out of phase component

Chapter 3

Flow in pipes and in conduits of non-circular cross-sections

3.1 Introduction

In the chemical and process industries, it is often required to pump fluids over long distances from storage to various processing units and/or from one plant site to another. There may be a substantial frictional pressure loss in both the pipe line and in the individual units themselves. It is thus often necessary to consider the problems of calculating the power requirements for pumping through a given pipe network, the selection of optimum pipe diameter, measurement and control of flow rate, etc. A knowledge of these factors also facilitates the optimal design and layout of flow networks which may represent a significant part of the total plant cost. Aside from circular pipes, one also encounters conduits of other cross-sections and may be concerned with axial flow in an annulus (as in a double pipe heat exchanger), rectangular, triangular and elliptic conduits as employed in nuclear reactors and for extrusion through dies. Furthermore, the velocity profile established in a given flow situation strongly influences the heat and mass transfer processes. For instance, the analysis and interpretation of data obtained in a standard falling-film absorber used for the determination of diffusion coefficients relies on the knowledge of flow kinematics. This chapter deals with engineering relationships describing flow in a variety of geometries. The treatment here is, however, restricted to the so-called purely viscous or time-independent type of fluids, for which the viscosity model describing the flow curve is already known. However, subsequently a generalised treatment for the laminar flow of time-independent fluids in circular tubes is presented. Notwithstanding the existence of time-dependent and visco-elastic fluid behaviour, experience has shown that the shear rate dependence of the viscosity is the most significant factor in most engineering applications which invariably operate under steady state conditions. Visco-elastic behaviour does not significantly influence laminar flow through circular tubes. Visco-elastic effects begin to manifest themselves for flow in non-circular conduits and/or in pipe fittings. Even in these circumstances, it is often possible to develop predictive expressions purely in terms of steady-shear viscous properties.

Many of the formulae to be developed here will relate the frictional pressure drop $(-\Delta p)$ to the volumetric flow rate (Q). The major application of such relationships is in the mechanical energy balance which is written to calculate the total head loss in a pipe network which in turn allows the estimation of the required power to be delivered by a pump. The total loss term, representing the conversion of the mechanical energy into thermal energy as a result of fluid friction, consists of two components: one due to the frictional pressure drop associated with fully developed flow through the conduit and the other due to the frictional losses associated with entrance effects, pipe fittings, etc. Because of their generally high viscosities, laminar flow is more commonly encountered in practice for non-Newtonian fluids, as opposed to Newtonian fluids.

3.2 Laminar flow in circular tubes

Consider the laminar, steady, incompressible fully-developed flow of a time-independent fluid in a circular tube of radius, R, as shown in Figure 3.1. Since there is no angular velocity, the force balance on a fluid element situated at distance r, can be written as:

$$p(\pi r^2) - (p + \Delta p)\pi r^2 = \tau_{rz} \cdot 2\pi r L \qquad (3.1)$$

i.e. $$\tau_{rz} = \left(\frac{-\Delta p}{L}\right)\frac{r}{2} \qquad (3.2)$$

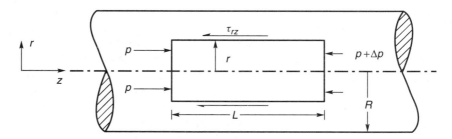

Figure 3.1   *Flow through a horizontal pipe*

This shows the familiar linear shear stress distribution across the pipe cross-section, the shear stress being zero at the axis of the tube, as shown in Figure 3.2. Note that equation (3.2) is applicable to both laminar and turbulent flow of any fluid since it is based on a simple force balance and no assumption has been made so far concerning the type of flow or fluid behaviour.

3.2.1 Power-law fluids

For a power-law fluid in a pipe, the shear stress is related to the shear rate by [Coulson and Richardson, 1999]:

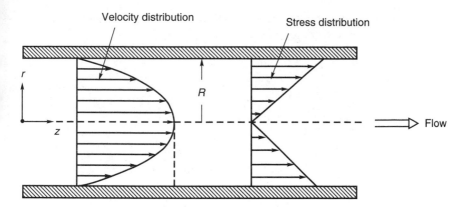

Figure 3.2   *Schematic representation of shear stress and velocity distribution in fully developed laminar flow in a pipe*

$$\tau_{rz} = m\left(-\frac{dV_z}{dr}\right)^n \tag{3.3}$$

where V_z is the velocity in the axial direction at radius r. Now combining equations (3.2) and (3.3) followed by integration yields the following expression for the velocity distribution:

$$V_z = -\left(\frac{n}{n+1}\right)\left\{\left(\frac{-\Delta p}{L}\right)\frac{1}{2m}\right\}^{1/n} r^{(n+1)/n} + \text{constant}$$

At the walls of the pipe (i.e. when $r = R$), the velocity V_z must be zero in order to satisfy the no-slip condition. Substituting the value $V_z = 0$, when $r = R$:

$$\text{constant} = \left(\frac{n}{n+1}\right)\left\{\left(\frac{-\Delta p}{L}\right)\frac{1}{2m}\right\}^{1/n} R^{(n+1)/n}$$

and therefore:

$$V_z = \left(\frac{n}{n+1}\right)\left(\frac{-\Delta p}{mL}\cdot\frac{R}{2}\right)^{1/n} R\left\{1 - \left(\frac{r}{R}\right)^{(n+1)/n}\right\} \tag{3.4}$$

The velocity profile may be expressed in terms of the average velocity, V, which is given by:

$$V = \frac{Q}{\pi R^2} = \frac{1}{\pi R^2}\int_0^R 2\pi r V_z\,dr \tag{3.5}$$

where Q is the volumetric flow rate of the liquid. On substitution for V_z from equation (3.4), and integration yields,

$$V = \frac{2\pi}{\pi R^2} \left(\frac{n}{n+1}\right) \left(\frac{-\Delta p}{L} \cdot \frac{R}{2m}\right)^{1/n} R \int_0^R r \left\{1 - \left(\frac{r}{R}\right)^{(n+1)/n}\right\} dr$$

$$= 2\left(\frac{n}{n+1}\right) \left(\frac{-\Delta p}{L} \cdot \frac{R}{2m}\right)^{1/n} R \int_0^R \frac{r}{R} \left\{1 - \left(\frac{r}{R}\right)^{(n+1)/n}\right\} d\left(\frac{r}{R}\right)$$

$$= 2\left(\frac{n}{n+1}\right) \left(\frac{-\Delta p}{L} \cdot \frac{R}{2m}\right)^{1/n} R \left[\frac{1}{2} - \frac{n}{3n+1}\right]$$

and therefore:

$$V = \left(\frac{n}{3n+1}\right) \left(\frac{-\Delta p \cdot R}{2mL}\right)^{1/n} R \qquad (3.6)$$

Equation (3.4) can now be re-written as:

$$\frac{V_z}{V} = \left(\frac{3n+1}{n+1}\right) \left\{1 - \left(\frac{r}{R}\right)^{(n+1)/n}\right\} \qquad (3.7a)$$

The velocity profiles calculated from equation (3.7a) are shown in Figure 3.3, for various values of n. Compared with the parabolic distribution for a Newtonian fluid ($n = 1$), the profile is flatter for a shear-thinning fluid and sharper for a shear-thickening fluid. The velocity is seen to be a maximum when $r = 0$, i.e. at the pipe axis. Thus the maximum velocity $V_{z\,max}$, at the pipe axis, is given by equation (3.7a) when $r = 0$ and:

$$V_{z\,max} = \left(\frac{3n+1}{n+1}\right) V$$

Hence:
$$\frac{V_{z\,max}}{V} = \left(\frac{3n+1}{n+1}\right) \qquad (3.7b)$$

Thus, it can be readily seen from equation (3.7b) that the value of the centre-line velocity drops from $2.33V$ to $1.18V$ as the value of the power-law index n decreases from 2 to 0.1.

Re-writing Equation (3.6) in terms of the volumetric flow rate and the pressure gradient:

$$Q = \pi R^2 V = \pi \left(\frac{n}{3n+1}\right) \left(\frac{-\Delta p}{2mL}\right)^{1/n} R^{(3n+1)/n} \qquad (3.8)$$

For a given power-law fluid and fixed pipe radius, $-\Delta p \propto Q^n$, i.e. for a shear-thinning fluid ($n < 1$), the pressure gradient is less sensitive than for a Newtonian fluid to changes in flow rate. The flow rate, on the other hand,

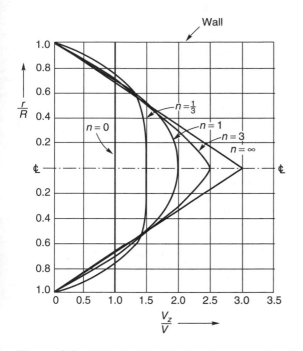

Figure 3.3   *Velocity distribution for power-law fluids in laminar flow in pipes*

does show a rather stronger dependence on the radius. For instance, for $n = 1$, $Q \propto R^4$ whereas for $n = 0.5$, $Q \propto R^5$.

It is useful to re-write equation (3.8) in dimensionless form by introducing a friction factor defined as $f = (\tau_w/(1/2)\rho V^2)$ where $\tau_w = (-\Delta p/L)(D/4)$ and defining a suitable Reynolds number which will yield the same relationship as that for Newtonian fluids, that is, $f = (16/\text{Re})$. Thus substitution for f in terms of the pressure gradient $(-\Delta p/L)$ in equation (3.8) gives:

$$f = \frac{16}{\text{Re}_{PL}} \tag{3.8a}$$

where the new Reynolds number Re_{PL} is defined by:

$$\text{Re}_{PL} = \frac{\rho V^{2-n} D^n}{8^{n-1} m \left(\dfrac{3n+1}{4n}\right)^n} \tag{3.8b}$$

It will be seen later (equations (3.28a) and (3.30b)) that this definition of the Reynolds number coincides with that of Metzner and Reed [1959], and hence hereafter it will be written as Re_{MR}, that is, $\text{Re}_{PL} = \text{Re}_{MR}$.

Example 3.1

A polymer solution (density $= 1075 \, \text{kg/m}^3$) is being pumped at a rate of $2500 \, \text{kg/h}$ through a 25 mm inside diameter pipe. The flow is known to be laminar and the power-law constants for the solution are $m = 3 \, \text{Pa·s}^n$ and $n = 0.5$. Estimate the pressure drop over a 10 m length of straight pipe and the centre-line velocity for these conditions. How does the value of pressure drop change if a pipe of 37 mm diameter is used?

Solution

Volumetric flow rate, $Q = \dfrac{2500}{3600} \times \dfrac{1}{1075} = 6.46 \times 10^{-4} \, \text{m}^3/\text{s}$

pipe radius, $R = \dfrac{25}{2} \times 10^{-3} = 0.0125 \, \text{m}$

Substitution in equation (3.8) and solving for pressure drop gives:

$-\Delta p = 110 \, \text{kPa}$

Average velocity in pipe, $V = \dfrac{Q}{\pi R^2} = \dfrac{6.46 \times 10^{-4}}{\pi (0.0125)^2}$

$= 1.32 \, \text{m/s}.$

The centre-line velocity is obtained by putting $r = 0$ in equation (3.7a),

$$V_z|_{r=0} = V_{z\,\text{max}} = V \left(\frac{3n + 1}{n + 1} \right) = 1.32 \times \frac{3 \times 0.5 + 1}{0.5 + 1}$$

$$= 2.2 \, \text{m/s}$$

For the pipe diameter of 37 mm, for the same value of the flow rate, equation (3.8) suggests that

$-\Delta p \propto R^{-(3n+1)}$

Hence, the pressure drop for the new pipe diameter can be estimated:

$$-\Delta p_{\text{new}} = -\Delta p_{\text{old}} \left(\frac{R_{\text{new}}}{R_{\text{old}}} \right)^{-(3n+1)}$$

$$= 110 \left(\frac{37/2}{25/2} \right)^{-2.5} = 41.3 \, \text{kPa}$$

Note that if the fluid were Newtonian ($n = 1$) of a viscosity of 3 Pa·s, the new value of the pressure drop would be 23 kPa, about half of that for the polymer solution.□

3.2.2 Bingham plastic and yield-pseudoplastic fluids

A fluid with a yield stress will flow only if the applied stress (proportional to pressure gradient) exceeds the yield stress. There will be a solid plug-like core flowing in the middle of the pipe where $|\tau_{rz}|$ is less than the yield stress, as shown schematically in Figure 3.4. Its radius, R_p, will depend upon the magnitude of the yield stress and on the wall shear stress. From equation (3.2),

$$\frac{\tau_0^B}{\tau_w} = \frac{R_p}{R} \tag{3.9}$$

where τ_w is the shear stress at the wall of the pipe.

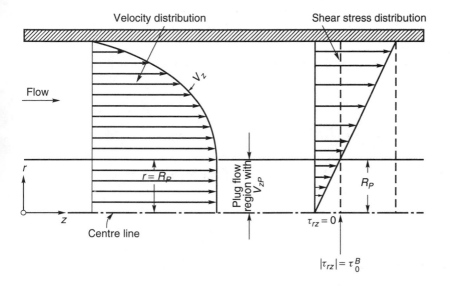

Figure 3.4 *Schematic velocity distribution for laminar flow of a Bingham plastic fluid in a pipe*

In the annular area $R_p < r < R$, the velocity will gradually decrease from the constant plug velocity to zero at the pipe wall. The expression for this velocity distribution will now be derived.

For the region $R_p < r < R$, the value of shear stress will be greater than the yield stress of the fluid, and the Bingham fluid model for pipe flow is given by (equation (1.16) in Chapter 1):

$$\tau_{rz} = \tau_0^B + \mu_B \left(-\frac{dV_z}{dr} \right) \tag{3.10}$$

Now combining equations (3.2) and (3.10) followed by integration yields the following expression for the velocity distribution

$$V_z = -\frac{1}{\mu_B} \left\{ \left(\frac{-\Delta p}{L} \right) \frac{r^2}{2} - r\tau_0^B \right\} + \text{constant}$$

At the walls of the pipe (i.e. when $r = R$), the velocity V_z must be zero to satisfy the condition of no-slip. Substituting the value $V_z = 0$, when $r = R$:

$$\text{constant} = \frac{1}{\mu_B} \left\{ \left(\frac{-\Delta p}{L} \right) \frac{R^2}{2} - R\tau_0^B \right\}$$

and, therefore:

$$V_z = \left(\frac{-\Delta p}{L} \right) \frac{R^2}{4\mu_B} \left(1 - \frac{r^2}{R^2} \right) - \frac{\tau_0^B}{\mu_B} R \left(1 - \frac{r}{R} \right) \qquad (3.11)$$

Clearly, equation (3.11) is applicable only when $|\tau_{rz}| > |\tau_0^B|$ and $r \geq R_p$. The corresponding velocity, V_{zp}, in the plug region $(0 \leq r \leq R_p)$ is obtained by substituting $r = R_p$ in equation (3.11) to give:

$$V_{zp} = \left(\frac{-\Delta p}{L} \right) \frac{R^2}{4\mu_B} \left(1 - \frac{R_p}{R} \right)^2 \qquad 0 \leq r \leq R_p \qquad (3.12)$$

The corresponding expression for the volumetric flow rate, Q, is obtained by evaluating the integral

$$Q = \int_0^R 2\pi r V_z \, \mathrm{d}r = \int_0^{R_p} 2\pi r V_{zp} \, \mathrm{d}r + \int_{R_p}^R 2\pi r V_z \, \mathrm{d}r$$

Substitution from equations (3.11) and (3.12), and integration yields for Q:

$$Q = \frac{\pi R^4}{8\mu_B} \left(\frac{-\Delta p}{L} \right) \left(1 - \frac{4}{3}\phi + \frac{1}{3}\phi^4 \right) \qquad (3.13)$$

where $\phi = \tau_0^B / \tau_w$.

It is useful to re-write equation (3.13) in a dimensionless form as:

$$f = \frac{16}{Re_B} \left(1 + \frac{1}{6} \frac{He}{Re_B} - \frac{1}{3} \frac{He^4}{f^3 Re_B^7} \right) \qquad (3.13a)$$

where f is the usual friction factor defined as $(-\Delta p / 2\rho u^2 \cdot D/L)$; Re_B is the Reynolds number $(= \rho V D / \mu_B)$ and He is the Hedström number defined as $(\rho D^2 \tau_0^B / \mu_B^2)$ [Hedström, 1952]. It should be noted that equation (3.13) is implicit in pressure gradient (because τ_w, and hence ϕ, is a function of pressure gradient) and therefore, for a specified flow rate, an iterative method is needed to evaluate the pressure drop.

This analysis can readily be extended to the laminar flow of Herschel–Bulkley model fluids (equation 1.17), and the resulting final expressions for the point velocity and volumetric flow rate are [Skelland, 1967; Govier and Aziz, 1982; Bird *et al.*, 1983, 1987]:

$$V_z = \frac{nR}{(n+1)} \left(\frac{\tau_w}{m}\right)^{1/n} \left\{(1-\phi)^{(n+1)/n} - \left(\frac{r}{R} - \phi\right)^{(n+1)/n}\right\} \tag{3.14a}$$

and

$$Q = \pi R^3 n \left(\frac{\tau_w}{m}\right)^{1/n} (1-\phi)^{(n+1)/n} \left\{\frac{(1-\phi)^2}{3n+1} + \frac{2\phi(1-\phi)}{2n+1} + \frac{\phi^2}{n+1}\right\}$$

$$\tag{3.14b}$$

where ϕ is now the ratio (τ_0^H/τ_w). These expressions are also implicit in pressure gradient.

Example 3.2

The rheological properties of a china clay suspension can be approximated by either a power-law or a Bingham plastic model over the shear rate range 10 to $100\,\mathrm{s}^{-1}$. If the yield stress is 15 Pa and the plastic viscosity is 150 mPa·s, what will be the approximate values of the power-law consistency coefficient and flow behaviour index?

Estimate the pressure drop when this suspension is flowing under laminar conditions in a pipe of 40 mm diameter and 200 m long, when the centre-line velocity is 0.6 m/s, according to the Bingham plastic model? Calculate the centre-line velocity for this pressure drop for the power-law model.

Solution

Using the Bingham plastic model,

$$\tau_{rz} = \tau_0^B + \mu_B(-dV_z/dr)$$

when $-dV_z/dr = 10\,\mathrm{s}^{-1}$, $\tau_{rz} = 15 + 150 \times 10^{-3} \times 10 = 16.5\,\mathrm{Pa}$

$-dV_z/dr = 100\,\mathrm{s}^{-1}$, $\tau_{rz} = 15 + 150 \times 10^{-3} \times 100 = 30\,\mathrm{Pa}$

Now using the power-law model,

$$\tau_{rz} = m\left(-\frac{dV_z}{dr}\right)^n$$

Substituting the values of τ_{rz} and $(-dV_z/dr)$:

$16.5 = m(10)^n$, and

$30 = m(100)^n$

Solving for m and n gives:

$n = 0.26$, $m = 9.08\,\mathrm{Pa\cdot s}^n$

For a Bingham plastic fluid, equation (3.12) gives:

$$V_{z\,\max} = V_{zp} = 0.6 = \left(\frac{-\Delta p}{L}\right)\left(\frac{(20 \times 10^{-3})^2}{4 \times 0.15}\right)\left(1 - \frac{R_p}{R}\right)^2$$

Substitution for (R_p/R) from equation (3.9) and writing the wall shear stress in terms of pressure gradient gives:

$$0.6 = \left(\frac{-\Delta p}{L}\right)\left(\frac{(20 \times 10^{-3})^2}{4 \times 0.15}\right)\left[1 - \frac{15}{\dfrac{20 \times 10^{-3}}{2}\left(\dfrac{-\Delta p}{L}\right)}\right]^2$$

A trial and error procedure leads to

$$\frac{-\Delta p}{L} = 3200\,\mathrm{Pa/m}$$

and therefore the total pressure drop over the pipe length, $-\Delta p = 3200 \times 200 = 640\,\mathrm{kPa}$

The centre-line velocity according to the power-law model is given by equation (3.4) with $r = 0$, i.e.

$$V_{z,\max} = \left(\frac{n}{n+1}\right)\left(-\frac{\Delta p}{mL} \cdot \frac{R}{2}\right)^{1/n} \cdot R$$

$$= \left(\frac{0.26}{0.26+1}\right)\left(\frac{3200 \times 20 \times 10^{-3}}{9.08 \times 2}\right)^{1/0.26} \times 20 \times 10^{-3}$$

$$= 0.52\,\mathrm{m/s}$$

One can easily see that the plug like motion occurs across a substantial portion of the cross-section as $R_p = 0.47\,R$.□

Before concluding this section, it is appropriate to mention here that one can establish similar $Q - \Delta p$ relations for the other commonly used viscosity models in an identical manner. A summary of such relations can be found in standard textbooks [Skelland, 1967; Govier and Aziz, 1982].

3.2.3 *Average kinetic energy of fluid*

In order to obtain the kinetic energy correction factor, α, for insertion in the mechanical energy balance, it is necessary to evaluate the average kinetic energy per unit mass in terms of the average velocity of flow. The calculation procedure is exactly similar to that used for Newtonian fluids, (e.g. see [Coulson and Richardson, 1999]).

Average kinetic energy/unit mass

$$= \frac{\int \frac{1}{2} V_z^2 \, d\dot{m}}{\int d\dot{m}} = \frac{\int_0^R \frac{1}{2} V_z^2 2\pi r V_z \rho dr}{\int_0^R 2\pi r V_z \rho dr} \qquad (3.15a)$$

$$= \frac{V^2}{2\alpha} \qquad (3.15b)$$

where α is a kinetic energy correction factor to take account of the non-uniform velocity over the cross-section. For power-law fluids, substitution for V_z from equation (3.7a) into equation (3.15a) and integration gives

$$\alpha = \frac{(2n+1)(5n+3)}{3(3n+1)^2} \qquad (3.16)$$

The corresponding expression for a Bingham plastic is cumbersome. However, Metzner [1956] gives a simple expression for α which is accurate to within 2.5%:

$$\alpha = \frac{1}{2 - \phi} \qquad (3.17)$$

Again, both equations (3.16) and (3.17) reduce to $\alpha = 1/2$ for Newtonian fluid behaviour. Note that as the degree of shear-thinning increases, i.e. the value of n decreases, the kinetic energy correction factor approaches unity at $n = 0$ as would be expected, as all the fluid is flowing at the same velocity (Figure 3.3). For shear-thickening fluids, on the other hand, it attains a limiting value of 0.37 for the infinite degree of shear-thickening behaviour ($n = \infty$).

3.2.4 Generalised approach for laminar flow of time-independent fluids

Approach used in section 3.2 for power-law and Bingham plastic model fluids can be extended to other fluid models. Even if the relationship between shear stress and shear rate is not known exactly, it is possible to use the following approach to the problem. It depends upon the fact that the shear stress distribution over the pipe cross-section is not a function of the fluid rheology and is given simply by equation (3.2), which can be re-written in terms of the wall shear stress, i.e.

$$\frac{\tau_{rz}}{\tau_w} = \frac{r}{R} \qquad (3.18)$$

The volumetric flow rate is given by

$$Q = \int_0^R 2\pi r V_z \, dr \qquad (3.19)$$

Integration by parts leads to:

$$Q = \pi r^2 V_z \Big|_0^R + \int_0^R \pi r^2 \left(-\frac{dV_z}{dr} \right) dr$$

For the no-slip boundary condition at the wall, the first term on the right hand side is identically zero and therefore:

$$Q = \int_0^R \pi r^2 \left(-\frac{dV_z}{dr} \right) dr \qquad (3.20)$$

$$\because V_z = 0 \quad \text{at} \quad r = R$$

Now changing the variable of integration from r to τ_{rz} using equation (3.18):

$$r = R \left(\frac{\tau_{rz}}{\tau_w} \right); \quad dr = R \frac{d\tau_{rz}}{\tau_w}$$

When $r = 0$, $\tau_{rz} = 0$ and at the walls of the pipe when $r = R$, $\tau_{+z} = \tau_w$. Substitution in equation (3.20) gives:

$$Q = \frac{\pi R^3}{\tau_w^3} \int_0^{\tau_w} \tau_{rz}^2 f(\tau_{rz}) \, d\tau_{rz} \qquad (3.21)$$

The velocity gradient (or the shear rate) term $(-dV_z/dr)$ has been replaced by a function of the corresponding shear stress via equation (1.10). The form of the function will therefore depend on the viscosity model chosen to describe the rheology of the fluid. Equation (3.21) can be used in two ways:

(i) to determine general non-Newtonian characteristics of a time-independent fluid, as demonstrated in Chapter 2 and in Section 3.2.5, or
(ii) to be integrated directly for a specific fluid model to obtain volumetric flow rate-pressure drop relationship. This is demonstrated for the flow of a power-law fluid, for which the shear rate is given by equation (3.3):

$$-\frac{dV_z}{dr} = f(\tau_{rz}) = \left(\frac{\tau_{rz}}{m} \right)^{1/n} \qquad (3.22)$$

Substitution of equation (3.22) into equation (3.21) followed by integration and re-arrangement gives:

$$V = \frac{Q}{\pi R^2} = \left(\frac{n}{3n+1} \right) \left(\frac{\tau_w}{m} \right)^{1/n} R \qquad (3.23)$$

Substituting for $\tau_w = (R/2)(-\Delta p/L)$ in equation (3.23) gives:

$$V = \left(\frac{n}{3n+1}\right)\left(\frac{(-\Delta p)R}{2mL}\right)^{1/n} R \tag{3.23a}$$

which is identical to equation (3.6).

A similar analysis for a Bingham plastic fluid will lead to the same expression for Q as equation (3.13). Thus, these are alternative methods of obtaining flow rate-pressure gradient relation for any specific model to describe the fluid rheology. The scheme given above provides a quicker method of obtaining the relation between pressure gradient and flow rate, but has the disadvantage that it does not provide a means of obtaining the velocity profile.

Example 3.3

The shear-dependent viscosity of a commercial grade of polypropylene at 403 K can satisfactorily be described using the three constant Ellis fluid model (equation 1.15), with the values of the constants: $\mu_0 = 1.25 \times 10^4$ Pa·s, $\tau_{1/2} = 6900$ Pa and $\alpha = 2.80$. Estimate the pressure drop required to maintain a volumetric flow rate of $4\,\text{cm}^3/\text{s}$ through a 50 mm diameter and 20 m long pipe. Assume the flow to be laminar.

Solution

Since we need the $Q - (-\Delta p)$ relation to solve this problem, such a relationship will be first derived using the generalised approach outlined in Section 3.2.4. For laminar flow in circular pipes, the Ellis fluid model is given as:

$$\tau_{rz} = \frac{\mu_0(-\mathrm{d}V_z/\mathrm{d}r)}{1+(\tau_{rz}/\tau_{1/2})^{\alpha-1}} \tag{1.15}$$

or

$$-\frac{\mathrm{d}V_z}{\mathrm{d}r} = f(\tau_{rz}) = \frac{1}{\mu_0}\left[\tau_{rz} + \frac{\tau_{rz}^{\alpha}}{\tau_{1/2}^{\alpha-1}}\right] \tag{3.24}$$

Substituting equation (3.24) into equation (3.21) followed by integration and rearrangement gives:

$$Q = \frac{\pi R^3 \tau_w}{4\mu_0}\left[1+\left(\frac{4}{\alpha+3}\right)\left(\frac{\tau_w}{\tau_{1/2}}\right)^{\alpha-1}\right]$$

Note that in the limit of $\tau_{1/2} \to \infty$, i.e. for Newtonian fluid behaviour, this equation reduces to the Hagen–Poiseuille equation.

Now substituting the numerical values:

$$4 \times 10^{-6} = \frac{3.14 \times (0.025)^3 \tau_w}{4 \times 1.25 \times 10^4}\left[1+\left(\frac{4}{2.8+3}\right)\left(\frac{\tau_w}{6900}\right)^{2.8-1}\right]$$

or $\quad 4074.37 = \tau_w(1+8.4 \times 10^{-8}\tau_w^{1.8})$

A trial and error procedure gives $\tau_w = 3412\,\text{Pa}$

$$\therefore\ -\Delta p = \frac{4\tau_w L}{D} = \frac{4 \times 3412 \times 20}{0.05} = 5.46 \times 10^6\,\text{Pa}$$

i.e. the pressure drop across the pipe will be 5.46 MPa.□

3.2.5 *Generalised Reynolds number for the flow of time-independent fluids*

It is useful to define an appropriate Reynolds number which will result in a unique friction factor-Reynolds number curve for all time-independent fluids in laminar flow in circular pipes. Metzner and Reed [1955] outlined a generalised approach obviating this difficulty. The starting point is equation 3.21:

$$Q = \frac{\pi R^3}{\tau_w^3} \int_0^{\tau_w} \tau_{rz}^2 f(\tau_{rz})\,d\tau_{rz} \tag{3.21}$$

Equation (3.21) embodies a definite integral, the value of which depends only on the values of the integral function at the limits, and not on the nature of the continuous function that is integrated. For this reason it is necessary to evaluate only the wall shear stress τ_w and the associated velocity gradient at the wall $(-dV_z/dr)$ at $r = R$ or $f(\tau_w)$. This is accomplished by the use of the Leibnitz rule which allows a differential of an integral of the form $(d/ds')\{\int_0^{s'} s^2 f(s)\,ds\}$ to be written as $(s')^2 f(s')$ where s is a dummy variable of integration (τ_{rz} here) and s' is identified as τ_w. First multiplying both sides of equation (3.21) by τ_w^3 and then differentiating with respect to τ_w gives:

$$\frac{d}{d\tau_w}\left\{\tau_w^3\left(\frac{Q}{\pi R^3}\right)\right\} = \frac{d}{d\tau_w} \int_0^{\tau_w} \tau_{rz}^2 f(\tau_{rz})\,d\tau_{rz}$$

Applying the Leibnitz rule to the integral on the right-hand side gives:

$$3\tau_w^2\left(\frac{Q}{\pi R^3}\right) + \tau_w^3 \frac{d}{d\tau_w}\left(\frac{Q}{\pi R^3}\right) = \tau_w^2 f(\tau_w)$$

Introducing a factor of 4 on both sides and further rearrangement of the terms on the left-hand side gives:

$$f(\tau_w) = \left(-\frac{dV_z}{dr}\right)_{\text{wall}} = \frac{4Q}{\pi R^3}\left\{\frac{3}{4} + \frac{1}{4} \cdot \frac{\dfrac{d(4Q/\pi R^3)}{(4Q/\pi R^3)}}{\dfrac{d\tau_w}{\tau_w}}\right\} \tag{3.25a}$$

or in terms of average velocity V and pipe diameter D,

$$\left(-\frac{dV_z}{dr}\right)_{wall} = \frac{8V}{D}\left\{\frac{3}{4} + \frac{1}{4}\frac{d\log(8V/D)}{d\log\tau_w}\right\}$$

(3.25b)

Here, $(8V/D)$ is the wall shear rate for a Newtonian fluid and is referred to as the nominal shear rate for a non-Newtonian fluid which is identical to equation (2.5) in Chapter 2. Alternatively, writing it in terms of the slope of $\log\tau_w - \log(8V/D)$ plot's,

$$\dot{\gamma}_w = \left(-\frac{dV_z}{dr}\right)_{wall} = \left(\frac{8V}{D}\right)\left(\frac{3n'+1}{4n'}\right)$$

(3.25c)

where $n' = (d\log\tau_w/d\log(8V/D))$ which is not necessarily constant at all shear rates. Equation (3.25c) is identical to equation (2.6) in Chapter 2.

Thus, the index n' is the slope of the log–log plots of the wall shear stress τ_w versus $(8V/D)$ in the laminar region (the limiting condition for laminar flow is discussed in Section 3.3). Plots of τ_w versus $(8V/D)$ thus describe the flow behaviour of time-independent non-Newtonian fluids and may be used directly for scale-up or process design calculations.

Over the range of shear rates over which n' is approximately constant, one may write a power-law type equation for this segment as

$$\tau_w = \frac{D}{4}\left(\frac{-\Delta p}{L}\right) = m'\left(\frac{8V}{D}\right)^{n'}$$

(3.26)

Substituting for τ_w in terms of the friction factor, f, $(= \tau_w/(1/2)\rho V^2)$, equation (3.26) becomes:

$$f = \frac{2}{\rho V^2}m'\left(\frac{8V}{D}\right)^{n'}$$

(3.27)

Now a Reynolds number may be defined so that in the laminar flow regime, it is related to f in the same way as is for Newtonian fluids, i.e.

$$f = \frac{16}{Re_{MR}}$$

(3.28a)

from which

$$Re_{MR} = \frac{\rho V^{2-n'} D^{n'}}{8^{n'-1} m'}$$

(3.28b)

Since Metzner and Reed [1955] seemingly were the first to propose this definition of the generalised Reynolds number, and hence the subscripts 'MR' in Re_{MR}.

It should be realised that by defining the Reynolds number in this way, the same friction factor chart can be used for Newtonian and time-independent non-Newtonian fluids in the laminar region. In effect, we are writing,

$$\text{Re}_{MR} = \frac{\rho V D}{\mu_{\text{eff}}} \tag{3.29}$$

Thus, the flow curve provides the value of the effective viscosity μ_{eff} where $\mu_{\text{eff}} = m'(8V/D)^{n'-1}$. It should be noted that the terms apparent and effective viscosity have been used to relate the behaviour of a non-Newtonian fluid to an equivalent property of a hypothetical Newtonian fluid. The apparent viscosity is the point value of the ratio of the shear stress to the shear rate. The effective viscosity is linked to the macroscopic behaviour $(Q - (-\Delta p))$ characteristics, for instance) and is equal to the Newtonian viscosity which would give the same relationship. It will be seen in Chapter 8 that this approach has also been quite successful in providing a reasonable basis for correlating much of the literature data on power consumption for the mixing of time-independent non-Newtonian fluids. The utility of this approach for reconciling the friction factor data for all time-independent fluids including shear-thinning and viscoplastic fluids, has been demonstrated by Metzner and Reed [1955] and subsequently by numerous other workers. Indeed, by writing a force balance on an element of fluid flowing in a circular pipe it can readily be shown that equation (3.28a) is also applicable for visco-elastic fluids. Figure 3.5 confirms this expectation for the flow of highly shear-thinning inelastic and visco-elastic polymer solutions in the range $0.28 \leq n' \leq 0.92$. [Chhabra *et al.*, 1984]. Griskey and Green [1971] have shown that the same approach may be adopted for the flow of shear-thickening materials, in the range $1.15 \leq n' \leq 2.50$. Experimental evidence suggests that stable laminar flow prevails for time-independent non-Newtonian fluids for Re_{MR} up to about 2000–2500; the transition from laminar to turbulent flow as well as the friction factor – Reynolds number characteristics beyond the laminar region are discussed in detail in the next section.

Before concluding this section, it is useful to link the apparent power-law index n' and consistency coefficient m' (equation 3.26) to the true power-law constants n and m, and to the Bingham plastic model constants τ_0^B and μ_B. This is accomplished by noting that $\tau_w = (D/4)(-\Delta p/L)$ always gives the wall shear stress and the corresponding value of the wall shear rate $\dot{\gamma}_w (= dV_z/dr)_w$ can be evaluated using the expressions for velocity distribution in a pipe presented in Sections 3.2.1 and 3.2.2.

For the laminar flow of a power-law fluid in a pipe, the velocity distribution over the pipe cross-section is given by equation (3.7a):

$$\frac{V_z}{V} = \left(\frac{3n+1}{n+1}\right) \left\{ 1 - \left(\frac{r}{R}\right)^{(n+1)/n} \right\} \tag{3.7a}$$

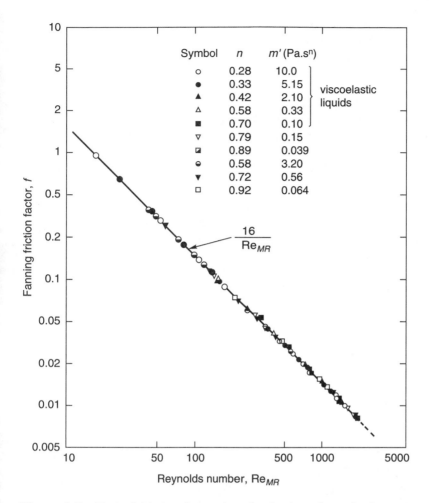

Figure 3.5 *Typical friction factor data for laminar flow of polymer solutions [Chhabra et al., 1984]*

Differentiating with respect to r and substituting $r = R$ gives the expression for the velocity gradient (or shear rate) at the wall as:

$$-\frac{dV_z}{dr} = \frac{3n+1}{n} \left(\frac{r}{R}\right)^{1/n} \frac{V}{R}$$

and:
$$\left(-\frac{dV_z}{dr}\right)_{r=R} = \dot{\gamma}_w = \left(\frac{3n+1}{n}\right)\frac{V}{R} = \left(\frac{3n+1}{4n}\right)\left(\frac{8V}{D}\right)$$

The corresponding value of the shear stress at the wall τ_w for a power-law fluid is obtained by substituting this value of $\dot{\gamma}_w$ in equation (3.3):

$$\tau_w = \tau_{rz}|_{r=R} = m \left[\left(-\frac{dV_z}{dr} \right)_{r=R} \right]^n$$

$$= m \left\{ \left(\frac{3n+1}{4n} \right) \left(\frac{8V}{D} \right) \right\}^n$$

which is identical to equation (2.3) presented in Chapter 2. Now comparing this equation with equation (3.26) for a power-law fluid gives:

$$n' = n; \quad m' = m \left(\frac{3n+1}{4n} \right)^n \tag{3.30a}$$

In this case therefore n' and m' are both constant and independent of shear rate. Similarly, it can be shown that the Bingham model parameters τ_0^B and μ_B are related to m' and n' as [Skelland, 1967]:

$$n' = \frac{1 - \frac{4}{3}\phi + \frac{\phi^4}{3}}{1 - \phi^4} \tag{3.30b}$$

and

$$m' = \tau_w \left[\frac{\mu_B}{\tau_w \left(1 - \frac{4}{3}\phi + \frac{\phi^4}{3} \right)} \right]^{n'} \tag{3.30c}$$

where $\phi = (\tau_0^B / \tau_w)$.

It should be noted that in this case, m' and n' are not constant and depend on the value of the wall shear stress τ_w.

3.3 Criteria for transition from laminar to turbulent flow

For all fluids, the nature of the flow is governed by the relative importance of the viscous and the inertial forces. For Newtonian fluids, the balance between these forces is characterised by the value of the Reynolds number. The generally accepted value of the Reynolds number above which stable laminar flow no longer occurs is 2100 for Newtonian fluids. For time-independent fluids, the critical value of the Reynolds number depends upon the type and the degree of non-Newtonian behaviour. For power-law fluids ($n = n'$), the criterion of Ryan and Johnson [1959] can be used,

$$Re_{MR} = \frac{6464n}{(3n+1)^2} (2+n)^{(2+n)/(1+n)} \tag{3.31}$$

While for Newtonian fluids equation (3.31) predicts the critical Reynolds number of 2100, the corresponding limiting values increase with decreasing values of the power-law index, reaching a maximum of about 2400 at $n = 0.4$ and then dropping to 1600 at $n = 0.1$. The latter behaviour is not in line with the experimental results of Dodge and Metzner [1959] who observed laminar flow conditions up to $\mathrm{Re}_{MR} \sim 3100$ for a fluid with $n' = 0.38$. Despite the complex dependence of the limiting Reynolds number on the flow behaviour index embodied in equation (3.31) and the conflicting experimental evidence, it is probably an acceptable approximation to assume that the laminar flow conditions cease to prevail at Reynolds numbers above ca. 2000–2500 and, for the purposes of process calculations, the widely accepted figure of 2100 can be used for time-independent fluids characterised in terms of n'. It is appropriate to add here that though the friction factor for visco-elastic fluids in the laminar regime is given by equation (3.28a), the limited experimental results available suggest much higher values for the critical Reynolds number. For instance, Metzner and Park [1964] reported that their friction factor data for visco-elastic polymer solutions were consistent with equation (3.28a) up to about $\mathrm{Re}_{MR} = 10\,000$. However, it is not yet possible to put forward a quantitative criterion for calculating the limiting value of Re_{MR} for visco-elastic fluids.

Several other criteria, depending upon the use of a specific fluid model, are also available in the literature [Hanks, 1963; Govier and Aziz, 1982; Wilson, 1996; Malin, 1997]. For instance, Hanks [1963] proposed the following criterion for Bingham plastic fluids:

$$(\mathrm{Re}_B)_c = \frac{\rho V D}{\mu_B} = \frac{1 - \dfrac{4}{3}\phi_c + \dfrac{\phi_c^4}{3}}{8\phi_c} He \qquad (3.32a)$$

where the ratio, $\phi_c = (\tau_0^B / \tau_{w_c})$, is given by:

$$\frac{\phi_c}{(1 - \phi_c)^3} = \frac{He}{16\,800} \qquad (3.32b)$$

The Hedström number, He, is defined as:

$$He = \frac{\rho D^2 \tau_0^B}{\mu_B^2} = \mathrm{Re}_B \times \mathrm{Bi} \qquad (3.33)$$

where $\mathrm{Bi} = (D\tau_0^B / \mu_B V)$ is the Bingham number. For a given pipe size (D) and Bingham plastic fluid behaviour (ρ, μ_B, τ_0^B), the Hedström number will be known and the value of ϕ_c can be obtained from equation (3.32b) which, in turn, facilitates the calculation of $(\mathrm{Re}_B)_c$ using equation (3.32a), as illustrated in example 3.4. More recent numerical calculations [Malin, 1997] lend further support to the validity of equations (3.32a,b).

Both Wilson [1996] and Slatter [1996] have also re-evaluated the available criteria for the laminar–turbulent transition, with particular reference to the flow of pseudoplastic and yield-pseudoplastic mineral slurries in circular pipes. Wilson [1996] has argued that the larger dissipative micro-eddies present in the wall region result in thicker viscous sub-layers in non-Newtonian fluids which, in turn, produce greater mean velocity, giving a friction factor lower than that for Newtonian fluids, for the same value of the pressure drop across the pipe. For power-law fluids, he was able to link the non-Newtonian apparent viscosity to the viscosity of a hypothetical Newtonian fluid simply through a function of n, the power-law flow behaviour index, such that the same $Q - (-\Delta p)$ relationship applies to both fluids. This, in turn, yields the criterion for laminar–turbulent transition in terms of the critical value of the friction factor as a function of n (power-law index) alone. Note that in this approach, the estimated value of the effective viscosity will naturally depend upon the type of fluid and pipe diameter, D. Similarly, Slatter [1996] has put forward a criterion in terms of a new Reynolds number for the flow of Herschel–Bulkley model fluids (equation (1.17)) to delineate the laminar–turbulent transition condition. His argument hinges on the fact that the inertial and viscous forces in the fluid are determined solely by that part of the fluid which is undergoing deformation (shearing), and hence he excluded that part of the volumetric flow rate attributable to the unsheared plug of material present in the middle of the pipe. These considerations lead to the following definition of the modified Reynolds number:

$$\text{Re}_{\text{mod}} = \frac{8\rho V_{\text{ann}}^2}{\tau_0^H + m \left(\dfrac{8V_{\text{ann}}}{D_{\text{shear}}} \right)^n} \tag{3.34}$$

where $V_{\text{ann}} = \dfrac{Q - Q_{\text{plug}}}{\pi(R^2 - R_p^2)}$, and $D_{\text{shear}} = 2(R - R_p)$

Laminar flow conditions cease to exist at $\text{Re}_{\text{mod}} = 2100$. The calculation of the critical velocity corresponding to $\text{Re}_{\text{mod}} = 2100$ requires an iterative procedure. For known rheology (ρ, m, n, τ_0^H) and pipe diameter (D), a value of the wall shear stress is assumed which, in turn, allows the calculation of R_p, from equation (3.9), and Q and Q_p from equations (3.14b) and (3.14a) respectively. Thus, all quanties are then known and the value of Re_{mod} can be calculated. The procedure is terminated when the value of τ_w has been found which makes $\text{Re}_{\text{mod}} = 2100$, as illustrated in example 3.4 for the special case of $n = 1$, i.e., for the Bingham plastic model, and in example 3.5 for a Herschel–Bulkley fluid. Detailed comparisons between the predictions of equation (3.34) and experimental data reveal an improvement in the predictions, though the values of the critical velocity obtained using the criterion $\text{Re}_{MR} = 2100$ are only 20–25% lower than those predicted by equation (3.34). Furthermore, the two criteria coincide for power-law model fluids. Subsequently, it has also been shown that while the laminar–turbulent transition in small diameter tubes is virtually unaffected by the value of the yield stress, both the flow behaviour index (n) and the yield stress play increasingly greater roles in determining the transition point with increasing pipe diameter. Finally, the scant results obtained with a kaolin slurry and a CMC solution seem to suggest that the laminar–turbulent transition is not influenced by the pipe roughness [Slatter, 1996, 1997].

Example 3.4

The rheological behaviour of a coal slurry ($1160\,\text{kg/m}^3$) can be approximated by the Bingham plastic model with $\tau_0^B = 0.5\,\text{Pa}$ and $\mu_B = 14\,\text{mPa·s}$. It is to be pumped through a 400 mm diameter pipe at the rate of 188 kg/s. Ascertain the nature of the flow by calculating the maximum permissible velocity for laminar flow conditions. Contrast the predictions of equations (3.33) and (3.34).

Solution

Here, the Hedström number, $\text{He} = \dfrac{\rho D^2 \tau_0^B}{\mu_B^2}$

$$= \frac{1160 \times 0.4^2 \times 0.5}{(14 \times 10^{-3})^2}$$

i.e. $He = 4.73 \times 10^5$ which when substituted in equation (3.32b) yields,

$$\frac{\phi_c}{(1 - \phi_c)^3} = \frac{4.73 \times 10^5}{16\,800} = 28.15$$

A trial and error procedure gives $\phi_c = 0.707$. Now substituting for He and ϕ_c in equation (3.32a):

$$(\text{Re}_B)_c = \frac{1 - \dfrac{4}{3} \times 0.707 + \dfrac{(0.707)^4}{3}}{8 \times 0.707} \times 4.73 \times 10^5$$

or $\quad (\text{Re}_B)_c = \dfrac{\rho V_c D}{\mu_B} = 11\,760$

and the maximum permissible velocity, V_c therefore is,

$$V_c = \frac{11\,760 \times 14 \times 10^{-3}}{1160 \times 0.4} = 0.354\,\text{m/s}.$$

The actual velocity in the pipe is

$$\left(\frac{188}{1160}\right)\left(\frac{4}{\pi \times 0.4^2}\right) = 1.29\,\text{m/s}$$

Thus, the flow in the pipe is not streamline.

Alternatively, one can use equation (3.34) to estimate the maximum permissible velocity for streamline flow in the pipe. In this example, $n = 1$, $m = 0.014$ Pa and $\tau_0^H = 0.5$ Pa. As mentioned previously the use of equation (3.34) requires an iterative procedure, and to initiate this method let us assume a value of $\tau_w = 0.6$ Pa.

$$\therefore \phi = \frac{\tau_0^H}{\tau_w} = \frac{0.5}{0.6} = \frac{R_p}{R}, \text{ i.e. } R_p = 0.166 \text{ m and}$$

$\phi = 0.833$. Now using equation (3.14b) for $n = 1$:

$$Q = \pi \times 0.2^3 \times 1 \left(\frac{0.6}{0.014} \right) (1 - 0.833)^2$$

$$\times \left[\frac{(1 - 0.833)^2}{4} + \frac{2 \times 0.833(1 - 0.833)}{3} + \frac{0.833^2}{2} \right]$$

$$= 0.0134 \text{ m}^3/\text{s}$$

The plug velocity, V_p, is calculated from equation (3.14a) by setting $r/R = R_p/R = \phi = 0.833$, i.e.

$$V_p = \frac{1 \times 0.2}{(1 + 1)} \left(\frac{0.6}{0.014} \right) (1 - 0.833)^2 = 0.1195 \text{ m/s}$$

$$Q_p = V_p \pi R_p^2 = 0.1195 \times \pi \times 0.166^2 = 0.01035 \text{ m}^3/\text{s}$$

$$\therefore \quad Q_{\text{ann}} = Q - Q_p = 0.0134 - 0.01035 = 0.00305 \text{ m}^3/\text{s}$$

$$V_{\text{ann}} = \frac{Q_{\text{ann}}}{\pi(R^2 - R_p^2)} = \frac{0.00305}{\pi(0.2^2 - 0.166^2)} = 0.078 \text{ m/s}$$

$$D_{\text{shear}} = 2(R - R_p) = 2(0.2 - 0.166) = 0.068 \text{ m}$$

$$\therefore \quad \text{Re}_{\text{mod}} = \frac{8\rho V_{\text{ann}}^2}{\tau_0^H + m \left(\frac{8 V_{\text{ann}}}{D_{\text{shear}}} \right)^n} = \frac{8 \times 1160 \times 0.078^2}{0.5 + 0.014 \left(\frac{8 \times 0.078}{0.068} \right)} = 90$$

which is too small for the flow to be turbulent. Thus, this procedure must be repeated for other values of τ_w to make $\text{Re}_{\text{mod}} = 2100$. A summary of calculations is presented in the table below.

τ_w (Pa)	Q (m³/s)	Q_p (m³/s)	Re_{mod}
0.6	0.0134	0.01035	90
0.70	0.0436	0.0268	890
0.73	0.0524	0.0305	1263
0.80	0.0781	0.0395	2700
0.77	0.0666	0.0358	1994
0.78	0.0706	0.0370	2233
0.775	0.0688	0.0365	2124

The last entry is sufficiently close to $\mathrm{Re_{mod}} = 2100$, and laminar flow will cease to exist at $\tau_w \geq 0.775\,\mathrm{Pa}$. Also, note that the use of equations (3.14a) and (3.14b) beyond this value of wall shear stress is incorrect.

$$\therefore \text{ maximum permissible velocity} = \frac{0.0688}{\dfrac{\pi}{4}(0.4)^2} = 0.55\,\mathrm{m/s}.$$

This value is some 40% higher than the previously calculated value of $0.35\,\mathrm{m/s}$. However, even on this count, the flow will be turbulent at the given velocity of $1.29\,\mathrm{m/s}.\square$

Example 3.5

Determine the critical velocity for the upper limit of laminar flow for a slurry with the following properties, flowing in a 150 mm diameter pipe.

$$\rho = 1150\,\mathrm{kg/m^3}; \quad \tau_0^H = 6\,\mathrm{Pa}; \quad m = 0.3\,\mathrm{Pa\cdot s^n} \quad \text{and} \quad n = 0.4$$

Solution

As in example 3.4, one needs to assume a value for τ_w, and then to calculate all other quantities using equations (3.14a) and (3.14b) which in turn allow the calculation of $\mathrm{Re_{mod}}$ using equation (3.34). A summary of the calculations is presented here in a tabular form.

τ_w (Pa)	$Q\,(\mathrm{m^3/s})$	$Q_p\,(\mathrm{m^3/s})$	$\mathrm{Re_{mod}}$
6.4	4.72×10^{-5}	4.27×10^{-5}	6.5×10^{-3}
7.4	3.097×10^{-3}	2.216×10^{-3}	26.6
8.4	0.01723	0.01	778
9.3	0.046	0.0224	5257
8.82	0.0287	0.0153	2100

Thus, the laminar–turbulent transition for this slurry in a 150 mm diameter pipe occurs when the wall shear stress is $8.82\,\mathrm{Pa}$ and the volumetric flowrate is $0.0287\,\mathrm{m^3/s}$.

$$\therefore \text{ mean velocity at this point} = \frac{Q}{(\pi/4)D^2} = \frac{0.0287}{(\pi/4)(0.15)^2} = 1.62\,\mathrm{m/s}$$

Hence, streamline flow will occur for this slurry in a 150 mm diameter pipe at velocities up to a value of $1.62\,\mathrm{m/s}.\square$

3.4 Friction factors for transitional and turbulent conditions

Though turbulent flow conditions are encountered less frequently with polymeric non-Newtonian substances, sewage sludges, coal and china clay suspensions are usually all transported in the turbulent flow regime in large diameter pipes. Therefore, considerable research efforts have been directed at developing a generalised approach for the prediction of the frictional pressure drop for turbulent flow in pipes, especially for purely viscous (power-law, Bingham plastic and Herschel–Bulkley models) fluids. Analogous studies for the flow of visco-elastic and the so-called drag-reducing fluids are somewhat inconclusive. Furthermore, the results obtained with drag-reducing polymer solutions also tend to be strongly dependent on the type and molecular weight of the polymers, the nature of the solvent and on the type of experimental set up used, and it is thus not yet possible to put forward generalised equations for the turbulent flow of such fluids. Therefore, the ensuing discussion is restricted primarily to the turbulent flow of time-independent fluids. However, excellent survey articles on the turbulent flow of visco-elastic and drag-reducing systems are available in the literature [Govier and Aziz, 1982; Cho and Hartnett, 1982; Sellin et al., 1982].

In the same way as there are many equations for predicting friction factor for turbulent Newtonian flow, there are numerous equations for time-independent non-Newtonian fluids; most of these are based on dimensional considerations combined with experimental observations [Govier and Aziz, 1982; Heywood and Cheng, 1984]. There is a preponderance of correlations based on the power-law fluid behaviour and additionally some expressions are available for Bingham plastic fluids [Tomita, 1959; Wilson and Thomas, 1985]. Here only a selection of widely used and proven methods is presented.

3.4.1 Power-law fluids

In a comprehensive study, Dodge and Metzner [1959] carried out a semi-empirical analysis of the fully developed turbulent flow of power-law fluids in smooth pipes. They used the same dimensional considerations for such fluids, as Millikan [1939] for incompressible Newtonian fluids, and obtained an expression which can be re-arranged in terms of the apparent power law index, n', (equation 3.26) as follows:

$$\frac{1}{\sqrt{f}} = A(n') \log[\text{Re}_{MR} f^{(2-n')/2}] + C(n') \tag{3.35}$$

where $A(n')$ and $C(n')$ are two unknown functions of n'. Based on extensive experimental results in the range $2900 \leq \text{Re}_{MR} \leq 36\,000$; $0.36 \leq n' \leq 1$ for polymer solutions and particulate suspensions, Dodge and Metzner [1959] obtained,

$$A(n') = 4(n')^{-0.75} \tag{3.36a}$$

$$C(n') = -0.4(n')^{-1.2} \tag{3.36b}$$

Incorporating these values in equation (3.35),

$$\frac{1}{\sqrt{f}} = \frac{4}{(n')^{0.75}} \log[\mathrm{Re}_{MR} f^{(2-n')/2}] - \frac{0.4}{(n')^{1.2}} \tag{3.37}$$

and this relation is shown graphically in Figure 3.6.

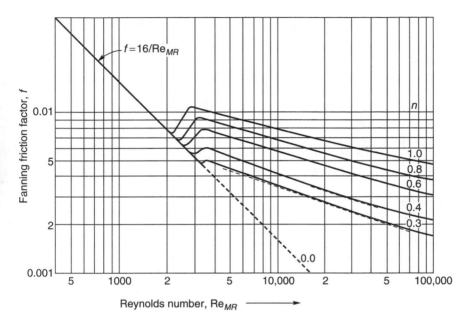

Figure 3.6 *Friction factor – Reynolds number behaviour for time-independent fluids [Dodge and Metzner, 1959]*

A more detailed derivation of equation (3.37) is available in their original paper and elsewhere [Skelland, 1967]. For Newtonian fluids ($n' = 1$), equation (3.37) reduces to the well-known Nikuradse equation. Dodge and Metzner [1959] also demonstrated that their data for clay suspensions which did not conform to power-law behaviour, were consistent with equation (3.37) provided that the slope of $\log \tau_w - \log(8V/D)$ plots was evaluated at the appropriate value of the wall shear stress. It is also important to point out here that equation (3.37) necessitates the values of n' and m' be evaluated from volumetric flow rate – pressure drop data for laminar flow conditions. Often, this requirement poses significant experimental difficulties. Finally, needless to say, this correlation is implicit in friction factor f (like the equation for Newtonian fluids) and hence an iterative technique is needed for its solution. The recent method of Irvine [1988] obviates this difficulty. Based on the Blasius expression for velocity profile for turbulent flow (discussed subsequently) together with modifications based on experimental results, Irvine [1988] proposed the following Blasius like expression for power-law fluids:

$$f = \{D(n)/\mathrm{Re}_{MR}\}^{1/(3n+1)} \tag{3.38}$$

$$\text{where} \quad D(n) = \frac{2^{n+4}}{7^{7n}} \left(\frac{4n}{3n+1}\right)^{3n^2}$$

Note that this cumbersome expression does reduce to the familiar Blasius expression for $n = 1$ and is explicit in friction factor, f. Equation (3.38) was stated to predict the values of friction factor with an average error of $\pm 8\%$ in the range of conditions: $0.35 \le n \le 0.89$ and $2000 \le \mathrm{Re}_{MR} \le 50\,000$. Though this approach has been quite successful in correlating most of the literature data, significant deviations from it have also been observed [Harris, 1968; Quader and Wilkinson, 1980; Heywood and Cheng, 1984]; though the reasons for such deviations are not immediately obvious but possible visco-elastic effects and erroneous values of the rheological parameters (e.g. n' and m') cannot be ruled out. Example 3.6 illustrates the application of these methods.

Example 3.6

A non-Newtonian polymer solution (density $1000\,\mathrm{kg/m^3}$) is in steady flow through a smooth 300 mm inside diameter 50 m long pipe at the mass flow rate of 300 kg/s. The following data have been obtained for the rheological behaviour of the solution using a tube viscometer. Two tubes, 4 mm and 6.35 mm in inside diameter and 2 m and 3.2 m long respectively were used to encompass a wide range of shear stress and shear rate.

Mass flow rate (kg/h)	Pressure drop (kPa)	Mass flow rate (kg/h)	Pressure drop (kPa)
$D = 4\,\mathrm{mm}, L = 2\,\mathrm{m}$		$D = 6.35\,\mathrm{mm}, L = 3.2\,\mathrm{m}$	
33.9	49	18.1	27
56.5	57.6	45.4	36
95	68.4	90.7	44
136	76.8	181	54
153.5	79.5	272	61

Determine the pump power required for this pipeline. How will the power requirement change if the flow rate is increased by 20%?

Solution

First, the tube viscometer data will be converted to give the wall shear stress, τ_w, and nominal shear rate, $(8V/D)$:

$$\tau_w = \frac{D}{4}\left(\frac{-\Delta p}{L}\right) = \frac{4 \times 10^{-3}}{4} \times \frac{49 \times 1000}{2} = 24.5\,\mathrm{Pa}$$

and $\dfrac{8V}{D} = \dfrac{8}{4 \times 10^{-3}} \times \dfrac{33.9}{1000 \times 3600} \times \dfrac{4}{\pi(4 \times 10^{-3})^2} = 1499\,\text{s}^{-1}$

Similarly the other mass flow rate–pressure drop data can be converted into τ_w – $(8V/D)$ form as are shown in the table.

τ_w (Pa)	$(8V/D)\,(\text{s}^{-1})$	τ_w (Pa)	$(8V/D)\,(\text{s}^{-1})$
	$D = 4\,\text{mm}$		$D = 6.35\,\text{mm}$
24.5	1499	13.4	200
28.8	2500	17.86	502
34.2	4200	21.83	1002
38.4	6000	26.8	2000
39.8	6800	30.26	3005

Note that since L/D for both tubes is 500, entrance effects are expected to be negligible. Figure 3.7 shows the $\tau_w - (8V/D)$ data on log-log coordinates. Obviously, n' is not constant, though there seem to be two distinct power-law regions with parameters:

$n' = 0.3 \quad m' = 2.74\,\text{Pa·s}^{n'} \quad (\tau_w < \sim 30\,\text{Pa})$

$n' = 0.35 \quad m' = 1.82\,\text{Pa·s}^{n'} \quad (\tau_w > \sim 30\,\text{Pa})$

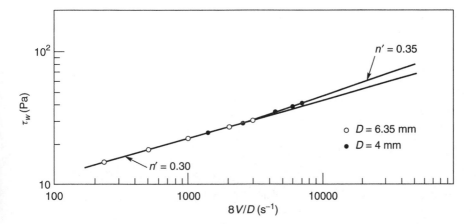

Figure 3.7 *Wall shear stress–apparent wall shear rate plot for data in example 3.6*

Also, the overlap in data obtained using tubes of two different diameters confirms the time-independent behaviour of the solution.

In the large pipe, the mean velocity of flow, $V = \dfrac{300}{1000} \times \dfrac{4}{\pi(0.3)^2}$

or $V = 4.1\,\text{m/s}$

Let us calculate the critical velocity, V_c, for the end of the streamline flow by setting $Re_{MR} = 2100$, i.e.

$$\frac{\rho V^{2-n'} D^{n'}}{8^{n'-1} m'} = 2100$$

Substituting values,

$$\frac{1000 \times V_c^{2-0.3} \times (0.3)^{0.3}}{8^{0.3-1} \times 2.74} = 2100$$

Solving, $V_c = 1.47$ m/s which is lower than the actual velocity of 4.1 m/s and hence, the flow in the 300 mm pipe is likely to be turbulent.

Initially, let us assume that the wall shear stress in the large pipe would be <30 Pa, i.e. $n' = 0.3$ and $m' = 2.74$ Pa·s^n can be used for calculating the value of the Re_{MR}, equation (3.28b),

$$Re_{MR} = \frac{\rho V^{2-n'} D^{n'}}{8^{n'-1} m'} = \frac{1000 \times 4.1^{2-0.3} \times (0.3)^{0.3}}{8^{0.3-1} \times 2.74}$$

$$= 12\,230$$

Now for $n' = 0.3$ and $Re_{MR} = 12\,230$, from Figure 3.6, $f \approx 0.0033$ (equation (3.38) gives $f = 0.0036$). The frictional pressure gradient $(-\Delta p/L)$, is calculated next:

$$\left(\frac{-\Delta p}{L}\right) = \frac{2 f \rho V^2}{D} = \frac{2 \times 0.0033 \times 1000 \times 4.1^2}{0.3} = 364 \text{ Pa/m}$$

The value of $\tau_w = (D/4)(-\Delta p/L) = (0.3 \times 364)/4 = 27.7$ Pa is within the range of the first power-law region and hence no further iteration is needed. The pump power is $Q \cdot \Delta p$, i.e. $(300/1000) \times 364 \times 50 = 5460$ W.

For the case when the flow rate has been increased by 20%, i.e. the new mass flow rate in the large pipe is 360 kg/s.

$$\text{mean velocity of flow, } V = \frac{360}{1000} \times \frac{4}{\pi (0.3)^2} = 4.92 \text{ m/s}$$

Based on the previous calculation, it is reasonable to assume that the new value of the wall shear stress would be greater than 30 Pa and therefore, one should use $n' = 0.35$ and $m' = 1.82$ Pa·s$^{n'}$.

$$\text{The Reynolds number, } Re_{MR} = \frac{1000 \times 4.92^{2-0.35} \times (0.3)^{0.35}}{8^{0.35-1} \times 1.82}$$

$$= 19\,410$$

For $n' = 0.35$ and $Re_{MR} = 19\,410$, from Figure 3.6, $f \approx 0.0032$ (while equation (3.38) also yields the same value). The frictional pressure gradient $(-\Delta p/L)$ is:

$$\frac{-\Delta p}{L} = \frac{2 f \rho V^2}{D} = \frac{2 \times 0.0032 \times 1000 \times 4.92^2}{(0.3)} = 511 \text{ Pa/m}$$

Checking: $\tau_w = \dfrac{D}{4}\left(\dfrac{-\Delta p}{L}\right) = \dfrac{0.3 \times 511}{4} = 39\,\text{Pa}.$

This value is just within the range of laminar flow data.

$$\text{pump power} = \dfrac{360}{1000} \times 511 \times 50 = 9200\,\text{W} \;\square$$

3.4.2 Viscoplastic fluids

Despite the fact that equation (3.37) is applicable to all kinds of time-independent fluids, numerous workers have presented expressions for turbulent flow friction factors for specific fluid models. For instance, Tomita [1959] applied the concept of the Prandtl mixing length and put forward modified definitions of the friction factor and Reynolds number for the turbulent flow of Bingham Plastic fluids in smooth pipes so that the Nikuradse equation, i.e. equation (3.37) with $n' = 1$, could be used. Though he tested the applicability of his method using his own data in the range $2000 \leq \text{Re}_B(1 - \phi)^2(3 - \phi) \leq 10^5$, the validity of this approach has not been established using independent experimental data.

In contrast, the semi-empirical equations due to Darby *et al.* (1992) obviate these difficulties due to their explicit form which is as follows:

$$f = (f_L{}^b + f_T{}^b)^{1/b} \tag{3.39}$$

where f_L is the solution of equation (3.13) and f_T is a function of the Reynolds number and Hedstrom number expressed as:

$$f_T = 10^{a_0}\text{Re}_B^{-0.193} \tag{3.40}$$

where $\quad a_0 = -1.47[1 + 0.146\exp(-2.9 \times 10^{-5}\text{He})]$

and $\quad b = 1.7 + \dfrac{40\,000}{\text{Re}_B}$

This method has been shown to yield satisfactory values of pressure drop under turbulent conditions for $D < 335\,\text{mm}$, $\text{Re}_B \leq 3.4 \times 10^5$ and $1000 \leq \text{He} \leq 6.6 \times 10^7$. Example 3.7 illustrates the application of this method.

Example 3.7

A 18% iron oxide slurry (density $1170\,\text{kg/m}^3$) behaves as a Bingham plastic fluid with $\tau_0^B = 0.78\,\text{Pa}$ and $\mu_B = 4.5\,\text{mPa·s}$. Estimate the wall shear stress as a function of the nominal wall shear rate $(8V/D)$ in the range $0.4 \leq V \leq 1.75\,\text{m/s}$ for flow in a 79 mm diameter pipeline.

Over the range of the wall shear stress encountered, the slurry can also be modelled as a power-law fluid with $m = 0.16\,\text{Pa·s}^n$ and $n = 0.48$. Contrast the predictions of the two rheological models.

Solution

(a) Bingham model

Here $\tau_0{}^B = 0.78\,\text{Pa};\quad \mu_B = 4.5\,\text{mPa·s}$

$$\rho = 1170\,\text{kg/m}^3 \quad \text{and} \quad D = 79 \times 10^{-3}\,\text{m}$$

The value of Hedström number is calculated first:

$$\text{He} = \frac{\rho D^2 \tau_0^B}{\mu_B{}^2} = \frac{1170 \times 0.079^2 \times 0.78}{(4.5 \times 10^{-3})^2} = 2.81 \times 10^5$$

From equation (3.40):

$$a_0 = -1.47[1 + 0.146\exp(-2.9 \times 10^{-5} \times 2.81 \times 10^5)] = -1.47$$

A sample calculation is shown for $V = 0.4\,\text{m/s}$.

$$\text{Apparent shear rate at wall,}\ \frac{8V}{D} = \frac{8 \times 0.4}{0.079} = 40.5\,\text{s}^{-1}$$

The Reynolds number of flow, Re_B is:

$$\text{Re}_B = \frac{\rho V D}{\mu_B} = \frac{1170 \times 0.4 \times 0.079}{4.5 \times 10^{-3}} = 8216$$

$\therefore$ the index b in equation (3.39) is given by:

$$b = 1.7 + \frac{40\,000}{\text{Re}_B} = 1.7 + \frac{40\,000}{8216} = 6.57$$

The value of f_L, i.e. the friction factor in streamline flow, is calculated using equation (3.13a) which can be re-cast in dimensionless form as:

$$f_L = \frac{16}{\text{Re}_B}\left[1 + \left(\frac{1}{6}\right)\left(\frac{\text{He}}{\text{Re}_B}\right) - \left(\frac{1}{3}\right)\frac{\text{He}^4}{f_L^3 \text{Re}_B^7}\right]$$

For $\text{Re}_B = 8216$, and $\text{He} = 2.81 \times 10^5$, this equation is solved to get $f_L = 0.0131$

The value of f_T is calculated using equation (3.40):

$$f_T = 10^{-1.47}(8216)^{-0.193} = 0.00595$$

Finally, the actual friction factor is estimated from equation (3.39):

$$f = (0.0131^{6.57} + 0.00595^{6.57})^{1/6.57} = 0.0131$$

and the wall shear stress, $\tau_w = \frac{1}{2}f\rho V^2 = \frac{1}{2} \times 0.0131 \times 1170 \times 0.4^2$, i.e. $\tau_w = 1.22\,\text{Pa}$.

This procedure is repeated for the other values of the average velocity, and a summary of results is shown in a tabular form here.

V (m/s)	$(8V/D)$ (s^{-1})	Re_B $(-)$	b $(-)$	f_T $(-)$	f_L $(-)$	f $(-)$	τ_w (Pa)
0.4	40.5	8216	6.57	0.00595	0.0131	0.0131	1.22
0.6	60.8	1.23×10^4	4.95	0.0055	0.00625	0.0068	1.44
0.8	81	1.64×10^4	4.14	0.00521	0.00377	0.00551	2.62
1.0	101.3	2.05×10^4	3.65	0.00499	0.00257	0.00510	2.99
1.25	126.6	2.57×10^4	3.26	0.00477	0.00176	0.00483	4.42
1.50	152	3.08×10^4	3.00	0.00461	0.00131	0.00465	6.12
1.75	177.2	3.6×10^4	2.81	0.00447	0.00102	0.0045	8.06

(b) Power-law model

Likewise, one sample calculation is shown here for $V = 0.4$ m/s.

The Metzner–Reed Reynolds number, Re_{MR} is given by equation (3.28b):

$$Re_{MR} = \frac{\rho V^{2-n} D^n}{8^{n-1} m \left(\dfrac{3n+1}{4n}\right)^n}$$

$$= \frac{1170 \times 0.4^{2-0.48} \times 0.079^{0.48}}{8^{0.48-1} \times 0.16 \times \left(\dfrac{3 \times 0.48 + 1}{4 \times 0.48}\right)^{0.48}} = 1407 < 2100$$

$\therefore$ the flow is streamline and $f = 16/Re_{MR}$,

$$f = 0.01136 \quad \text{and} \quad \tau_w = \tfrac{1}{2} f \rho V^2 = \tfrac{1}{2} \times 0.01136 \times 1170 \times 0.4^2$$

or $\tau_w = 1.06$ Pa

For $V \geq 0.6$ m/s, the value of Re_{MR} ranges from 2607 for $V = 0.6$ m/s to 1.33×10^4 for $V = 1.75$ m/s. Therefore, one must use either equation (3.37) or (3.38) under these condition; the latter is used here owing to its explicit form. A summary of calculations is tabulated below:

V (m/s)	$(8V/D)$ (s^{-1})	Re_{MR} $(-)$	f $(-)$	τ_w (Pa)
0.4	40.5	1407	0.0114	1.06
0.6	60.8	2607	0.00911	1.92
0.8	81	4037	0.00761	2.85
1.0	101.3	5667	0.00662	3.87
1.25	126.6	7956	0.00577	5.27
1.50	152	1.05×10^4	0.00515	6.77
1.75	177.2	1.33×10^4	0.00468	8.37

The predicted values of τ_w according to power-law and Bingham plastic models are plotted in Figure 3.8 together with the experimental values. While the maximum

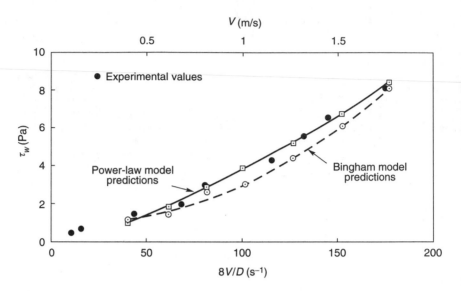

Figure 3.8 $\tau_w - (8V/D)$ *plot for example 3.7*

discrepancy between the two predictions is of the order of 20%, in this particular example the actual values are seen to be closer to the predictions of the power-law model than that of the Bingham model.□

3.4.3 Bowen's general scale-up method

Bowen [1961], on the other hand, proposed that for turbulent flow of a *particular* fluid (exhibiting time independent behaviour), the wall shear stress, τ_w, could be expressed as:

$$\tau_w = A \frac{V^b}{D^c} \tag{3.41}$$

where A, b and c are constants for the fluid and may be determined from experimental measurements in small diameter tubes. For laminar flow, wall shear stress (τ_w) – apparent wall shear stress $(8V/D)$ data for different diameter tubes, coincide. However, for turbulent conditions, as shown schematically in Figure 3.9, diameter appears to be an additional independent parameter. Because of the inclusion of a diameter effect explicitly, this method is to be preferred for scaling up the results of small scale pipe experiments to predict frictional pressure drops for the same fluid in large diameter pipes. Note that this method does not offer a generalised scheme of calculating pressure drop in contrast to the method of Metzner and Reed [1955] and Dodge and Metzner [1959].

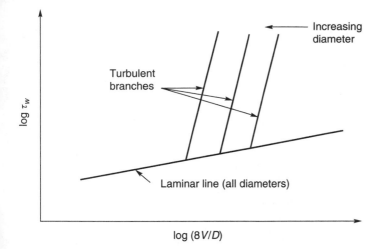

Figure 3.9 *Schematics of $\tau_w - (8V/D)$ behaviour for different diameter tubes in laminar and turbulent regions*

The three turbulent branches shown in Figure 3.9, which all have the same slope, correspond to three different pipe diameters, whilst a single line suffices for all diameters for laminar flow in the absence of wall-slip effects. The Reynolds number at the transition can be calculated from the value of $(8V/D)$ at which the turbulent branch deviates from the laminar line. In practice, however, the laminar–turbulent transition occurs over a range of conditions rather than abruptly as shown in Figure 3.9. Strictly speaking, only turbulent flow data for small scale pipe should be used to scale up to turbulent flow in larger pipes. However, this is likely to pose problems because of the extremely high velocities and pressure drops which would frequently be required to achieve turbulence in small diameter pipes. But nevertheless measurements in the streamline and transitional regimes facilitate the delineation of the critical Reynolds number. Due to the fluctuating nature of flow, the operation of pipe lines in the transition region should be avoided as far as possible. Finally, this method can be summarised as follows:

(i) Obtain $\tau_w - (8V/D)$ data using at least two pipe sizes under both laminar and turbulent flow conditions if possible. Plot these data on log-log coordinates as shown schematically in Figure 3.9. The laminar flow data should collapse on to one line and there should be one branching for turbulent flow line for each pipe diameter.

(ii) The index of the velocity term, b, in equation (3.41) is the slope of the turbulent branch. Ideally, all branches would have identical slopes but in practice b should be evaluated for each value of D and the mean value used.

(iii) Next, plot $LV^b/(-\Delta p)$ against D on log-log coordinates for each turbulent flow data point; the slope of this line will be $(1+c)$ and hence c can be evaluated.
(iv) The remaining constant, A, is evaluated by calculating its value for each turbulent flow data point as $\tau_w D^c/V^b$ and again the mean value of A should be used.

Equation (3.41) can now be used directly to give the pressure drop for any pipe diameter if the flow is turbulent. Alternatively, this approach can be used to construct a wall shear stress (τ_w)–apparent wall shear rate $(8V/D)$ turbulent flow line for any pipe diameter. This method is particularly suitable when either the basic rheological measurements for laminar flow are not available or it is not possible to obtain a satisfactory fit of such data. The application of this method is illustrated in example 3.8.

Example 3.8

The following flow rate – pressure drop data for a 0.2% aqueous carbopol solution (density $1000\,\text{kg/m}^3$) have been reported for two capillary tubes and three pipes of different diameters. (Data from D.W. Dodge, Ph D Thesis, University of Delaware, Newark, 1958).

Capillary Data				Pipeline Data					
$D = 0.84\,\text{mm}$ $L = 155.5\,\text{mm}$		$D = 0.614\,\text{mm}$ $L = 203.8\,\text{mm}$		$D = 12.7\,\text{mm}$		$D = 25.4\,\text{mm}$		$D = 50.8\,\text{mm}$	
$\left(\dfrac{8V}{D}\right)$ (s^{-1})	τ_w (Pa)	$\left(\dfrac{8V}{D}\right)$ (s^{-1})	τ_w (Pa)	$\left(\dfrac{8V}{D}\right)$ (s^{-1})	τ_w (Pa)	$\left(\dfrac{8V}{D}\right)$ (s^{-1})	τ_w (Pa)	$\left(\dfrac{8V}{D}\right)$ (s^{-1})	τ_w (Pa)
213.2	4.73	820.5	12.85	472.5	7.76	205	4.53	74	2.19
397.1	7.59	1456	19.50	753.2	10.93	301	6.16	114	2.90
762.8	12.16	2584	29.61	1121	14.94	418	7.88	151	3.87
1472	19.67	4691	46.24	1518	19.68	504	9.42	177	4.86
1837	23.27	8293	69.71	1715	26.45	562	11.83	194*	5.64
2822	31.39	14 420	104.50	1849*	34.45	602	14.24	212	7.61
4000	41.90	25 140	157	1989	40.12	691*	19.93	236.8	9.33
5237	49.08	43 310	231.7	2250	49.41	825	27.03	264	11.40
9520	77.33	76 280	343.4	2642	64.11	1004	37.21	303	14.14
10 300	81.1	111 900	496	3043	81.11	1188	49.08	348	17.61
19 010	125.35	–	–	3485	101	1473	69.76	395	21.62
22 400	142.54	–	–	4047	128.6	1727	90.30	453	27
35 490	191	–	–	4610	160.6	2071	120.40	517	33.52

Capillary Data		Pipeline Data					
$D = 0.84$ mm $L = 155.5$ mm		$D = 12.7$ mm		$D = 25.4$ mm		$D = 50.8$ mm	
$\left(\dfrac{8V}{D}\right)$ (s^{-1})	τ_w (Pa)	$\left(\dfrac{8V}{D}\right)$ (s^{-1})	τ_w (Pa)	$\left(\dfrac{8V}{D}\right)$ (s^{-1})	τ_w (Pa)	$\left(\dfrac{8V}{D}\right)$ (s^{-1})	τ_w (Pa)
50 400	246	5192	196	2482	162.1	582	40.64
59 170	272	6106	260.5	2990	222.7	671	51.33
94 260	365	6899	320	3482	286.3	776	64.78
95 600	358	7853	399	4261	407	907	82.64

Using the methods of Bowen [1961] and of Dodge and Metzner [1959], construct the wall shear stress-apparent shear rate plots for turbulent flow of this material in 101.6 mm and 203.2 mm diameter pipes. Also, calculate the velocity marking the end of the streamline flow.

Solution

(a) Bowen's method

Figure 3.10 shows the flow behaviour data tabulated. As expected, all the laminar flow data are independent of tube diameter and collapse on to one line, with $n' = 0.726$ and $m' = 0.0981$ Pa·s$^{n'}$. Also, the same value of n' is seen to cover the entire range of measurements.

On the other hand, three turbulent branches are obtained with, each corresponding to a particular pipe diameter. The slopes of these lines are remarkably similar at 1.66, 1.66 and 1.65 for the 12.7 mm, 25.4 mm and 50.8 mm diameter pipes respectively.

$$\therefore b = 1.66$$

Note that the first few data points in each case seem to lie in laminar and transitional range as can be seen in Figure 3.10. These are also indicated in the tabulated data by an asterisk.

Now using this value of b, the quantity $(LV^b/-\Delta p)$ is evaluated for each turbulent data point, and the resulting mean values for $(LV^b)/(-\Delta p)$ are 0.000535, 0.001183 and 0.002713 for the 12.7 mm, 25.4 mm and 50.8 mm diameter pipes respectively. These values are plotted against pipe diameter in Figure 3.11. Again, as expected, a straightline results with a slope of 1.18.

$$\therefore 1 + c = 1.18, \text{ i.e. } c = 0.18$$

Finally, the value of A is calculated by evaluating the quantity $(\tau_w D^c / V^b)$ for each turbulent flow data point.

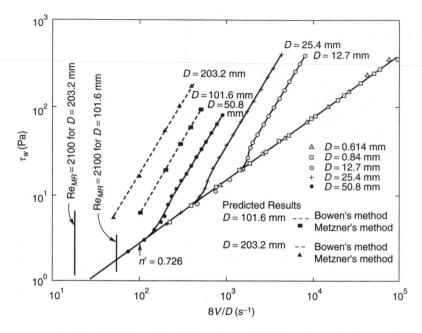

Figure 3.10   *Turbulent flow behaviour of 0.2% carbopol solution (example 3.8)*

For   12.7 mm diameter pipe, $2.67 \leq (\tau_w D^c / V^b) \leq 2.8$

25.4 mm diameter pipe, $2.75 \leq (\tau_w D^c / V^b) \leq 2.86$

50.8 mm diameter pipe, $2.75 \leq (\tau_w D^c / V^b) \leq 2.86$

The resulting mean value of A is 2.78. Therefore, for this carbopol solution in turbulent flow, the wall shear stress is given as:

$$\tau_w = 2.78 \frac{V^{1.66}}{D^{0.18}}$$

where all quantities are in S.I. units. Now, the predicted values for $D = 101.6$ mm and 203.2 mm pipes are shown in the following tables, and are also included in Figure 3.10 as broken lines.

Since the values of n' and m' remain constant over the entire range of wall shear stress, one can readily apply the method of Metzner and Reed [1955] to calculate the wall shear stress in large diameter pipes here. For instance, for $V = 1.27$ m/s, $D = 101.6$ mm

$$\text{Re}_{MR} = \frac{\rho V^{2-n'} D^{n'}}{8^{n'-1} m'} = \frac{1000 \times 1.27^{2-0.726} \times 0.1016^{0.726}}{8^{0.726-1} \times 0.0981}$$

$$= 4645$$

From Figure 3.6, $f \approx 0.0077$.

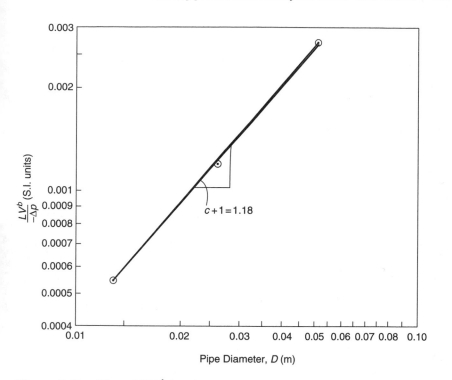

Figure 3.11   *Plot of $(LV^b/\Delta p)$ versus pipe diameter for evaluating c*

Predicted values of $\tau_w - (8V/D)$ using the methods of Bowen and of Metzner and co-workers

V (m/s)	$\left(\dfrac{8V}{D}\right)$ s^{-1}	Re$_{MR}$	f	τ_w (Pa)	
				Bowen method	Metzner et al.
			$D = 101.6$ mm		
1.27	100	4645	0.0077	6.2	6.21
2.54	200	11 234	6.012×10^{-3}	19.65	19.39
3.81	300	18 831	5.348×10^{-3}	38.51	38.82
5.08	400	27 168	4.695×10^{-3}	62	60.58
6.35	500	36 100	4.5×10^{-3}	90	90.73
			$D = 203.2$ mm		
1.27	50	7684	0.00668	5.5	5.39
2.54	100	18 582	0.00513	17.34	16.55
5.08	200	44 937	0.0043	54.8	55.5
7.62	300	75 325	0.0037	107.5	107.4
10.16	400	1 08 670	0.0035	173	180.6

Alternatively equation (3.38) may be used with $n = n' = 0.726$

Then: $\quad D(n) = \dfrac{2^{4.726}}{7^{7 \times 0.726}} \left(\dfrac{4 \times 0.726}{3 \times 0.726 + 1} \right)^{3 \times 0.726^2} = 0.001164$

$\quad \therefore f = \left(\dfrac{0.001164}{4645} \right)^{1/(3 \times 0.726 + 1)} \approx 0.00837$

The two values differ by about 8%. Using the value of $f = 0.0077$:

$$\tau_w = f \cdot 1/2 \rho V^2 = 0.0077 \times \dfrac{1000}{2} \times 1.27^2 = 6.21 \, \text{Pa}.$$

Similarly, one can calculate the values of the wall shear stress for the other values of the average velocity. These results are also summarised in the above Table where a satisfactory correspondence is seen to exist between the two predicted values of the wall shear stress.

Finally, the critical values of the velocity marking the limit of streamline flow in 101.6 mm and 203.2 mm diameter pipes are calculated by setting $Re_{MR} = 2100$. These values are found to be $V = 0.68$ m/s in the 101.6 mm diameter pipe and $V = 0.37$ m/s for the other pipe. These values are also shown in Figure 3.10.□

Many other correlations for power-law, Bingham plastic and yield-pseudo-plastic fluid models are available in the literature [Govier and Aziz, 1982; Hanks, 1986; Thomas and Wilson, 1987; Darby, 1988] but unfortunately their validity has been established only for a limited range of data and hence they are too tentative to be included here. In an exhaustive comparative study, Heywood and Cheng [1984] evaluated the relative performance of seven such correlations. They noted that the predictions differed widely and that the uncertainty was further compounded by the inherent difficulty in unambiguously evaluating the rheological properties to be used under turbulent flow conditions. Indeed in some cases, the estimated values of friction factor varied by up to ±50%, which is in stark contrast to Newtonian turbulent flow where most predictive expressions yield values of the friction factor within ±5% of each other. Thus, they recommended that as many predictive methods as possible should be used to calculate the value of f, so that an engineering judgement can be made taking into account the upper and lower bounds on the value of f for the particular application. However, in more recent as well as more extensive evaluations of turbulent and transitional flow data [Garcia and Steffe, 1987; Hartnett and Kostic, 1990], it is recommended that the method of Metzner and Reed [1955] be used to calculate frictional losses in straight sections of smooth pipes, i.e. Figure 3.6 or equations (3.37) or (3.38). The method of Bowen [1961] is recommended for scaling up small-scale turbulent flow data to turbulent flow in large diameter pipes. However, it should be re-iterated here that the method of Metzner and Reed [1955] necessitates using the values of m' and n' evaluated at the prevailing wall shear stress; this usually means relying on $\tau_w - (8V/D)$ data obtained in laminar region. For a true power law fluid, $n' = n$ and there is no difficulty provided this value of n is applicable all the way up to the wall shear stress levels under turbulent flow conditions. However, since the pressure drop (hence wall shear stress) is usually unknown, a trial and error solution is required, except for the case when the pressure loss drop is known and the flow rate is to be calculated. Thus, first a value of $(-\Delta p/L)$ is assumed and based on this, m' and n' are evaluated from the plot of τ_w versus $(8V/D)$ obtained for laminar conditions. From these parameters, the value of Re_{MR} is calculated and f is found either from Figure 3.6 or equation (3.37) or (3.38). If the value of $(-\Delta p/L)$ calculated using this value of f is appreciably different from that obtained using the assumed value of $(-\Delta p/L)$, other iterations are carried out till the two values are fairly close to each other. The calculation usually converges quickly. However, there does not appear to be a simple way of determining *a priori* the shear stress range which must be covered in the laminar flow rheological tests to ensure that these will include the values of τ_w likely to be encountered under turbulent flow conditions. However it is usually recommended that the approximate $(8V/D)$ range over which the pipeline is to operate be used as a guide for performing laminar flow tests. On the other hand, the method of Bowen [1961], while free from all these complications, relies on the use of the same fluid in turbulent flow in both small and large scale pipes.

3.4.4 Effect of pipe roughness

Considerable confusion exists regarding the effect of the pipe wall roughness on the value of friction factor in the turbulent flow region, though the effect is qualitatively similar to that for Newtonian fluids [Slatter, 1997]. Thus, Torrance [1963] and Szilas *et al.* [1981] have incorporated a small correction to account for pipe roughness in their expressions for the turbulent flow of power-law and Bingham plastic fluids; under fully turbulent conditions, it is assumed that the value of the friction factor is determined only by pipe roughness and rheological properties, and is independent of the Reynolds number. However, in view of the fact that laminar sub-layers tend to be somewhat thicker for non-Newtonian fluids than that for Newtonian liquids, the effect of pipe roughness is likely to be smaller for time-independent fluids [Bowen, 1961; Wójs, 1993]. Despite this, Govier and Aziz [1982] recommend the use of the same function of relative roughness for time-independent fluids as for Newtonian systems. In any case, this will yield conservative estimates.

3.4.5 Velocity profiles in turbulent flow of power-law fluids

No exact mathematical analysis of turbulent flow has yet been developed for power-law fluids, though a number of semi-theoretical modifications of the expressions for the shear stress in Newtonian fluids at the walls of a pipe have been proposed.

The shear stresses within the fluid are responsible for the frictional force at the walls and the velocity distribution over the cross-section of the pipe. A given assumption for the shear stress at the walls therefore implies some particular velocity distribution. In line with the traditional concepts that have proved of value for Newtonian fluids, the turbulent flow of power-law fluids in smooth pipes can be considered by dividing the flow into three zones, as shown schematically in Figure 3.12.

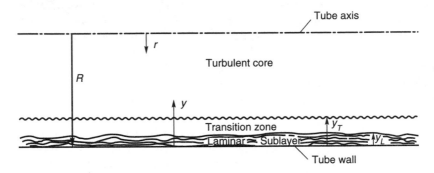

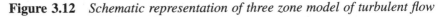

Figure 3.12   *Schematic representation of three zone model of turbulent flow*

(i) Laminar sub-layer

This represents a thin layer $(0 \leq y \leq y_L)$ next to the pipe wall in which the effects of turbulence are assumed to be negligible. Assuming the no-slip boundary condition at the wall, $y = 0$, the fluid in contact with the surface is at rest. Furthermore, all the fluid close to the surface is moving at a very low velocity and therefore any changes in its momentum as it flows in the z direction must be extremely small. Consequently, the net shear force acting on any element of fluid in this zone must be negligible, the retarding force at its lower boundary being balanced by the accelerating force at its upper boundary. Thus, the shear stress in the fluid near the surface must approach a constant value which implies that the shear rate in this layer must also be constant or conversely, the velocity variation must be linear in the laminar sub-layer.

(ii) Transition Zone

This region separates the so-called viscous or laminar sub-layer and the fully turbulent core prevailing in the middle portion of the pipe, and it extends over $y_L \leq y \leq y_T$, and finally

(iii) Turbulent Core $(y_T \leq y \leq R)$

A fully turbulent region comprising the bulk of the fluid stream where momentum transfer is attributable virtually entirely to random eddies and the effects of viscosity are negligible.

A brief derivation of the turbulent velocity profile for Newtonian fluids in smooth pipes will first be presented and then extended to power-law fluids. The shear stress at any point in the fluid, at a distance y from the wall, is made up of 'viscous' and 'turbulent' contributions, the magnitudes of which vary with distance from the wall. Expressing shear stress in terms of a dynamic viscosity and an eddy momentum diffusivity (or eddy kinematic viscosity), E,

$$\tau_{yz} = \left(\frac{\mu}{\rho} + E\right) \frac{\mathrm{d}}{\mathrm{d}y}(\rho V_z) \tag{3.42}$$

Prandtl postulated that E could be expressed as

$$E = l^2 \left|\frac{\mathrm{d}V_z}{\mathrm{d}y}\right| \tag{3.43}$$

where the so-called 'mixing' length l (analogous to the mean free path of the molecules) is assumed to be directly proportional to the distance from the wall, i.e. $l = ky$. Thus, equation (3.42) can be re-written as:

$$\tau_w \left(\frac{R - y}{R}\right) = \left\{\frac{\mu}{\rho} + k^2 y^2 \frac{\mathrm{d}V_z}{\mathrm{d}y}\right\} \frac{\mathrm{d}}{\mathrm{d}y}(\rho V_z) \tag{3.44}$$

In the laminar sub-layer (small values of y), $E \simeq 0$ and:

$$\tau_w \simeq \frac{\mu}{\rho} \frac{\mathrm{d}}{\mathrm{d}y}(\rho V_z) \tag{3.45}$$

which upon integration with $V_z = 0$ at $y = 0$ to V_z at y yields the expected linear velocity profile (discussed earlier):

$$V_z = \frac{\tau_w}{\mu} y \tag{3.46}$$

It is customary to introduce the friction velocity $V^* = \sqrt{\tau_w/\rho}$ and to express equation (3.46) in dimensionless form:

$$V^+ = y^+ \tag{3.47}$$

where

$$V^+ = V_z/V^*; \quad y^+ = \frac{yV^*\rho}{\mu} \tag{3.48}$$

In the turbulent core, but yet close to the wall $y/R \ll 1$, (μ/ρ) is small compared with E. In addition, $(\mathrm{d}V_z/\mathrm{d}y)$ will be positive close to the wall and therefore the modulus signs can be omitted in equation (3.43) and equation (3.44) becomes:

$$\tau_w \simeq \rho k^2 y^2 \left(\frac{\mathrm{d}V_z}{\mathrm{d}y}\right)^2 \tag{3.49}$$

Substitution for $V^* = \sqrt{(\tau_w/\rho)}$ and integration leads to:

$$V_z = \frac{V^*}{k} \ln y + B_0 \tag{3.50}$$

Now introducing the non-dimensional velocity and distance, V^+ and y^+,

$$V^+ = A \ln y^+ + B \tag{3.51}$$

where all constants have been absorbed in A and B. It might be expected that since equation (3.51) has been based on the approximation that $y/R \ll 1$, it should be valid only near the wall. In fact, it has been found in pipe flow to correlate experimental data well over most of the turbulent core, except close to the centre of the pipe. Experimental values of $A = 2.5$ and $B = 5.5$ have been obtained from a very wide range of experimental data. B is a function of the relative roughness for rough pipes and A which is independent of roughness is readily seen to be equal to $1/k$.

No such simple analysis is possible in the transition zone nor is it possible to delineate the transition boundaries for the three regions of flow. Based on experimental results, it is now generally believed that the laminar sub-layer extends up to $y^+ \approx 5$ and the turbulent core begins at $y^+ \simeq 30$. The following empirical correlation provides an adequately approximate velocity distribution in the transition layer ($5 \le y^+ \le 30$) in smooth pipes:

$$V^+ = 5 \ln y^+ - 3.05 \tag{3.52}$$

Equation (3.52) is a straight line on semi-logarithmic coordinates joining the laminar sub-layer values at $y^+ = 5$ and equation (3.51) for the turbulent core at $y^+ = 30$. Finally, it should be noted that equation (3.51) does not predict the expected zero velocity gradient at the centre of the pipe but this deficiency has little influence on the volumetric flow rate – pressure drop relationship.

Numerous attempts [Dodge and Metzner, 1959; Bogue and Metzner, 1963; Wilson and Thomas, 1985; Shenoy and Talathi, 1985; Shenoy, 1988] have been made at developing analogous expressions for velocity profiles for the steady state turbulent flow of power-law fluids in smooth pipes; most workers have modified the definitions of y^+ and V^+, but Brodkey *et al.* [1961] used a polynomial approximation for the velocity profile. Figure 3.13 shows the velocity profiles derived on this basis for power-law fluids; the transition region, shown as dotted lines, is least understood.

The velocity distribution within the laminar sub-layer can be derived by the same reasoning as for a Newtonian fluid that the velocity varies linearly with distance from the wall (equation 3.46),

$$\frac{dV_z}{dy} = \frac{\tau_w}{\mu} = \frac{V_z}{y} \tag{3.53}$$

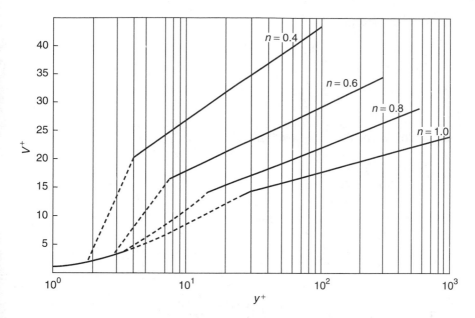

Figure 3.13 *Typical velocity profiles in the three zone representation of turbulent flow*

For a power-law fluid, the shear stress in this layer can be written as,

$$\tau_w \left(1 - \frac{y}{R}\right) = m \left(\frac{\mathrm{d}V_z}{\mathrm{d}y}\right)^n$$

or
$$\tau_w \simeq m \left(\frac{\mathrm{d}V_z}{\mathrm{d}y}\right)^n \tag{3.54}$$

which upon integration with the no-slip condition at the wall ($V_z = 0$ when $y = 0$) gives:

$$V_z = \left(\frac{\tau_w}{m}\right)^{1/n} y \tag{3.55}$$

Introducing the friction velocity, V^*, and re-arrangement leads to

$$V^+ = (y^+)^{1/n} \tag{3.56}$$

where
$$y^+ = \frac{\rho (V^*)^{2-n} y^n}{m} \tag{3.57}$$

It should be noted that for $n = 1$, equations (3.56) and (3.57) reduce to equations (3.47) and (3.48) respectively.

For the turbulent core, Dodge and Metzner [1959] used a similar approach to that given above for Newtonian fluids and proposed the following expression for power-law fluids:

$$V^+ = \frac{5.66}{n^{0.75}} \log y^+ - \frac{0.566}{n^{1.2}} + \frac{3.475}{(n)^{0.75}}$$

$$\times \left\{ 1.96 + 0.815\, n - 1.628\, n \log \left(\frac{3n+1}{n} \right) \right\} \tag{3.58}$$

It should be noted that in the limit of $n = 1$, equation (3.58) also reduces to equation (3.51) with $A = 2.47$ and $B = 5.7$; the slight discrepancy in the values of the constants arises from the fact that experimental $Q - (-\Delta p)$ data have been used to obtain the values of the constants rather than velocity measurements. The detailed derivation of equation (3.58) has been given by Skelland [1967]. Subsequently, Bogue and Metzner [1963] used point velocity measurements to modify equation (3.58) to give:

$$V^+ = \frac{5.57}{n} \log y^+ + C(y^*, f) + I(n, \mathrm{Re}_{MR}) \tag{3.59}$$

where $C(y^*, f) = 0.05 \frac{\sqrt{2}}{f} \exp \left\{ \frac{-(y^* - 0.8)^2}{0.15} \right\}$

The friction factor, f, is calculated using equation (3.37) and the typical values of $I(n, \mathrm{Re}_{MR})$ are presented in Table 3.1. Bogue and Metzner [1963] also noted that the velocity distributions for Newtonian and power-law fluids were virtually indistinguishable from each other if plotted in terms of (V_z/V) versus y/R instead of $V^+ - y^+$ coordinates, as illustrated in Figure 3.14. Finally, attention is drawn to the fact that both Dodge and Metzner [1959] and Bogue and Metzner [1963] have implicitly neglected the transition layer. Clapp [1961], on the other hand, has combined the Prandtl and von Karman approaches to put forward the following expressions for velocity distribution in the transition and turbulent zones:

$$V^+ = \frac{5}{n} \ln y^+ - 3.05 \qquad (5^n \le y^+ \le y_T^+) \tag{3.60a}$$

$$V^+ = \frac{2.78}{n} \ln y^+ + \frac{3.8}{n} \qquad (y^+ > y_T^+) \tag{3.60b}$$

where y_T^+ is evaluated as the intersection point of equations (3.60a) and (3.60b). The laminar sub-layer region, equation (3.56), extends up to a value of 5^n. The numerical constants in these equations were evaluated using point velocity measurements in the range $0.7 \le n \le 0.81$. For $n = 1$, these equations yield $y_T^+ = 22$ which is somewhat lower than the generally accepted figure of 30 for Newtonian fluids. Finally, all the above-mentioned velocity distributions fail to predict $(dV_z/dy) = 0$ at $y = R$.

Table 3.1   *Values of $I(n, Re_{MR})$ in equation (3.59)*

n ↓	$Re_{MR} \rightarrow$	5000	10 000	50 000	10^5
1		5.57	5.57	5.57	5.57
0.8		6.01	5.92	5.69	5.58
0.6		6.78	5.51	5.89	5.60
0.4		8.39	7.70	6.27	5.60

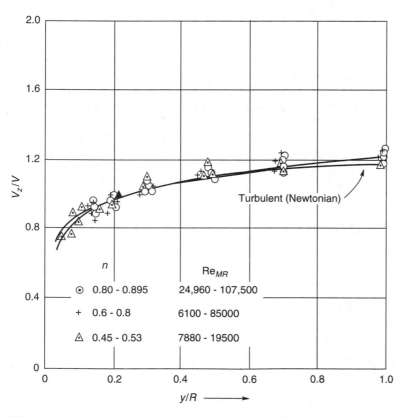

Figure 3.14   *Measured velocity profiles in circular pipes for turbulent flow of power-law fluids (data from Bogue and Metzner [1963])*

Extensive experimental work indicates that the form of the turbulent velocity profiles established for non-elastic fluids without yield stress is very similar to that found for Newtonian fluids. This supports the contention, implicit in the discussion on friction factors, that it is the properties of the fluid in the wall region which are most important. This is particularly so with shear-thinning materials, for which the apparent viscosity in this region is lower than that in the bulk.

A potential and interesting application of this idea is to use pulsating flow in pipelines transporting shear-thinning materials. The super-imposition of a small oscillating component on the bulk velocity has the effect of raising the average shear rate, thereby lowering the apparent viscosity. It appears advantageous to use some of the pumping energy in this way enabling the use of smaller pumps than would otherwise be required [Edwards *et al.*, 1972].

Before concluding this section, it is appropriate to mention briefly that some work is available on velocity distributions in turbulently flowing visco-elastic and drag-reducing polymer solutions in circular pipes. Qualitatively similar velocity distributions have been recorded for such systems as those mentioned in the preceding section for Newtonian and power-law fluids. However, the resulting equations tend to be more complex owing to the additional effects arising from visco-elasticity. Most of these studies have been critically reviewed by Shenoy and Talathi [1985] and Tam *et al.* [1992].

Aside from the foregoing discussion for the fully established flow, the problems involving transient flows of non-Newtonian fluids is circular pipes have also been investigated [Edwards *et al.*, 1972; Brown and Heywood, 1991]. Similarly, some guidelines for handling time-dependent thixotropic materials are also available in the literature [Govier and Aziz, 1982; Wardhaugh and Boger, 1987, 1991; Brown and Heywood, 1991].

3.5 Laminar flow between two infinite parallel plates

The steady flow of an incompressible power-law fluid between two parallel plates extending to infinity in x- and z-directions, as shown schematically in Figure 3.15 will now be considered. The mid-plane between the plates will be taken as the origin with the flow domain extending from $y = -b$ to $y = +b$. The force balance on the fluid element ABCD situated at distance $\pm y$ from the mid-plane, can be set up in a similar manner to that for flow through pipes.

$$p \cdot 2Wy - (p + \Delta p)2Wy = \tau_{yz} \cdot 2Wdz. \tag{3.61}$$

i.e.
$$\tau_{yz} = \left(\frac{-\Delta p}{L}\right) y \tag{3.62}$$

The shear stress is thus seen to vary linearly, from zero at the mid-plane to a maximum value at the plate surface, as in the case of pipe flow. The system is symmetrical about the mid-plane ($y = 0$) and equation (3.62) needs to be solved only for $0 < y < b$. Because dV_z/dy is negative in this region, the shear stress for a power-law fluid is given by:

$$\tau_{yz} = m \left(-\frac{dV_z}{dy}\right)^n \tag{3.63}$$

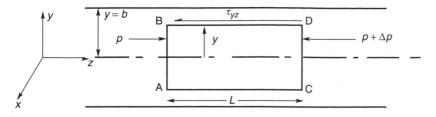

Figure 3.15   *Laminar flow between parallel plates*

Now combining equations (3.62) and (3.63) followed by integration yields for the velocity distribution:

$$V_z = -\left(\frac{n}{n+1}\right)\left\{\frac{1}{m}\left(\frac{-\Delta p}{L}\right)\right\}^{1/n} y^{(n+1)/n} + \text{constant}$$

At the walls of the channel (i.e. when $y = \pm b$), the velocity V_z must be zero to satisfy the condition of no-slip. Substituting the value $V_z = 0$, when $y = b$:

$$\text{constant} = \left(\frac{n}{n+1}\right)\left\{\frac{1}{m}\left(\frac{-\Delta p}{L}\right)\right\}^{1/n} b^{(n+1)/n}$$

and therefore:

$$V_z = \left(\frac{n}{n+1}\right)\left\{\frac{1}{m}\left(\frac{-\Delta p}{L}\right)\right\}^{1/n} b^{(n+1)/n}\left\{1 - \left(\frac{y}{b}\right)^{(n+1)/n}\right\} \qquad (3.64)$$

The velocity distribution is seen to be parabolic for a Newtonian fluid and becomes progressively blunter as the value of n decreases below unity, and sharper for shear-thickening fluids. The maximum velocity occurs at the centre-plane and its value is obtained by putting $y = 0$ in equation (3.64):

$$\text{Maximum velocity} = V_{\max} = \left(\frac{n}{n+1}\right) b^{(n+1)/n}\left\{\frac{1}{m}\left(\frac{-\Delta p}{L}\right)\right\}^{1/n} \qquad (3.65)$$

The total rate of flow of fluid between the plates is obtained by calculating the flow through two laminae of thickness dy and located at a distance y from the centre-plane and then integrating. Flow through laminae:

$$dQ = 2W \, dy \, V_z$$

$$= 2W \left(\frac{n}{n+1}\right) b^{(n+1)/n}\left\{\frac{1}{m}\left(\frac{-\Delta p}{L}\right)\right\}^{1/n}\left\{1 - \left(\frac{y}{b}\right)^{(n+1)/n}\right\} dy$$

Then, on integrating between the limits of y from 0 to b:

$$Q = 2bW \left(\frac{n}{2n+1}\right) \left\{\frac{1}{m}\left(\frac{-\Delta p}{L}\right)\right\}^{1/n} b^{(n+1)/n} \tag{3.66}$$

The average velocity of the fluid:

$$V = \frac{Q}{2bW}$$

$$= \left(\frac{n}{2n+1}\right) \left\{\frac{1}{m}\left(\frac{-\Delta p}{L}\right)\right\}^{1/n} b^{(n+1)/n} \tag{3.67}$$

A similar procedure can, in principle, be used for other rheological models by inserting an appropriate expression for shear stress in equation (3.62). The analogous result for the laminar flow of Bingham plastic fluids in this geometry is given here:

$$Q = \frac{2Wb^2}{3\mu_B} \left(\frac{-\Delta p}{L} \cdot b\right) \left\{1 - \frac{3}{2}\phi + \frac{1}{2}\phi^3\right\} \tag{3.68}$$

where $\phi = \tau_0^B \Big/ \left(\frac{-\Delta p}{L} \cdot b\right) = \tau_0^B/\tau_w$.

Example 3.9

Calculate the volumetric flow rate per unit width at which a 0.5% polyacrylamide solution will flow down a wide inclined surface (30° from horizontal) as a 3 mm thick film. The shear stress–shear rate behaviour of this polymer solution may be approximated by the Ellis fluid model, with the following values of the model parameters: $\mu_0 = 9\,\text{Pa·s}$; $\tau_{1/2} = 1.32\,\text{Pa}$; $\alpha = 3.22$ and the solution has a density of $1000\,\text{kg/m}^3$. Assume the flow to be laminar.

Solution

A general equation will be derived first for the flow configuration shown in Figure 3.16 by writing a force balance on a differential element of the fluid.

In a liquid flowing down a surface, a velocity distribution will be established with the velocity increasing from zero at the surface itself ($y = 0$) to a maximum at the free surface ($y = H$). For viscoplastic fluids, it can be expected that plug-like motion may occur near the free surface. The velocity distribution in the film can be obtained in a manner similar to that used previously for pipe flow, bearing in mind that the driving force here is that due to gravity rather than a pressure gradient; which is absent everywhere in the film.

In an element of fluid of length dz, the gravitational force acting on that part of the liquid which is at a distance greater than y from the surface is given by:

$$(H - y)W\,dz\rho g \sin \beta$$

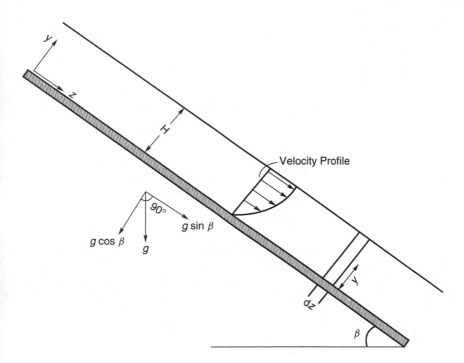

Figure 3.16   *Schematics of flow on an inclined plate*

If the drag force at the free surface is negligible, the retarding force for flow will be attributable to the shear stress prevailing in the liquid at the distance y from the surface and this will be given by:

$$\tau_{yz} \cdot W \, dz$$

At equilibrium therefore:

$$\tau_{yz} \cdot W \, dz = (H - y)W \, dz \rho g \sin \beta$$

or $\qquad \tau_{yz} = \rho g (H - y) \sin \beta$ $\qquad\qquad\qquad\qquad\qquad$ (3.69)

The shear stress is seen to vary linearly from a maximum at the solid surface to zero at the free surface.

Since dV_z/dy is positive here, the shear stress for an Ellis model fluid is given by:

$$\tau_{yz} = \frac{\mu_0 \left(\dfrac{dV_z}{dy} \right)}{1 + \left(\dfrac{\tau_{yz}}{\tau_{1/2}} \right)^{\alpha - 1}}$$

or $\quad \mu_0 \dfrac{dV_z}{dy} = \tau_{yz} \left\{ 1 + \left(\dfrac{\tau_{yz}}{\tau_{1/2}} \right)^{\alpha - 1} \right\}$ $\qquad\qquad\qquad\qquad$ (3.70)

Substitution of equation (3.69) into equation (3.70), followed by integration and using the no-slip boundary condition at the solid surface ($y = 0$) yields:

$$V_z = \frac{\rho g \sin \beta}{\mu_0}\left(Hy - \frac{y^2}{2}\right) + \frac{(\rho g \sin \beta)^\alpha H^{\alpha+1}}{\mu_0(\alpha+1)\tau_{1/2}^{\alpha-1}}\left\{1 - \left(1 - \frac{y}{H}\right)^{\alpha+1}\right\} \tag{3.71}$$

The volumetric flow rate of liquid down the surface can now be calculated:

$$Q = W \int_0^H V_z \, dy \tag{3.72}$$

Substituting equation (3.71) in equation (3.72) and integrating:

$$Q = \frac{\rho g H^3 W \sin \beta}{3\mu_0} + \frac{W(\rho g \sin \beta)^\alpha H^{\alpha+2}}{\mu_0(\alpha+2)(\tau_{1/2})^{\alpha-1}} \tag{3.73}$$

For a Newtonian fluid, $\tau_{1/2} \to \infty$, and both equations (3.71) and (3.73) reduce to the corresponding Newtonian expressions.

The maximum velocity occurs at the free surface, and its value is obtained by putting $y = H$ in equation (3.71):

$$\text{Maximum velocity} = V_{max} = \frac{\rho g \sin \beta H^2}{2\mu_0} + \frac{(\rho g \sin \beta)^\alpha H^{\alpha+1}}{\mu_0(\alpha+1)(\tau_{1/2})^{\alpha-1}} \tag{3.74}$$

For a vertical surface, $\sin \beta = 1$.

For the numerical example, $\beta = 30°$; $H = 3$ mm and substituting the other values in equation (3.73):

$$\frac{Q}{W} = \frac{1000 \times 9.81 \times (3 \times 10^{-3})^3 \left(\frac{1}{2}\right)}{3 \times 9}$$

$$+ \frac{\left(1000 \times 9.81 \times \frac{1}{2}\right)^{3.22}(3 \times 10^{-3})^{5.22}}{9(3.22 + 2)(1.32)^{3.22-1}}$$

$$= 4.91 \times 10^{-6} + 5.95 \times 10^{-4}$$

$$= 6 \times 10^{-4} \text{m}^3/\text{s per metre width.} \square$$

3.6 Laminar flow in a concentric annulus

The flow of non-Newtonian fluids through concentric and eccentric annuli represents an idealisation of several industrially important processes. One important example is in oil well drilling where a heavy drilling mud is circulated through the annular space around the drill pipe in order to carry the drilling debris to the surface. These drilling muds are typically either Bingham plastic or power-law type fluids. Other examples include the extrusion of plastic tubes and pipes in which the molten polymer is forced through an annular die, and the flow in double-pipe heat exchangers. In all these applications, it is often required to predict the frictional pressure gradient to sustain a fixed flow rate or vice versa. In this section, the isothermal, steady and fully-developed flow of power-law and Bingham plastic fluids in concentric annulus is analysed and appropriate expressions and/or charts are presented which permit the calculation of pressure gradient for a given application.

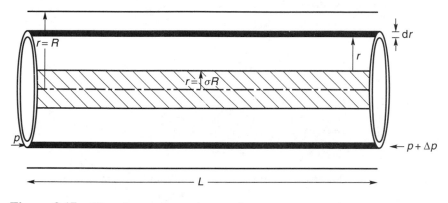

Figure 3.17   *Flow in a concentric annulus*

The calculation of the velocity distribution and the mean velocity of a fluid flowing through an annulus of outer radius R and inner radius σR is more complex than that for flow in a pipe or between two parallel planes (Figure 3.17), though the force balance on an element of fluid can be written in a manner similar to that used in previous sections. If the pressure changes by an amount Δp as a consequence of friction in a length L of annulus, the resulting force can be equated to the shearing force acting on the fluid. Consider the flow of the fluid situated at a distance not greater than r from the centreline of the pipe. The shear force acting on this fluid comprises two parts: one is the drag on its outer surface ($r = R$) which can be expressed in terms of the shear stress in the fluid at that location; the other contribution is the drag occurring at the inner (solid) boundary of the annulus, i.e. at $r = \sigma R$. This component cannot be estimated at present, however. Alternatively, this difficulty can be obviated by considering the equilibrium of a thin ring of fluid of radius r and thickness dr (Figure 3.17). The pressure force acting on this fluid element is:

$$2\pi r dr\{p - (p + \Delta p)\}$$

The only other force acting on the fluid element in the z-direction is that arising from the shearing on both surfaces of the element. Note that, not only will the shear stress change from r to $r + dr$ but the surface area over which shearing occurs will also depend upon the value of r. The net force can be written as:

$$\left(2\pi r \, L \cdot \tau_{rz}\big|_{r+dr} - 2\pi r \, L \, \tau_{rz}\big|_r \right)$$

At equilibrium therefore:

$$2\pi r dr \left(\frac{-\Delta p}{L} \right) L = 2\pi L \left\{ r \, \tau_{rz}\big|_{r+dr} - r\tau_{rz}\big|_r \right\}$$

or $$\frac{r\tau_{rz}\big|_{r+dr} - r\tau_{rz}\big|_r}{dr} = r \left(\frac{-\Delta p}{L} \right)$$

Now taking limits as $dr \to 0$, it becomes

$$\frac{d}{dr}(r\tau_{rz}) = r \left(\frac{-\Delta p}{L} \right) \tag{3.75}$$

The shear stress distribution across the gap is obtained by integration:

$$\tau_{rz} = \frac{r}{2} \left(\frac{-\Delta p}{L} \right) + \frac{C_1}{r} \tag{3.76}$$

Because of the no-slip boundary condition at both solid walls, i.e. at $r = \sigma R$ and $r = R$, the velocity must be maximum at some intermediate point, say at $r = \lambda R$. Then, for a fluid without a yield stress, the shear stress must be zero at this position and for a viscoplastic fluid, there will be a plug moving *en masse*. Equation (3.76) can therefore be re-written:

$$\tau_{rz} = \left(\frac{-\Delta p}{L} \right) \frac{R}{2} \left(\xi - \frac{\lambda^2}{\xi} \right) \tag{3.77}$$

where $\xi = r/R$, the dimensionless radial coordinate.

3.6.1 Power-law fluids

For this flow, the power-law fluid can be written as:

$$\tau_{rz} = -m \left| \frac{dV_z}{dr} \right|^{n-1} \left(\frac{dV_z}{dr} \right) \tag{3.78}$$

It is important to write the equation in this form whenever the sign of the velocity gradient changes within the flow field. In this case, (dV_z/dr) is positive for $\sigma \leq \xi \leq \lambda$ and negative for $\lambda \leq \xi \leq 1$. Now equation (3.78) can be substituted in equation (3.77) and integrated to obtain

$$V_{z_i} = R \left(\frac{-\Delta p}{L} \cdot \frac{R}{2m} \right)^{1/n} \int_{\sigma}^{\xi} \left(\frac{\lambda^2}{x} - x \right)^{1/n} dx \qquad (3.79a)$$

$$\sigma \leq \xi \leq \lambda$$

$$V_{z_o} = R \left(\frac{-\Delta p}{L} \cdot \frac{R}{2m} \right)^{1/n} \int_{\xi}^{1} \left(x - \frac{\lambda^2}{x} \right)^{1/n} dx \qquad (3.79b)$$

$$\lambda \leq \xi \leq 1$$

where subscripts '*i*' and '*o*' denote the inner ($\sigma \leq \xi \leq \lambda$) and outer ($\lambda \leq \xi \leq 1$) regions respectively and x is a dummy variable of integration. The no-slip boundary conditions at $\xi = \sigma$ and $\xi = 1$ have been incorporated in equation (3.79). Clearly, the value of λ is evaluated by setting $V_{z_i} = V_{z_o}$ at $\xi = \lambda$, i.e.

$$\int_{\sigma}^{\lambda} \left(\frac{\lambda^2}{x} - x \right)^{1/n} dx = \int_{\lambda}^{1} \left(x - \frac{\lambda^2}{x} \right)^{1/n} dx \qquad (3.80)$$

The volumetric flow rate of the fluid, Q, is obtained as:

$$Q = 2\pi \int_{\sigma R}^{R} r V_z \, dr = 2\pi R^2 \int_{\sigma}^{1} \xi V_z \, d\xi$$

$$= 2\pi R^3 \left(\frac{-\Delta p}{L} \cdot \frac{R}{2m} \right)^{1/n} \left[\int_{\sigma}^{\lambda} \xi V_{z_i} d\xi + \int_{\lambda}^{1} \xi V_{z_o} \, d\xi \right] \qquad (3.81)$$

Clearly, equations (3.80) and (3.81) must be solved and integrated simultaneously to eliminate λ and to evaluate the volumetric rate of flow of liquid, Q. Analytical solutions are possible only for integral values of $(1/n)$, i.e. for $n = 1, 0.5, 0.33, 0.25$, etc. Thus, Fredrickson and Bird [1958] evaluated the integral in equation (3.81) for such values of n and, by interpolating the results for the intermediate values of power law index, they presented a chart relating non-dimensional flowrate, pressure drop, σ and n. However, the accuracy of their results deteriorates rapidly with decreasing values of n and/or $(1 - \sigma) \ll 1$, i.e. with narrowing annular region. Subsequently, however, Hanks and Larsen [1979] were able to evaluate the volumetric flow rate, Q, analytically and their final expression is:

$$Q = \frac{n\pi R^3}{(3n + 1)} \left(\frac{-\Delta p}{L} \cdot \frac{R}{2m} \right)^{1/n} \{ (1 - \lambda^2)^{(n+1)/n} - \sigma^{(n-1)/n} (\lambda^2 - \sigma^2)^{(n+1)/n} \}$$

$$(3.82)$$

Table 3.2 *Values of λ computed from equation (3.80) [Hanks and Larson, 1979]*

n	σ								
	0.10	0.20	0.30	0.40	0.50	0.60	0.70	0.80	0.90
0.10	0.3442	0.4687	0.5632	0.6431	0.7140	0.7788	0.8389	0.8954	0.9489
0.20	0.3682	0.4856	0.5749	0.6509	0.7191	0.7818	0.8404	0.8960	0.9491
0.30	0.3884	0.4991	0.5840	0.6570	0.7229	0.7840	0.8416	0.8965	0.9492
0.40	0.4052	0.5100	0.5912	0.6617	0.7259	0.7858	0.8426	0.8969	0.9493
0.50	0.4193	0.5189	0.5970	0.6655	0.7283	0.7872	0.8433	0.8972	0.9493
0.60	0.4312	0.5262	0.6018	0.6686	0.7303	0.7884	0.8439	0.8975	0.9494
0.70	0.4412	0.5324	0.6059	0.6713	0.7319	0.7893	0.8444	0.8977	0.9495
0.80	0.4498	0.5377	0.6093	0.6735	0.7333	0.7902	0.8449	0.8979	0.9495
0.90	0.4572	0.5422	0.6122	0.6754	0.7345	0.7909	0.8452	0.8980	0.9495
1.00	0.4637	0.5461	0.6147	0.6770	0.7355	0.7915	0.8455	0.8981	0.9496

The only unknown now remaining is λ, which locates the position where the velocity is maximum. Table 3.2 presents the values of λ for a range of values of σ and n.

Example 3.10

A polymer solution exhibits power-law behaviour with $n = 0.5$ and $m = 3.2\,\mathrm{Pa\cdot s^{0.5}}$. Estimate the pressure gradient required to maintain a steady flow of $0.3\,\mathrm{m^3/min}$ of this polymer solution through the annulus between a 10 mm and a 20 mm diameter tube.

Solution

Here, $R = \dfrac{20}{2} \times 10^{-3} = 0.01\,\mathrm{m}$

$\sigma R = \dfrac{10}{2} \times 10^{-3} = 0.005\,\mathrm{m}$

or $\sigma = 0.5$

From Table 3.2, for $\sigma = 0.5$ and $n = 0.5$, $\lambda = 0.728$
Substituting these values in equation (3.82)

$$\left(\frac{0.3}{60}\right) = \frac{(0.5)(3.14)(0.01)^3}{(3 \times 0.5 + 1)} \left(\frac{-\Delta p}{L}\right)^2 \left(\frac{0.01}{2 \times 3.2}\right)^2$$
$$\times \{(1 - 0.728^2)^{(0.5+1)/0.5} - 0.5^{(0.5-1)/0.5}(0.728^2 - 0.5^2)^{(0.5+1)/0.5}\}$$

and solving: $\dfrac{-\Delta p}{L} = 169\,\mathrm{kPa/m}$ □

3.6.2 *Bingham plastic fluids*

The laminar axial flow of Bingham plastic fluids through a concentric annulus has generated even more interest than that for the power-law fluids, e.g. see refs. [Laird, 1957; Fredrickson and Bird, 1958; Bird *et al.*, 1983; Fordham *et al.*, 1991]. The main feature which distinguishes the flow of a Bingham plastic fluid from that of a power-law fluid is the existence of a plug region in which the shear stress is less than the yield stress. Figure 3.18 shows qualitatively the salient features of the velocity distribution in an annulus; the corresponding profile for a fluid without the yield stress (e.g. power-law fluid) is also shown for the sake of comparison.

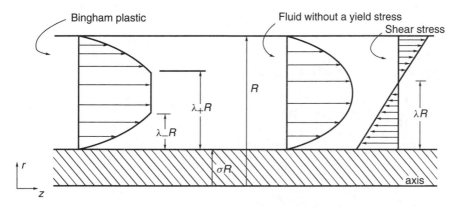

Figure 3.18   *Schematics of velocity profiles for Bingham plastic and power-law fluids in an concentric annulus*

In principle, the velocity distribution and the mean velocity of a Bingham plastic fluid flowing through an annulus can be deduced by substituting for the shear stress in equation (3.76) in terms of the Bingham plastic model, equation (3.10). However, the signs of the shear stress (considered positive in the same sense as the flow) and the velocity gradients in the two flow regions need to be treated with special care. With reference to the sketch shown in Figure 3.18, the shearing force on the fluid is positive ($\sigma R \leq r \leq \lambda_- R$) where the velocity gradient is also positive. Thus, in this region:

$$\tau_{rz} = \tau_0^B + \mu_B \left(\frac{dV_z}{dr} \right) \tag{3.83}$$

On the otherhand, in the region $\lambda_+ R \leq r \leq R$, the velocity gradient is negative and the shearing force is also in the negative r-direction and hence

$$-\tau_{rz} = \tau_0^B + \mu_B \left(-\frac{dV_z}{dr} \right) \tag{3.84}$$

Equations (3.83) and (3.84) can now be substituted in equation (3.76) and integrated to deduce the velocity distributions. The constants of integration can be evaluated by using the no-slip boundary condition at both $r = \sigma R$ and $r = R$. However, the boundaries of the plug existing in the middle of the annulus are not yet known; nor is the plug velocity known. These unknowns are evaluated by applying the following three conditions, namely, the continuity of velocity at $r = \lambda_- R$ and $r = \lambda_+ R$, the velocity gradient is also zero at these boundaries and finally, the force balance on the plug of fluid:

$$2\pi R(\lambda_+ + \lambda_-)\tau_0^B = \left(\frac{-\Delta p}{L}\right) \cdot \pi((\lambda_+ R)^2 - (\lambda_- R)^2) \tag{3.85}$$

Unfortunately, the algebraic steps required to carry out the necessary integrations and the evaluation of the constants are quite involved and tedious. Thus, these are not presented here and readers are referred to the original papers [Laird, 1957] or to the book by Skelland [1967] for detailed derivations. Instead consideration is given here to the practical problem of

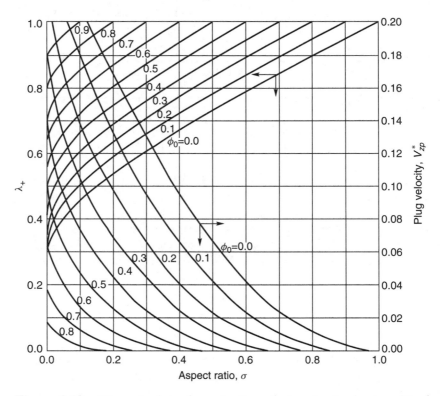

Figure 3.19 *Dimensionless plug velocity and plug size for laminar Bingham plastic flow in an annulus*

estimating the necessary pressure gradient to maintain a fixed flow rate of a Bingham plastic fluid or vice versa. Fredrickson and Bird [1958] organised their numerical solutions of the equations presented above in terms of the following dimensionless parameters:

$$\text{dimensionless velocity: } V_z^* = \frac{2\mu_B V_z}{R^2 \left(\dfrac{-\Delta p}{L} \right)}$$

$$\text{dimensionless yield stress: } \phi_0 = \frac{2\tau_0^B}{R \left(\dfrac{-\Delta p}{L} \right)}$$

$$\text{dimensionless flowrate: } \Omega = \frac{Q}{Q_N}$$

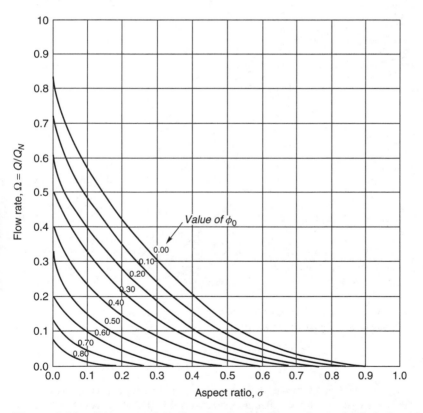

Figure 3.20 *Dimensionless flowrate for Bingham plastic fluids in laminar flow through an annulus*

where Q_N is the flow rate of a Newtonian liquid of viscosity, μ_B. Thus,

$$Q_N = \frac{\pi R^4}{8\mu_B} \left(\frac{-\Delta p}{L} \right)$$

Fredrickson and Bird [1958] presented three charts (Figures 3.19–3.21) showing relationships between V_z^*, ϕ_0, Ω, σ and λ_+.

For given values of the rheological constants (μ_B, τ_0^B), pressure gradient $(-\Delta p/L)$ and the dimensions of the annulus (σ, R), the values of λ_+ and the plug velocity V_{zp}^* can be read from Figure 3.19 and the value of Ω from Figure 3.20 from which the volumetric rate of flow, Q, can be estimated. For the reverse calculation, the group (Ω/ϕ_0) is independent of the pressure gradient and one must use Figure 3.21 to obtain the value of ϕ_0 and thus evaluate the required pressure gradient $(-\Delta p/L)$.

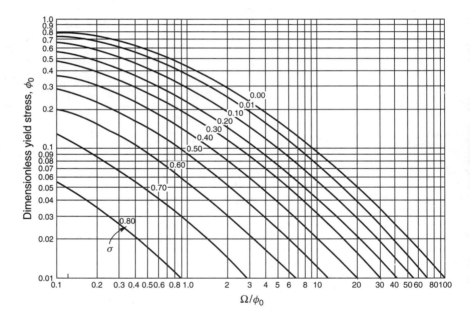

Figure 3.21   *Chart for the estimation of pressure gradient for laminar flow of Bingham plastic fluids in an annulus*

Example 3.11

A molten chocolate (density = 1500 kg/m³) flows through a concentric annulus of inner and outer radii 10 mm and 20 mm, respectively, at 30°C at the constant flow rate of 0.03 m³/min. The steady-shear behaviour of the chocolate can be approximated by a Bingham plastic model with $\tau_0^B = 35$ Pa and $\mu_B = 1$ Pa·s.

(a) Estimate the required pressure gradient to maintain the flow, and determine the velocity and the size of the plug.
(b) Owing to a pump malfunction, the available pressure gradient drops by 25% of the value calculated in (a), what will be the new flow rate?

Solution

Part (a):

In this case, $\tau_0^B = 35\,\text{Pa}$, $\mu_B = 1\,\text{Pa·s}$

$$Q = 0.03\,\text{m}^3/\text{min} = \frac{0.03}{60}\,\text{m}^3/\text{s}$$

$$\sigma = \frac{10}{20} = 0.5;\ R = 20 \times 10^{-3}\,\text{m}$$

Since the pressure gradient $(-\Delta p/L)$ is unknown, one must use Figure 3.21.

$$\therefore\ \frac{\Omega}{\phi_0} = \frac{4\mu_B Q}{\pi R^3 \tau_0^B} = \frac{4 \times 1 \times (0.03/60)}{3.14 \times (20 \times 10^{-3})^3 \times 35} = 2.28$$

For $\dfrac{\Omega}{\phi_0} = 2.28$ and $\sigma = 0.5$, Figure 3.21 gives:

$$\phi_0 = \sim 0.048$$

$$\therefore\ \left(\frac{-\Delta p}{L}\right) = \frac{2\tau_0^B}{R\phi_0} = \frac{2 \times 35}{20 \times 10^{-3} \times 0.048}$$

$$= 73\,000\,\text{Pa/m}$$

$$= 73\,\text{kPa/m}$$

Now from Figure 3.19, $\phi_0 = 0.048$ and $\sigma = 0.5$,

$$V_{z,p}^* = \sim 0.05$$

$$\lambda_+ = 0.76$$

From the definition of V_z^*, we have

$$V_{zp}^* = \frac{2\mu_B V_{zp}}{R^2 \left(\dfrac{-\Delta p}{L}\right)}$$

$$0.05 = \frac{2 \times 1 \times V_{zp}}{(20 \times 10^{-3})^2 (73\,000)}$$

$$\therefore\ V_{zp} = 0.73\,\text{m/s}$$

i.e. the plug in the central region has a velocity of 0.73 m/s (compared with the mean velocity of $Q/\pi R^2(1 - \sigma^2)$, i.e. 0.53 m/s).

From equation (3.85):

$$(\lambda_+ - \lambda_-)\frac{R}{2}\left(\frac{-\Delta p}{L}\right) = \tau_0^B$$

Substitution of values gives $\lambda_- = 0.71$. Thus the plug region extends from $\lambda_- R$ to $\lambda_+ R$, i.e. from 14.2 to 15.2 mm. These calculations assume the flow to be laminar. As a first approximation, one can define the corresponding Reynolds number based on the hydraulic diameter, D_h.

$$D_h = \frac{4 \times \text{Flow area}}{\text{wetted perimeter}} = \frac{4\pi R^2(1 - \sigma^2)}{2\pi R(1 + \sigma)} = 2R(1 - \sigma)$$

$$= 2 \times 20 \times 10^{-3}(1 - 0.5) = 0.02\,\text{m}$$

Reynolds number, $\text{Re} = \dfrac{\rho V D_h}{\mu_B} = \dfrac{1500 \times 0.53 \times 0.02}{1} = 16$

The flow is thus likely to be streamline.

Part (b): In this case, the available pressure gradient is only 75% of the value calculated above,

$$\frac{-\Delta p}{L} = 73 \times 0.75 = 54.75\,\text{kPa/m}$$

We can now evaluate ϕ_0:

$$\phi_0 = \frac{2\tau_0^B}{R\left(\dfrac{-\Delta p}{L}\right)} = \frac{2 \times 35}{20 \times 10^{-3} \times 54.75 \times 1000} = 0.064$$

From Figure 3.20, for $\phi_0 = 0.064$ and $\sigma = 0.5$,

$$\Omega = \frac{Q}{Q_N} = \sim 0.1$$

$$\therefore Q = 0.1 \times Q_N = 0.1 \times \frac{\pi R^4}{8\mu_B}\left(\frac{-\Delta p}{L}\right)$$

$$= \frac{0.1 \times 3.14 \times (20 \times 10^{-3})^4}{8 \times 1} \times 54.75 \times 1000\,\text{m}^3/\text{s}$$

$$= 0.000344\,\text{m}^3/\text{s or } 0.0206\,\text{m}^3/\text{min}$$

Two observations can be made here. The 25% reduction in the available pressure gradient has lowered the flow rate by 31%. Secondly, in this case the flow rate is only one tenth of that of a Newtonian fluid of the same viscosity as the plastic viscosity of the molten chocolate!□

This section is concluded by noting that analogous treatments for the concentric and eccentric annular flow of Herschel–Bulkley and other viscosity models are also available in the literature [Hanks, 1979; Uner *et al.*, 1988; Walton and Bittleston, 1991; Fordham *et al.*, 1991; Gücüyener and Mehmeteoglu, 1992].

3.7 Laminar flow of inelastic fluids in non-circular ducts

Analytical solutions for the laminar flow of time-independent fluids in non-axisymmetric conduits are not possible. Numerous workers have obtained approximate numerical solutions for specific flow geometries including rectangular and triangular pipes [Schechter, 1961; Wheeler and Wissler, 1965; Miller, 1972; Mitsuishi and Aoyagi, 1973]. On the other hand, semi-empirical attempts have also been made to develop methods for predicting pressure drop for time-independent fluids in ducts of non-circular cross-section. Perhaps the most systematic and successful friction factor analysis is that provided by Kozicki *et al.* [1966, 1967]. By noting the similarity between the form of the Rabinowitsch–Mooney equation for the flow of time-independent fluids in circular pipes (equation 3.25) and that in between two plates, they suggested that it could be extended to ducts having a constant cross-section of arbitrary shape as follows:

$$\left(-\frac{dV_z}{dr}\right)_w = f(\overline{\tau}_w) = a\overline{\tau}_w \frac{d\left(\dfrac{8V}{D_h}\right)}{d\overline{\tau}_w} + b\left(\frac{8V}{D_h}\right) \tag{3.86}$$

where a and b are two geometric parameters characterising the cross-section of the duct, D_h is the hydraulic diameter (= 4 times flow area/wetted perimeter), and $\overline{\tau}_w$ is the mean value of shear stress at the wall, and is related to the pressure gradient as:

$$\overline{\tau}_w = \frac{D_h}{4}\left(\frac{-\Delta p}{L}\right) \tag{3.87}$$

For constant values of a and b, equation (3.86) can be integrated to obtain:

$$\left(\frac{8V}{D_h}\right) = \frac{1}{a}(\overline{\tau}_w)^{-b/a}\int_0^{\overline{\tau}_w} \tau^{(b/a)-1} f(\tau)\,d\tau \tag{3.88}$$

It should be noted that for a circular pipe of diameter D, $D_h = D$; $a = 1/4$ and $b = 3/4$; equation (3.88) then reduces to equation (3.21). For the flow of a power-law fluid, $f(\tau) = (\tau/m)^{1/n}$ and integration of (3.88) yields:

$$\overline{\tau}_w = m\left\{\frac{8V}{D_h}\left(b + \frac{a}{n}\right)\right\}^n \tag{3.89}$$

which can be re-written in terms of the friction factor, $f = 2\overline{\tau}_w/\rho V^2$ as:

$$f = \frac{16}{\mathrm{Re}_g} \tag{3.90}$$

Table 3.3   *Values of a and b to be used in equations (3.90) and (3.91)*

1. *Concentric annuli*

$\sigma = R_i/R_o$	a	b
0.00	0.2500	0.7500
0.01	0.3768	0.8751
0.03	0.4056	0.9085
0.05	0.4217	0.9263
0.07	0.4331	0.9383
0.10	0.4455	0.9510
0.20	0.4693	0.9737
0.30	0.4817	0.9847
0.40	0.4890	0.9911
0.50	0.4935	0.9946
0.60	0.4965	0.9972
0.70	0.4983	0.9987
0.80	0.4992	0.9994
0.90	0.4997	1.0000
1.00	0.5000	1.0000

2. *Elliptical ducts*

$\beta = b'/a'$	a	b
0.00	0.3084	0.9253
0.10	0.3018	0.9053
0.20	0.2907	0.8720
0.30	0.2796	0.8389
0.40	0.2702	0.8107
0.50	0.2629	0.7886
0.60	0.2575	0.7725
0.70	0.2538	0.7614
0.80	0.2515	0.7546
0.90	0.2504	0.7510
1.00	0.2500	0.7500

3. *Rectangular ducts*

$E = H/W$	a	b
0.00	0.5000	1.0000
0.25	0.3212	0.8182
0.50	0.2440	0.7276
0.75	0.2178	0.6866
1.00	0.2121	0.6766

Table 3.3 *(continued)*

4. *Isosceles triangular ducts*

2α	a	b
10°	0.1547	0.6278
20°	0.1693	0.6332
40°	0.1840	0.6422
60°	0.1875	0.6462
80°	0.1849	0.6438
90°	0.1830	0.6395

5. *Regular polygonal ducts*

N	a	b
4	0.2121	0.6771
5	0.2245	0.6966
6	0.2316	0.7092
8	0.2391	0.7241

where the generalised Reynolds number,

$$\text{Re}_g = \frac{\rho V^{2-n} D_h^n}{8^{n-1} m \left(b + \dfrac{a}{n}\right)^n}.$$
(3.90a)

The main virtue of this approach lies in its simplicity and the fact that the geometric parameters a and b can be deduced from the behaviour of Newtonian fluids in the same flow geometry. Table 3.3 lists values of a and b for a range of flow geometries commonly encountered in process applications. A typical comparison between predicted and experimental values of friction factor for rectangular ducts is shown in Figure 3.22. Similar agreement has been reported by, among others, Mitsuishi *et al.* [1972] and, more recently, by Xie and Hartnett [1992] for visco-elastic fluids in rectangular ducts. Kozicki *et al.* [1966] argued that equation (3.37) can be generalised to include turbulent flow in non-circular ducts by re-casting it in the form:

$$\frac{1}{\sqrt{f}} = \frac{4}{n^{0.75}} \log_{10}(\text{Re}_g f^{(2-n)/2}) - \frac{0.4}{n^{1.2}} + 4n^{0.25} \log\left[\frac{4(a+bn)}{(3n+1)}\right]$$
(3.91)

Note that since for a circular tube, $a = 1/4$ and $b = 3/4$, equation (3.91) is consistent with that for circular pipes, equation (3.37). The limited data available on turbulent flow in triangular [Irvine, Jr., 1988] and rectangular ducts [Kostic and Hartnett, 1984] conforms to equation (3.91). In the absence

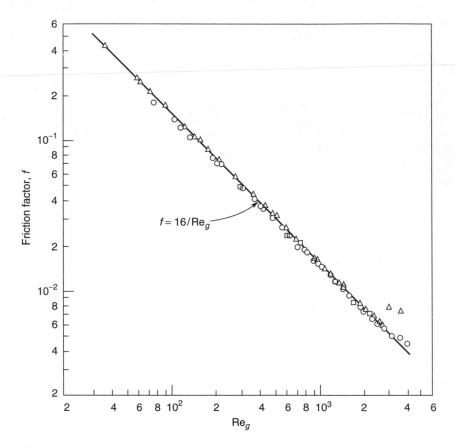

Figure 3.22 *Experimental friction factor values for power-law fluids in laminar regime in rectangular channels.* ○, *Wheeler and Wissler (1965);* △, *Hartnett et al. (1986);* □, *Hartnett and Kostic (1985)*

of any definite information, Kozicki and Tiu [1988] suggested that the Dodge–Metzner criterion, $Re_{gen} \le 2100$, can be used for predicting the limit of laminar flow in non-circular ducts.

Scant analytical and experimental results suggest that visco-elasticity in a fluid may induce secondary motion in non-circular conduits, even under laminar conditions. However, measurements reported to date indicate that the friction factor – Reynolds number behaviour is little influenced by such secondary flows [Hartnett and Kostic, 1989].

Example 3.12

A power-law fluid ($m = 0.3\,Pa \cdot s^n$ and $n = 0.72$) of density $1000\,kg/m^3$ is flowing in a series of ducts of the same flow area but different cross-sections as listed below:

(i) concentric annulus with $R = 37\,\mathrm{mm}$ and $\sigma = 0.40$
(ii) circular pipe
(iii) rectangular, $(H/W) = 0.5$
(iv) elliptical $b'/a' = 0.5$
(v) isosceles triangular with half-apex angle, $\alpha = 20°$.

Estimate the pressure gradient required to maintain an average velocity of $1.25\,\mathrm{m/s}$ in each of these channels. Use the geometric parameter method. Also, calculate the value of the generalised Reynolds number as a guide to the nature of the flow.

Solution

(i) For a concentric annulus, for $\sigma = 0.4$, and from Table 3.3:

$$a = 0.489; \quad b = 0.991$$

The hydraulic diameter, $D_h = 2R(1 - \sigma)$

$$= 2 \times 37 \times 10^{-3} \times (1 - 0.4)$$

$$= 0.044\,\mathrm{m}$$

Reynolds number, $\mathrm{Re}_g = \dfrac{\rho V^{2-n} D_h^n}{8^{n-1} m \left(b + \dfrac{a}{n}\right)^n}$ (3.90a)

$$= \frac{1000 \times 1.25^{2-0.72} \times 0.044^{0.72}}{8^{0.72-1} \times 0.3 \times \left(0.991 + \dfrac{0.489}{0.72}\right)^{0.72}}$$

$$= 579$$

Thus, the flow is laminar.

$$\therefore \qquad f = \frac{16}{579} = 0.0276$$

$$\text{and} \quad \frac{-\Delta p}{L} = \frac{2f\rho V^2}{D_h} = \frac{2 \times 0.0276 \times 1000 \times 1.25^2}{0.044} = 1963\,\mathrm{Pa/m}$$

For the sake of comparison, equation (3.82) yields a value of $1928\,\mathrm{Pa/m}$ which is remarkably close.

(ii) For a circular tube, the area of flow

$$= \pi R^2 (1 - \sigma^2) = 3.14 \times 0.037^2 (1 - 0.4^2) = 0.003613\,\mathrm{m}^2$$

which corresponds to the pipe radius of $0.0339\,\mathrm{m}$ or diameter of $0.0678\,\mathrm{m}$.

For a circular pipe, $a = 0.25$, $b = 0.75$, $D_h = D = 0.0678$ m.

$$\therefore Re_g = \frac{\rho V^{2-n} D_h^n}{8^{n-1} m \left(b + \dfrac{a}{n}\right)^n} = \frac{(1000)(1.25)^{2-0.72}(0.0678)^{0.72}}{8^{0.72-1}(0.3)\left(0.75 + \dfrac{0.25}{0.72}\right)^{0.72}}$$

$$= 1070$$

Thus, the flow is laminar and $f = 16/1070 = 0.01495$ and pressure gradient,

$$\frac{-\Delta p}{L} = \frac{2f\rho V^2}{D_h} = \frac{2 \times 0.01495 \times 1000 \times 1.25^2}{0.0678}$$

$$= 689\,\text{Pa/m}$$

(iii) For a rectangular duct with $H/W = 0.5$, $H = 0.0425\,m$ and $W = 0.085$ m (for the same area of flow), and from Table 3.3:
$a = 0.244$, $b = 0.728$

$$\text{The hydraulic diameter, } D_h = \frac{4 \times \text{Flow area}}{\text{wetted perimeter}}$$

$$= \frac{4 \times A}{2(H + W)} = \frac{4 \times 0.003613}{2(0.0425 + 0.085)}$$

$$= 0.0567\,\text{m}$$

$$\therefore Re_g = \frac{\rho V^{2-n} D_h^n}{8^{n-1} \cdot m \left(b + \dfrac{a}{n}\right)^n} = \frac{(1000) \times (1.25)^{2-0.72}(0.0567)^{0.72}}{8^{0.72-1} \times 0.3 \times \left(0.728 + \dfrac{0.244}{0.72}\right)^{0.72}}$$

$$= 960$$

Again, the flow is laminar and $f = 16/960 = 0.0167$

$$\text{The pressure gradient, } \frac{-\Delta p}{L} = \frac{2f\rho V^2}{D_h}$$

$$= \frac{2 \times 0.0167 \times 1000 \times 1.25^2}{0.0567} = 919\,\text{Pa/m}$$

(iv) The cross-sectional area of an elliptic pipe with semi-axes a' and b' is $\pi a'b'$, while $a = 0.2629$ and $b = 0.7886$ from Table 3.3 corresponding to $b'/a' = 0.5$

$$\therefore \pi a'b' = 0.003613$$

Solving with $b'/a' = 0.5$, $a' = 0.04795\,\text{m}$ and $b' = 0.024\,\text{m}$. The hydraulic diameter D_h is calculated next as:

$$D_h = \frac{4 \times \text{Flow area}}{\text{Wetted perimeter}}$$

No analytical expression is available for the perimeter of an ellipse; however, it can be approximated by $2\pi((a'^2 + b'^2)/2)^{1/2}$

$$\therefore \qquad D_h = \frac{4 \times 0.003613}{2\pi\left(\dfrac{0.04795^2 + 0.024^2}{2}\right)^{1/2}} = 0.0607\,\text{m}$$

$$Re_g = \frac{1000 \times 1.25^{1.28} \times 0.0607^{0.72}}{8^{-0.28} \times 0.3 \times \left(0.7886 + \dfrac{0.2629}{0.72}\right)^{0.72}} = 953$$

and $\qquad f = 16/Re_g = 16/953 = 0.0168$

$$\therefore \qquad \frac{-\Delta p}{L} = \frac{2f\rho V^2}{D_h} = \frac{2 \times 0.0168 \times 1000 \times 1.25^2}{0.0607} = 864\,\text{Pa/m}$$

(v) For the pipe of triangular cross-section, with $\alpha = 20°$ $a = 0.184$ and $b = 0.6422$

Let the base of the triangle be x

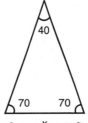

$$\therefore \text{ height of the triangle} = \frac{x}{2\tan\alpha} = \frac{x}{2\tan 20}$$

$$= 1.37x$$

$$\therefore \text{ Area for flow} = \tfrac{1}{2} \times x \times 1.37x = 0.003613$$

$$\text{or } x = 0.0725\,\text{m}$$

The hydraulic diameter is calculated next:

$$D_h = \frac{4 \times 0.003613}{x + 2\left(\dfrac{x}{2\sin\alpha}\right)} = \frac{4 \times 0.003613}{0.0725\left(1 + \dfrac{1}{\sin 20}\right)}$$

$$= 0.0508\,\text{m}$$

The Reynolds number of flow,

$$Re_g = \frac{1000 \times 1.25^{1.28} \times 0.0508^{0.72}}{8^{-0.28} \times 0.3 \times \left(0.6422 + \dfrac{0.184}{0.72}\right)^{0.72}}$$

$$= 1006$$

$$\therefore \qquad f = \frac{16}{484} = 0.016$$

and $\qquad \dfrac{-\Delta p}{L} = \dfrac{2f\rho V^2}{D_h} = \dfrac{2 \times 0.016 \times 1000 \times 1.25^2}{0.0508}$

$$= 980\,\text{Pa/m}$$

This example clearly shows that the pressure drop is a minimum in the case of circular pipes, followed by the elliptic, rectangular and triangular cross-sections, and the concentric annulus for flow. On the other hand, if one were to maintain the same hydraulic diameter in each case, the corresponding pressure gradients range from 2500 Pa/m to 4000 Pa/m.□

3.8 Miscellaneous frictional losses

In the analysis of pipe networks, one is usually concerned either with how much power is required to deliver a set flow rate through an existing flow system or with the optimum pipe diameter for a given pump and duty. All such calculations involve determining the frictional pressure losses in the systems, both in the region of fully established flow (as has been assumed so far), and in the associated sudden changes in cross-section (expansions and contractions) and other fittings such as bends, elbows, valves, etc. It is also necessary to establish whether the flow is laminar or turbulent. For Newtonian flow, the magnitudes of these losses are well known, and, although the theory is far from complete, the established design procedures are usually quite satisfactory. The analogous results for non-Newtonian (mainly time-independent type) are presented here, although the experimental results are scant in this field.

3.8.1 Sudden enlargement

When the cross-section of a pipe enlarges gradually, the streamlines follow closely the contours of the duct and virtually no extra frictional losses are incurred. On the other hand, whenever the change is sudden, additional losses arise due to the eddies formed as the fluid enters the enlarged cross-section. The resulting head loss for laminar flow can be evaluated by applying the mechanical energy balance in conjunction with the integral momentum balance. Consider the flow configuration shown in Figure 3.23 in which the section '2' is located immediately after the end of the smaller pipe. By suitable choice of plane '1', the frictional pressure loss may be assumed to be negligible between planes '1' and '2' and hence $p_1 \sim p_2$; the latter acts over the whole cross-section (πR_2^2). Also, immediately following the expansion at section '2', the streamlines will be nearly parallel to the axis and the velocity profile at section '2' will be similar to the fully developed profile at section '1'. If section '3' is sufficiently far down-stream, the velocity profile again be fully established. On applying the integral momentum balance over the control volume as shown in Figure 3.23:

$$\Sigma F_z = (p_2 - p_3)\pi R_2^2 = -\int_0^{R_1} 2\pi r\rho V_z^2 \, dr + \int_0^{R_2} 2\pi r\rho V_z^2 \, dr \qquad (3.92)$$

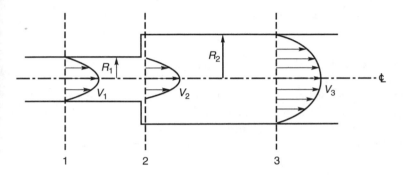

Figure 3.23   *Schematics of laminar flow through a sudden expansion in a tube*

The integration in equation (3.92) can be carried out after insertion of the velocity profiles for the appropriate viscosity model to obtain the pressure loss ($p_2 - p_3$). For power-law fluids, this procedure leads to:

$$\frac{p_2 - p_3}{\rho} = \left(\frac{3n+1}{2n+1}\right) \frac{Q^2}{A_1^2} \left[\left(\frac{A_1}{A_2}\right)^2 - \frac{A_1}{A_2}\right] \tag{3.93}$$

where $A_1 = \pi R_1^2$ and $A_2 = \pi R_2^2$. Applying the mechanical energy balance equation between points '1' and '3':

$$\frac{p_1}{\rho} + \frac{V_1^2}{2\alpha} + gz_1 = \frac{p_3}{\rho} + \frac{V_3^2}{2\alpha} + gz_3 + \Sigma F_{exp}$$

or
$$\Sigma F_{exp} = \frac{p_1 - p_3}{\rho} + (z_1 - z_3)g + \frac{V_1^2 - V_3^2}{2\alpha} \tag{3.94}$$

For a horizontal system, $z_1 = z_3$, putting $p_1 \approx p_2$ and substituting for α from equation (3.16), equations (3.93) and (3.94) yield the following expression for the head loss, h_e:

$$h_e = \frac{\Sigma F_{exp}}{g} = \frac{1}{g} \left(\frac{Q}{A_1}\right)^2 \left(\frac{3n+1}{2n+1}\right)$$

$$\times \left[\frac{(n+3)}{2(5n+3)} \left(\frac{A_1}{A_2}\right)^2 - \left(\frac{A_1}{A_2}\right) + \frac{3(3n+1)}{2(5n+3)}\right] \tag{3.95}$$

If n were equal to zero, the velocity would be uniform across the pipe cross-section ($\alpha = 1$) and equation (3.95) would reduce to

$$h_e = \frac{V_1^2}{2g} \left(1 - \frac{A_1}{A_2}\right)^2 \tag{3.96}$$

This agrees with the expression for turbulent Newtonian flow when the velocity profile is assumed to be approximately flat.

3.8.2 Entrance effects for flow in tubes

The previous discussion on flow in pipes has been restricted to fully-developed flow where the velocity at any position in the cross-section is independent of distance along the pipe. In the entrance and exit sections of the pipe this will no longer be true. Since exit effects are much less significant than entrance effects, only the latter are dealt with in detail here.

For all fluids entering a small pipe from either a very much larger one or from a reservoir, the initial velocity profile will be approximately flat, and will then undergo a progressive change until fully developed flow is established, as shown schematically in Figure 3.24.

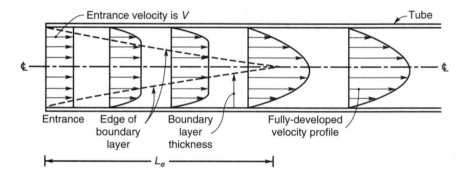

Figure 3.24   *Development of the boundary layer and velocity profile for laminar flow in the entrance region of a pipe*

The thickness of the boundary layer is theoretically zero at the entrance and increases progressively along the tube. The retardation of the fluid in the wall region must be accompanied by a concomitant acceleration in the central region in order to maintain continuity. When the velocity profile has reached its final shape, the flow is fully developed and the boundary layers may be considered to have converged at the centre line. It is customary to define an entry length, L_e, as the distance from the inlet at which the centreline velocity is 99% of that for the fully-developed flow. The pressure gradient in this entry region is different from that for fully developed flow and is a function of the initial velocity profile. There are two factors influencing the pressure gradient in the entry region: firstly, some pressure energy is converted into kinetic energy as the fluid in the central core accelerates, and secondly, the higher velocity gradients in the wall region result in greater frictional losses. It is important to estimate both the pressure drop occurring in the region before flow has been fully developed and the extent of this entrance length. This situation is amenable to analysis by repeated use of the mechanical energy balance equation. Consider the schematics of the flow shown in Figure 3.25. The stations '1' and '3' are well removed from the tube entrance, '2' is in the plane of entrance while '3' is situated in the fully developed region. The frictional pressure loss between points '1' and '3' can be expressed as:

$$\rho \Sigma F = \Delta p_{fd(1-2)} + \Delta p_{fd(2-3)} + \Delta p_{ex(1-2)} + \Delta p_{ex(2-3)} \tag{3.97}$$

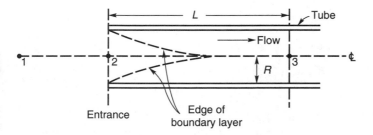

Figure 3.25 *Schematics for calculation of entrance effects*

where the subscripts 'fd' and 'ex' respectively denote the pressure drops over the regions of fully-developed flow and the additional pressure drop due to the acceleration of the fluid. Because $V_1 \ll V_3$, the fully developed pressure loss between points '1' and '2' is assumed to be negligible and that between '2' and '3' can be expressed in terms of the wall shear stress in the smaller tube as:

$$\Delta p_{fd(2-3)} = 2\tau_w \left(\frac{L}{R}\right) \tag{3.98}$$

where L is the length of the pipe between '2' and '3'. Thus, the extra frictional loss between '1' and '3' arising from the fact that flow is developing, $\Delta p_{\text{entrance}}$, can be written as:

$$\Delta p_{\text{entrance}} = (p_1 - p_3) - \Delta p_{fd(2-3)} = p_1 - p_3 - 2\tau_w \left(\frac{L}{R}\right) \tag{3.99}$$

Applying the mechanical energy balance between points '1' and '3', noting $z_1 = z_3$ and $V_1 \ll V_3$, and substituting for $\rho \Sigma F$ from equations (3.97) and (3.98):

$$(p_1 - p_3) = \frac{\rho V_3^2}{2\alpha_3} + \rho \Sigma F = \frac{\rho V_3^2}{2\alpha_3} + 2\tau_w \left(\frac{L}{R}\right) + \Delta p_{ex(1-3)} + \Delta p_{ex(2-3)}$$

or

$$\Delta p_{\text{entrance}} = p_1 - p_3 - 2\tau_w \left(\frac{L}{R}\right) = \frac{\rho V_3^2}{2\alpha_3} + \Delta p_{ex(1-2)} + \Delta p_{ex(2-3)}$$

(3.100)

Noting $(1/2)\rho V_3^3 = \tau_w/f$ and that $f = 16/\text{Re}_{MR}$ in laminar region, it is customary to re-arrange equation (3.100) as:

$$\frac{\Delta p_{\text{ent}}}{2\tau_w} = C_1(n)\frac{\text{Re}_{MR}}{32} + C_2(n)$$

(3.101)

where $$C_1(n) = \left(\frac{1}{\alpha_3} + \frac{\Delta p_{ex(2-3)}}{(1/2)\rho V_3^2}\right)$$

(3.102)

and $$C_2(n) = \frac{\Delta p_{ex(1-2)}}{2\tau_w}$$

(3.103)

For laminar flow of power-law fluids, it has been found that both C_1 and C_2 (also known as Couette correction) are functions of n alone. Obviously, C_2 representing the loss between points '1' and '2' would be strongly dependent on the geometrical details of the system, more gradual or smooth the entrance, smaller will be the value of C_2. However, to date, its values have been computed only for an abrupt change. Table 3.4 summarises the predicted values of C_1 and C_2 for a range of values of n. It should be noted that $C_1(n)$ decreases with the increasing degree of shear-thinning behaviour while $C_2(n)$ shows the exactly opposite type of dependence on n. That is, the contribution of the excess pressure drop between points '1' and '2' increases with decreasing value of n. Based on extensive comparisons

Table 3.4   *Values of $C_1(n)$ and $C_2(n)$ [Boger, 1987]*

n	$C_1(n)$	$C_2(n)$
1	2.33	0.58
0.9	2.25	0.64
0.8	2.17	0.70
0.7	2.08	0.79
0.6	1.97	0.89
0.5	1.85	0.99
0.4	1.70	1.15
0.3	1.53	1.33

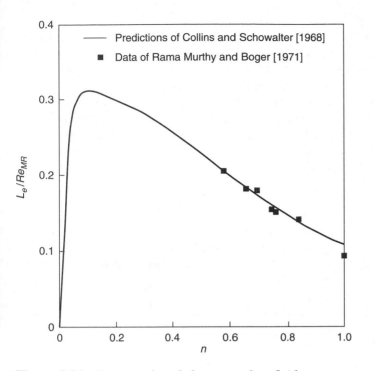

Figure 3.26   *Entrance length for power-law fluids*

between predictions and experimental data, this approach has been found to be reliable for estimating the value of Δp_{en} for linear contraction ratios greater than 2 and downstream Reynolds number $(\rho V^{2-n} D^n / m) > 5$ [Boger, 1987]. Figure 3.26 shows the entry length L_e, required to attain fully developed flow in tubes; excellent agreement is seen to exist between predictions and limited experimental data. From a practical standpoint, the currently available body of information suggests that the entrance length, L_e, is of the order of forty pipe diameters for inelastic fluids and about $110D$ for visco-elastic fluids [Cho and Hartnett, 1982] in streamline flow. The literature on this subject has been critically reviewed by Boger [1987].

There is little information on either the entrance length or the additional pressure drop for fully developed turbulent flow. Dodge and Metzner [1959] indicated that both the entrance length and the extra pressure loss for inelastic fluids were similar to those for Newtonian fluids.

3.8.3 Minor losses in fittings

Little reliable information is available on the pressure drop for the flow of non-Newtonian fluids through pipe fittings. The limited work carried out on the laminar flow of inelastic polymer solutions and of suspensions [Urey, 1966; Steffe *et al.*, 1984; Edwards *et al.*, 1985; Banerjee *et al.*, 1994] and on the turbulent flow of magnesia and titania slurries [Weltmann and Keller, 1957; Cheng, 1970] through a range of fittings including elbows, tees, bends and valves suggests that the shear-dependence of viscosity exerts little influence on such minor losses and therefore values for Newtonian fluids can be used. However, more recent work [Slatter, 1997] on the flow of a kaolin slurry and a polymer solution through sudden contractions and expansions and 90° elbows suggests that pressure drops are much larger than for Newtonian fluids. Therefore, great care must be exercised in estimating such losses, though it is always preferable to carry out some tests with the fluid under consideration. The corresponding losses for visco-elastic fluids are likely to be greater if their extensional viscosities are high. Additional uncertainties arise because the critical Reynolds number for the laminar–turbulent transition is not well defined and is strongly influenced by the design of the particular fitting. Thus, Edwards *et al.* [1985] reported the critical value of Re_{MR} as 900 for an elbow and 12 for globe valves! Overall there appears to be little definitive information for the minor losses in various fittings.

Some guidelines for the design and selection of valves for pseudoplastic materials have been developed by De Haven [1959] and Beasley [1992].

3.8.4 Flow measurement

Little information is available on suitable flow measurement devices. It has already been shown that pressure drop across a straight length of pipe is relatively insensitive to flow rate in laminar region of highly shear-thinning materials and therefore it is an unsuitable parameter for flow measurement. In principle, the devices which depend upon the conversion of pressure into kinetic energy (e.g. orifice and venturi meters, rotameters) can be used but they need calibration for each fluid. The problem is compounded in practice because the fluid may not have a constant rheology and/or may display visco-elastic or time-dependent behaviour. Though many investigators have studied the flow of polymer solutions through orifices, most of this work has not been directed towards flow measurement but has related either to the measurement of extensional viscosity or to extrusion behaviour. Both Harris and Magnall [1972] and Edwards *et al.* [1985] have examined the suitability of orifice and venturi meters for measuring the flow rates of inelastic polymer solutions and kaolin suspensions. While Edwards *et al.* [1985] were unable to bring together data for different aspect ratios, Harris and Magnall [1972] reported satisfactory correlation for the discharge coefficient in the turbulent regime. Despite all these difficulties, the measurement of the flow of shear-thinning oil-water emulsions using orifice and venturi meters has been the subject of a recent study [Pal, 1993]. The discharge coefficients were successfully correlated with the generalised Reynolds number (Re_{MR}). However, extrapolation to other types of non-Newtonian materials, e.g. polymer solutions, slurries, etc., must be treated with reserve. In view of these difficulties, indirect methods of flow measurement are generally preferred, the electromagnetic flow meter being the most common choice [Dodge and Metzner, 1959; Quader and Wilkinson, 1980; Chhabra et al., 1984] covering a wide variety of materials including polymer solutions, kaolin, anthracite and titania slurries. This method is, however, limited to substances possessing some degree of electrical conductivity. In addition, both Skelland [1967] and Liptak [1967] have suggested rotary type volumetric displacement flow meters for non-Newtonian viscous fluids but have given virtually no practical information on their performance. Ginesi [1991] has found magnetic and coriolis type flow meters to be reliable for slurries and viscous fluids. Brown and Heywood (1992) have discussed instrumentation for the measurement of both flow rate and density for non-Newtonian slurries.

Example 3.13

A kaolin slurry (density $= 1200\,\text{kg/m}^3$; $n = 0.2$ and $m = 25\,\text{Pa·s}^n$) flows under gravity from reservoir A to B both of which are of large diameter, as shown in Figure 3.27. Estimate the flow rate of slurry through the 50 mm diameter connecting pipe of total length of 75 m.

Solution

The mechanical energy balance can be applied between points '1' and '2' shown in Figure 3.27.

$$\frac{V_1^2}{2\alpha_1 g} + \frac{p_1}{\rho g} + z_1 = \frac{V_2^2}{2\alpha_2 g} + \frac{p_2}{3g} + z_2 + \Sigma h_L$$

Here $p_1 = p_2 = p_{\text{atm}}$. Because the two reservoirs are of large diameter, V_1 and V_2 are both approximately zero. With these simplifications, the equation simplifies to:

$$\Sigma h_L = z_1 - z_2 = 40 - 5 = 35\,\text{m}$$

The friction losses are associated with the flow in a 75 m length of 50 mm pipe including one gate valve, one globe valve, two short curvature elbows, a contraction (at A) and an expansion at B. At this stage, neglecting the contraction and expansion losses, one can express the loss term in terms of the unknown velocity V as:

$$\Sigma h_L = \frac{2fLV^2}{gD} + K\frac{V^2}{2g} + K\frac{V^2}{2g} + 2K\frac{V^2}{2g} = 35\,\text{m}$$

$$\text{(gate} \quad \text{(globe} \quad \text{(elbows)}$$
$$\text{valve)} \quad \text{valve)}$$

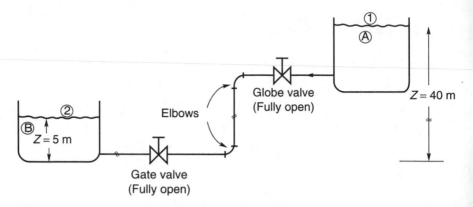

Figure 3.27   *Schematics of flow for example 3.13*

Edwards *et al.* [1985] have published the following values of K for various fittings:

For a 2 inch (50 mm) fully-open gate valve, $K = \dfrac{273}{\mathrm{Re}_{MR}}$ $\mathrm{Re}_{MR} < 120$

For a 2 inch (50 mm) fully-open globe valve, $K = \dfrac{384}{\mathrm{Re}_{MR}}$ $\mathrm{Re}_{MR} < 15$

$K = 25.4$ $\mathrm{Re}_{MR} > 15$

For a 2 inch (50 mm) short-curvature elbow, $K = \dfrac{842}{\mathrm{Re}_{MR}}$ $\mathrm{Re}_{MR} < 900$

$K = 0.9$ $\mathrm{Re}_{MR} > 900$

Assuming the flow to be laminar and $\mathrm{Re}_{MR} > 15$. The total head loss is then:

$$\frac{V^2}{2g}\left(\underbrace{\frac{64L}{D\,\mathrm{Re}_{MR}}}_{\substack{\text{(straight pipe}\\\text{length)}}} + \underbrace{\frac{273}{\mathrm{Re}_{MR}}}_{\substack{\text{(1 gate}\\\text{valve)}}} + \underbrace{25.4}_{\substack{\text{(1 globe}\\\text{valve)}}} + \underbrace{2 \times \frac{842}{\mathrm{Re}_{MR}}}_{\text{(2 elbows)}}\right) = 35\,\mathrm{m}$$

where $\mathrm{Re}_{MR} = \dfrac{\rho V^{2-n} D^n}{8^{n-1} m \left(\dfrac{3n+1}{4n}\right)^n}$ (3.8b)

Here, $\rho = 1200\,\mathrm{kg/m^3}$; $D = 50 \times 10^{-3}\,\mathrm{m}$; $m = 25\,\mathrm{Pa\cdot s^n}$ $L = 75\,\mathrm{m}$; $n = 0.2$

$$\therefore \mathrm{Re}_{MR} = \frac{(1200)V^{2-0.2}(50 \times 10^{-3})^{0.2}}{(8^{0.2-1})(25)\left(\dfrac{3 \times 0.2 + 1}{4 \times 0.2}\right)^{0.2}} = 121.13\,V^{1.8}$$

Substituting for Re_{MR} and other values:

$$V^2 \left\{ \frac{64 \times 75}{50 \times 10^{-3} \times 121.13V^{1.8}} + \frac{273}{121.13V^{1.8}} + 25.4 + \frac{2 \times 842}{121.13V^{1.8}} \right\}$$
$$= 35 \times 9.81 \times 2$$

or $808.69V^{0.2} + 25.4V^2 = 686.7$

A trial and error method gives, $V = 0.43\,\mathrm{m/s}$. Check: the value of $\mathrm{Re}_{MR} = 121.13 \times 0.43^{1.8} = 26.52$ which is in laminar range and is also larger than the value of 15 assumed above.$\square$

3.9 Selection of pumps

Non-Newtonian characteristics, notably shear-dependent viscosity and yield stress, strongly influence the choice of a suitable pump and its performance. While no definite quantitative information is available on this subject, general features of a range of pumps commonly used in industry are briefly described here. In particular, consideration is given to positive-displacement, centrifugal, and screw pumps.

3.9.1 Positive displacement pumps

Reciprocating pumps

Difficulties experienced in initiating the flow of pseudoplastic materials (owing to their high apparent viscosities and/or the yield stress effects) are frequently countered by the use of one of the various types of positive-displacement pumps, see e.g. Coulson and Richardson [1999]. Non-Newtonian fluids which are sensitive to breakdown, particularly agglomerates in suspensions, are best handled with pumps which subject the liquid to a minimum of shearing. Diaphragm pumps are then frequently used, but care must be taken that the safe working pressure for the pump and associated pipe network is not exceeded. The use of a hydraulic drive and a pressure-limiting relief valve fitted to the pump ensures that the system is protected from damage. Such an arrangement is shown schematically in Figure 3.28, which depicts general features of all systems using positive displacement pumps.

Rotary pumps

The selection of a suitable rotary positive-displacement pump for a viscoplastic material has been discussed by Steffe and Morgan [1986].

Such pumps operate on the principle of using mechanical means to transfer small elements or "packages" of fluid from the low pressure (inlet) side to the high pressure (delivery) side. There is a wide range of designs available

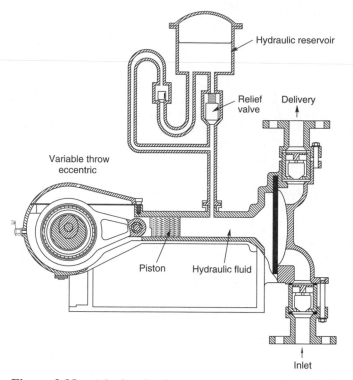

Figure 3.28   *A hydraulic drive to protect a positive displacement pump*

for achieving this end. The general characteristics of the pumps are similar to those of reciprocating piston pumps but the delivery is more even because of the fluid stream is broken down into much smaller elements. The pumps are capable of delivering to a high pressure, and the pumping rate is approximately proportional to the speed of the pump and is not greatly influenced by the pressure against which it is delivering. Again, it is necessary to provide a pressure relief system to ensure that the safe operating pressure is not exceeded.

One of the commonest forms of the pump is the gear pump in which one of the gear wheels is driven and the other turns as the teeth engage; two versions are illustrated in Figures 3.29 and 3.30. The liquid is carried round in the spaces between consecutive gear teeth and the outer casing of the pump, and the seal between the high and low pressure sides of the pump is formed as the gears come into mesh and the elements of fluid are squeezed out. Gear pumps are extensively used for both high-visosity Newtonian liquids and non-Newtonian fluids. The lobe-pump (Figure 3.31) is similar, but the gear teeth are replaced by two or three lobes and both axles are driven; it is therefore possible to maintain a small clearance between the lobes, and wear is reduced.

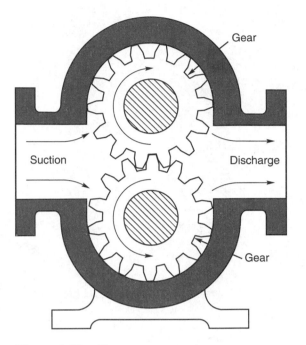

Figure 3.29 *Gear pump*

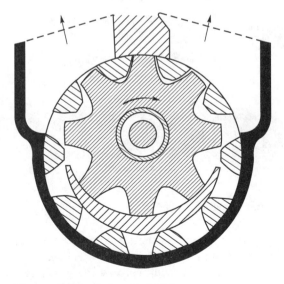

Figure 3.30 *Internal gear pump*

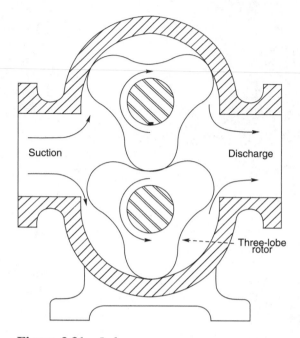

Figure 3.31 *Lobe pump*

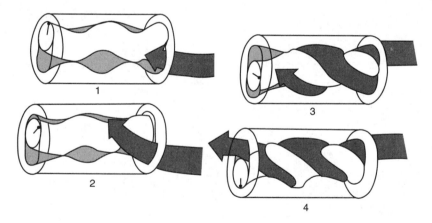

Figure 3.32 *The mode of operation of a Mono pump*

Another form of positive-acting rotary pump is the single-screw extruder pump typified by the Mono-pump, illustrated in Figures 3.32 and 3.33. A specially shaped metal helical-rotor revolves eccentrically within a resilient rubber or plastic double-helix, thus creating a continuous forming cavity which progresses towards the discharge of the pump. A continuous seal is created

Figure 3.33   *Section of Mono pump*

and, the higher the delivery pressure, the greater is the required number of turns, and hence the longer the stator and rotor. This type of pump is suitable for pumping slurries and pastes, whether Newtonian or non-Newtonian in character.

3.9.2 Centrifugal pumps

The most common type of pump used in the chemical industry is the centrifugal pump, though its performance deteriorates rapidly with increasing viscosity of fluids even with Newtonian fluids. The underlying principle is the conversion of kinetic energy into a static pressure head. For a pump of this type, the distribution of shear within the pump will vary with throughput. Considering Figure 3.34 where the discharge is completely closed off, the

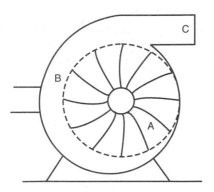

Figure 3.34   *Zones of differing shear in a centrifugal pump*

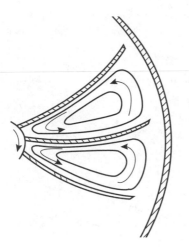

Figure 3.35   *Circulation within a centrifugal pump impeller*

highest degree of shearing is in the gap between the rotor and shell, i.e. at point B. Within the vanes of the rotor (region A) there will be some circulation as sketched in Figure 3.35, but in the discharge line C, the fluid will be essentially static. If the fluid is moving through the pump, there will still be differences between these shear rates but they will be less extreme. For a pseudoplastic material, the effective viscosity will vary in these different regions, being less at B than at A and C, while a shear-thickening material will exhibit the opposite behaviour. Under steady conditions, the pressure developed in the rotor produces a uniform flow through the pump. However, there may be problems on starting, when the very high effective viscosities of the fluid as the system starts from rest might result in the overloading of the motor. At this time too, the apparent viscosity of the liquid in the delivery line is at its maximum value, and the pump may take an inordinately long time to establish the required flow. Many pseudoplastic materials (such as food stuffs, pharmaceutical formulations) are damaged and degraded by prolonged shearing, and such a pump would be unsuitable.

As mentioned previously, it is generally accepted that the performance of a centrifugal pump deteriorates increasingly as the extent of non-Newtonian characteristics increases. Both head and particullarly efficiency, are adversely affected, and the performance of small pumps is impaired to the greatest extent. Severe erosion of the impeller and pump casing is also encountered especially with particulate suspensions. Some guidelines for using charts based on water for non-Newtonian materials by selecting a suitable value of the effective viscosity are, however, available [Duckham, 1971; Walker and Goulas, 1984]. Carter and Lambert [1972], on the other hand, have found helical gear pumps more suitable for viscous Newtonian and non-Newtonian fluids.

3.9.3 Screw pumps

Screw extruders, as used in the polymer processing and food industries, form a most important class of pumps for handling highly viscous non-Newtonian materials. Extruders are used for forming simple and complex sections (rods, tubes, etc). The shape of section produced for a given material is dependent only on the profile of the dies through which the fluid is forced just before it cools and solidifies, though additional complications may arise due to die-swell whereby the diameter of the extrudate may be larger than that of the die.

The basic function of the screw pump or extruder is to shear the fluid in the channel between the screw and the wall of the barrel, as shown schematically in Figure 3.36. The mechanism that generates the pressure can be visualised in terms of a model consisting of an open channel covered by a moving plane surface (Figure 3.37).

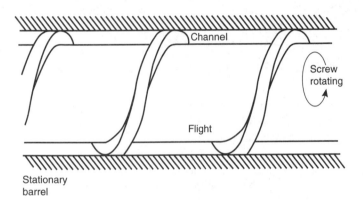

Figure 3.36   *Section of a screw pump*

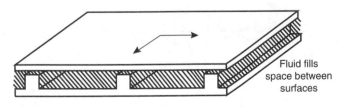

Figure 3.37   *Planar model of part of a screw pump*

This planar simplification of a stationary screw with rotating barrel is not unreasonable, provided that the depth of the screw channel is small compared with the barrel diameter. The distribution of centrifugal forces will, of course, be different according to whether the rotating member is the wall or the screw; this distinction must be drawn before a detailed force balance can be undertaken, but in any event the centrifugal (inertial) forces are generally far smaller than the viscous forces.

If the upper plate is moved along in the direction of the channel then a velocity profile is established, giving an approximately linear velocity distribution between two walls which are moving parallel to each other. If it moves at right angles to the channel axis, however, a circulation pattern is developed in the gap, as shown in Figure 3.38. In fact, the relative movement is somewhere between these two extremes and is determined by the pitch of the screw.

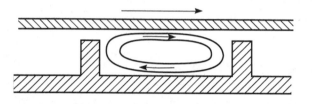

Figure 3.38   *Fluid displacement resulting from movement of plane surface*

The fluid path in a screw pump is therefore of a complex helical form within the channel section. The velocity components along the channel depend on the pressure generated and the resistance at the discharge end. If there is no resistance, the velocity profile in the direction of the channel will be of the Couette type, as depicted in Figure 3.39a. With a totally closed discharge end, the net flow would be zero, but the velocity components at the walls would not be affected. As a result, the flow field necessarily would be of the form shown in Figure 3.39b.

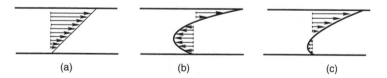

| (a) | (b) | (c) |

Figure 3.39   *Velocity profile produced between screw pump surfaces with (a) no resistance on fluid flow (b) no net flow (c) partially restricted discharge*

Viscous forces within the fluid will always prevent a completely unhindered discharge, but in extrusion-practice the die-head provides additional resistance which generates back-flow and mixing, thus creating a more uniform product. Under these conditions, the flow profile along the channel is of some intermediate form such as that shown in Figure 3.39c.

It must be stressed here that flow in a screw pump is produced as a result of viscous forces and hence the pressures achieved at the outlet with low viscosity materials are small. The screw pump is thus not a modification of the Archimedes screw used in antiquity to raise water–that was essentially a positive-displacement device using a deep cut helix, not running full, mounted at an angle to the horizontal. In any detailed analysis of the flow in a screw pump, it is also necessary to consider the small leakage flow that will occur between the flight and the wall. With the large pressures generated in a polymer extruder ($\sim$100 bar), the flow through this gap (typically 2 per cent of the barrel internal diameter) can be significant because the pressure drop over a single pitch length may be of the order of 10 bar. Once in this region, the viscous fluid is subject to a high rate of shear (the rotational speed of about 2 Hz) and an appreciable part of the fine-scale mixing and viscous heat generation occurs in this part of the extruder. It is thus important to bear in mind that with the very high viscosity materials generally involved, heat generation can be very large and so the temperature of the fluid (and hence its rheological properties) may be a strong function of the power input to the extruder. One must thus solve the coupled momentum and energy equations, including the viscous dissipation term. General descriptions of extrusion technology are available in several books, e.g. McKelvey [1962], Tadmor and Gogos [1979].

3.10 Further reading

Bird, R.B., Armstrong, R.C. and Hassager, O., *Dynamics of Polymeric Liquids. Vol 1. Fluid Dynamics*, 2nd edn. Wiley, New York (1987).

Carreau, P.J., Dekee, D and Chhabra, R.P., *Rheology of Polymeric Systems: Principles and Applications*. Hanser, Munich (1997).

Darby, R., *Visco-elastic Fluids*. Marcel–Dekker, New York (1976).

Govier, G.W. and Aziz, K., *The Flow of Complex Mixtures in Pipes*. Krieger, Malabar, FL (1982).

3.11 References

Banerjee, T.K., Das, M and Das, S.K., *Can. J. Chem. Eng.* **72** (1994) 207.

Beasley, M.E., *Chem. Eng.* **99** (September, 1992) 112.

Bird, R.B., Armstrong, R.C. and Hassager, O., *Dynamics of Polymeric Liquids*. Vol. 1 *Fluid Dynamics*, 2nd edn. Wiley, New York (1987).

Bird, R.B., Dai, G.C. and Yarusso, B.J., *Rev. Chem. Eng.* **1** (1983) 1.

Boger, D.V., *Ann. Rev. Fluid Mech.* **19** (1987) 157.

Bogue, D.C. and Metzner, A.B., *Ind. Eng. Chem. Fundam.* **2** (1963) 143.

Bowen, R.L., *Chem. Eng.* **68** (July, 1961) 143.

Brodkey, R.S., Lee, J. and Chase, R.C., *AIChEJ.* **7** (1961) 392.

Brown, N.P. and Heywood, N.I., (editors), *Slurry Handling: Design of Solid-Liquid Systems*. Elsevier, Amsterdam (1991).

Brown, N.P. and Heywood, N.I., *Chem. Eng.* **99** (Sep, 1992) 106.

Carter, G. and Lambert, D.J., *The Chem. Engr.* (Sep, 1972) 355.

Cheng, D.C.-H., *Proc. Hydrotransport* **1**, (1970) 77.

Chhabra, R.P., Farooqi, S.I. and Richardson, J.F., *Chem. Eng. Res. Des.* **62** (1984) 22.

Cho, Y.I. and Hartnett, J.P., *Adv. Heat Transf.* **15** (1982) 59.

Clapp, R.M., *Int. Dev. Heat Transf.* Part III, 625, D159 & D211–215, ASME, New York (1961).

Collins, M. and Schowalter, W.R., *AIChEJ.* **9** (1968) 98.

Coulson, J.M. and Richardson, J.F., *Chemical Engineering*. Vol. 1 6th edn. Butter-worth–Heinemann, Oxford (1999).

Darby, R., *in Encyclopedia of Fluid Mechanics*. (edited by Cheremisinoff, N.P., Gulf, Houston) **7** (1988) 19.

Darby, R., Mun, R. and Boger, D.V., *Chem. Eng.* **99** (1992) 116.

De Haven, E.S., *Ind. Eng. Chem.* **51** (July, 1959) 59A.

Dodge, D.W. and Metzner, A.B., *AIChEJ.* **5** (1959) 189. (corrections ibid 8 (1962) 143).

Duckham, C.B., *Chem. & Proc. Eng.* **52** (July, 1971) 66.

Edwards, M.F., Jadallah, M.S.M. and Smith, R., *Chem. Eng. Res. Des.* **63** (1985) 43.

Edwards, M.F., Nellist, D.A. and Wilkinson, W.L., *Chem. Eng. Sci.* **27** (1972) 545. *Also see ibid* **27** (1972) 295.

Fordham, E.J., Bittleston, S.H. and Tehrani, M.A., *Ind. Eng. Chem. Res.* **30** (1991) 517.

Fredrickson, A.G. and Bird, R.B., *Ind. Eng. Chem.* **50** (1958) 347. Also see *ibid* **50** (1958) 1599 & *Ind. Eng. Chem. Fundam.* **3** (1964) 383.

Garcia, E.J. and Steffe, J.F., *J. Food Process Eng.* **9** (1987) 93.

Ginesi, D., *Chem. Eng.* **98** (May, 1991) 146.

Govier, G.W. and Aziz, K., *The Flow of Complex Mixtures in Pipes*. Krieger, Malabar, FL (1982).

Griskey, R.G. and Green, R.G., *AIChEJ.* **17** (1971) 725.

Gücüyener, H.I. and Mehmeteoglu, T., *AIChEJ.* **38** (1992) 1139.

Hanks, R.W., *AIChEJ.* **9** (1963) 306.

Hanks, R.W., *Ind. Eng. Chem., Proc. Des. Dev.* **18** (1979) 488.

Hanks, R.W., *in Encyclopedia of Fluid Mechanics* (edited by Cheremisinoff, N.P., Gulf, Houston), **5** (1986) 213.

Hanks, R.W. and Larsen, K.M., *Ind. Eng. Chem. Fundam.* **18** (1979) 33.

Harris, J., *Rheol. Acta*, **7** (1968) 228.

Harris, J. and Magnall, A.N., *Trans Inst. Chem. Engrs.* **50** (1972) 61.

Hartnett, J.P. and Kostic, M., *Int. J. Heat Mass Transf.* **28** (1985) 1147.

Hartnett, J.P. and Kostic, M., *Adv. Heat Transf.* **19** (1989) 247.

Hartnett, J.P. and Kostic, M., *Int. Commun. Heat Mass Transf.* **17** (1990) 59.

Hartnett, J.P., Kwack, E.Y. and Rao, B.K., *J. Rheol.* **30** (1986) S45.

Hedström, B.O.A., *Ind. Eng. Chem.* **44** (1952) 651.

Heywood, N.I. and Cheng, D.C.-H., *Trans. Inst. Meas. & Control* **6** (1984) 33.

Irvine, Jr., T.F., *Chem. Eng. Commun.* **65** (1988) 39.

Kostic, M. and Hartnett, J.P., *Int. Comm. Heat Mass Transf.* **11** (1984) 345.

Kozicki, W., Chou, C.H. and Tiu, C., *Chem. Eng. Sci.* **21** (1966) 665.

Kozicki, W. and Tiu, C. Can, J., *Chem. Eng.* **45** (1967) 127.

Kozicki, W. and Tiu, C., *in Encyclopedia of Fluid Mechanics* (edited by Cheremisinoff, N.P., Gulf, Houston) **7** (1988) 199.

Laird, W.M., *Ind. Eng. Chem.* **49** (1957) 138.

Liptik, B.G., *Chem. Eng.* **70** (Jan 30, 1967) **133** & **70** (Feb 13, 1967) 151.

Malin, M.R., *Int. Comm. Heat Mass Transf.* **24** (1997) 793.

McKelvey, J.M., *Polymer Processing*, Wiley, New York (1962).

Metzner, A.B., *Adv. Chem. Eng.* **1** (1956) 79.

Metzner, A.B. and Park, M.G., *J. Fluid Mech.* **20** (1964) 291.

Metzner, A.B. and Reed, J.C., *AIChEJ.* **1** (1955) 434.

Miller, C., *Ind. Eng. Chem. Fundam.* **11** (1972) 524.

Millikan, C.B., *Proc. 5th Int. Cong. Applied Mech.* (1939) 386.

Mitsuishi, N. and Aoyagi, Y., *Chem. Eng. Sci.* **24** (1969) 309.

Mitsuishi, N. and Aoyagi, Y., *J. Chem. Eng. Jpn.* **6** (1973) 402.

Mitsuishi, N., Aoyagi, Y. and Soeda, H., *Kagaku–Kogaku.* **36** (1972) 182.

Pal, R., *Ind. Eng. Chem. Res.* **32** (1993) 1212.

Quader, A.K.M.A. and Wilkinson, W.L., *Int. J. Multiphase Flow* **6** (1980) 553.

Ramamurthy, A.V. and Boger, D.V., *Trans. Soc. Rheol.* **15** (1971) 709.

Ryan, N.W. and Johnson, M.M., *AIChEJ.* **5** (1959) 433.

Schechter, R.S., *AIChEJ.* **7** (1961) 445.

Sellin, R.H.J., Hoyt, J.W. and Scrivener, O., *J. Hyd. Res.* **20** (1982) 29 & 235.

Shenoy, A.V., *in Encyclopedia of Fluid Mechanics* (edited by Cheremisinoff, N.P., Gulf, Houston), **7** (1988) 479.

Shenoy, A.V. and Talathi, M.M., *AIChEJ.* **31** (1985) 520.

Skelland, A.H.P., *Non-Newtonian Flow and Heat Transfer* Wiley, New York (1967).

Slatter, P.T., *Hydrotransport* **13**, (1996) 97.

Slatter, P.T., *9th Int. Conf. Transport & Sedimentation of Solid Particles*, Cracow (1997) p. 547.

Slatter, P.T. and van Sittert, F.P., *9th Int. Conf. on Transport & Sedimentation of Solid Particles*, Cracow (1997) p. 621.

Slatter, P.T., Pienaar, V.G. and Petersen, F.W., *9th Int. Conf. on Transport & Sedimentation of Solid Particles*, Cracow (1997) p. 585.

Steffe, J.F., Mohamed, I.O. and Ford, E.W., *Trans. Amer. Soc. Agri. Engrs.* **27** (1984) 616.

Steffe, J.F. and Morgan, R.G., *Food Technol.* **40** (Dec 1986) 78.

Szilas, A.P., Bobok, E. and Navratil, L., *Rheol. Acta* **20** (1981) 487.

Tadmor, Z. and Gogos, G., *Principles of Polymer Processing*, Wiley, New York (1979).

Tam, K.C., Tiu, C. and Keller, R.J., *J. Hyd. Res.* **30** (1992) 117.

Thomas, A.D. and Wilson, K.C., *Can. J. Chem. Eng.* **65** (1987) 335.

Tomita, Y., *Bull. Jap. Soc. Mech. Engrs.* **2** (1959) 10.

Torrance, B. McK., *South African Mech. Engr.* **13** (1963) 89.

Uner, D., Ozgen, C. and Tosun, I., *Ind. Eng. Chem. Res.* **27** (1988) 698.

Urey, J.F., *J. Mech. Eng. Sci.* **8** (1966) 226.

Walker, C.I. and Goulas, A., *Proc. Inst. Mech. Engrs*, Part A, **198** (1984) 41.

Walton, I.C. and Bittleston, S.H., *J. Fluid Mech.* **222** (1991) 39.

Wardhaugh, L.T. and Boger, D.V., *Chem. Eng. Res. Des.* **65** (1987) 74.

Wardhaugh, L.T. and Boger, D.V., *AIChEJ.* **37** (1991) 871.

Weltmann, R.N. and Keller, T.A., *NACA Technical Note* 3889 (1957).

Wheeler, J.A. and Wissler, E.H., *AIChEJ.* **11** (1965) 207.

Wilson, K.C., *Hydrotransport* **13**, pp. 61–74 (1996). *BHR Group* Cranfield, U.K.

Wilson, K.C. and Thomas, A.D., *Can. J. Chem. Eng.* **63** (1985) 539.

Wojs, K., *J. Non-Newt. Fluid Mech.* **48** (1993) 337.

Xie, C. and Hartnett, J.P., *Ind. Eng. Chem. Res.* **31** (1992) 727.

3.12 Nomenclature

		Dimensions in **M, L, T**
a, b	geometric parameters for non-circular ducts (–)	$\mathbf{M^0L^0T^0}$
b	half gap between two parallel plates (m)	$\mathbf{L}$
$\mathrm{Bi} = \dfrac{D\tau_0^B}{V\,\mu_B}$	Bingham number (–)	$\mathbf{M^0L^0T^0}$
D	pipe diameter (m)	$\mathbf{L}$

D_h	hydraulic equivalent diameter (m)	**L** Dimensions in **M, L, T**
E	eddy diffusivity (m^2/s)	$\mathbf{L^2T^{-1}}$
f	friction factor (–)	$\mathbf{M^0L^0T^0}$
g	acceleration due to gravity (m/s^2)	$\mathbf{LT^{-2}}$
H	film thickness in flow over inclined surface (m)	**L**
$\text{He} = \dfrac{\rho D^2 \tau_0^B}{\mu_B^2}$	Hedström number (–)	$\mathbf{M^0L^0T^0}$
h_l	frictional loss in head (m)	**L**
L	pipe length (m)	**L**
l	Prandtl mixing length (m)	**L**
m	power law consistency coefficient (Pa · s^n)	$\mathbf{ML^{-1}T^{n-2}}$
m'	apparent power law consistency coefficient (Pa · s$^{n'}$)	$\mathbf{ML^{-1}T^{n'-2}}$
n	power-law flow behaviour index (–)	$\mathbf{M^0L^0T^0}$
n'	apparent power-law flow behaviour index (–)	$\mathbf{M^0L^0T^0}$
p	pressure (Pa)	$\mathbf{ML^{-1}T^{-2}}$
$-\Delta p$	pressure drop (Pa)	$\mathbf{ML^{-1}T^{-2}}$
Q	volumetric flow rate (m^3/s)	$\mathbf{L^3T^{-1}}$
r	radial coordinate (m)	**L**
R	pipe radius/annulus outer radius (m)	**L**
R_p	radius of plug in centre of pipe (m)	**L**
Re_B	Reynolds number based on Bingham plastic viscosity (–)	$\mathbf{M^0L^0T^0}$
Re_{MR}	Metzner–Reed definition of Reynolds number (–)	$\mathbf{M^0L^0T^0}$
Re_g	generalised Reynolds number for non-circular conduits, eq. 3.90 (–).	$\mathbf{M^0L^0T^0}$
Re_T	modified Reynolds number, equation 3.39 (–)	$\mathbf{M^0L^0T^0}$
V	mean velocity of flow (m/s)	$\mathbf{LT^{-1}}$
V_z	point velocity of flow in z-direction (m/s)	$\mathbf{LT^{-1}}$
V_z^*	non-dimensional point velocity (–)	$\mathbf{M^0L^0T^0}$
$V^* = \sqrt{\tau_w/\rho}$	friction velocity (m/s)	$\mathbf{LT^{-1}}$
V^+	non-dimensional velocity, equation 3.47. (–)	$\mathbf{M^0L^0T^0}$
W	width (m)	**L**
y	distance from wall (m)	**L**
y^+	non-dimensional distance from wall, equation 3.47 (–).	$\mathbf{M^0L^0T^0}$
z	axial coordinate in the flow direction (m).	**L**

Greek letters

α	parameter in Ellis fluid model or kinetic energy correction factor (–)	$\mathbf{M^0L^0T^0}$
λ	radial coordinate corresponding to zero shear stress in annulus (–)	$\mathbf{M^0L^0T^0}$
λ_+, λ_-	boundaries of plug of Bingham plastic fluid in annular flow (–)	$\mathbf{M^0L^0T^0}$
μ	apparent viscosity (Pa · s)	$\mathbf{ML^{-1}T^{-1}}$
μ_0	zero shear viscosity (Pa · s)	$\mathbf{ML^{-1}T^{-1}}$
μ_B	Bingham plastic viscosity (Pa · s)	$\mathbf{ML^{-1}T^{-1}}$
$\zeta = r/R$	non-dimensional radial coordinate (–)	$\mathbf{M^0L^0T^0}$

		Dimensions in **M, L, T**
ρ	density (kg/m^3)	$\mathbf{ML}^{-3}$
σ	non-dimensional inner radius of annulus (–)	$\mathbf{M^0L^0T^0}$
τ_{rz}	shear stress in fluid (Pa)	$\mathbf{ML}^{-1}\mathbf{T}^{-2}$
τ_w	shear stress at pipe wall (Pa)	$\mathbf{ML}^{-1}\mathbf{T}^{-2}$
$\tau_{1/2}$	parameter in Ellis fluid model (Pa)	$\mathbf{ML}^{-1}\mathbf{T}^{-2}$
τ_0^B	yield stress in Bingham plastic model (Pa)	$\mathbf{ML}^{-1}\mathbf{T}^{-2}$
$\phi = \tau_0^B/\tau_w$	non-dimensional ratio (–)	$\mathbf{M^0L^0T^0}$
Ω	non-dimensional flow rate (–).	$\mathbf{M^0L^0T^0}$

Chapter 4
Flow of multi-phase mixtures in pipes

4.1 Introduction

The flow problems considered in the previous chapter have concerned either single phases or pseudo-homogeneous fluids such as emulsions and suspensions of fine particles in which little or no separation occurs. Attention will now be focussed on the far more complex problem of the flow of multi-phase systems in which the composition of the mixture may show spatial variation over the cross-section of the pipe or channel. Furthermore, the two components may have different in-situ velocities as a result of which there is 'slip' between the two phases and in-situ holdups which are different from those in the feed or exit stream. Furthermore, the residence times of the two phases will be different.

Multiphase flow is encountered in many chemical and process engineering applications, and the behaviour of the material is influenced by the properties of the components, such as their Newtonian or non-Newtonian characteristics or the size, shape and concentration of particulates, the flowrate of the two components and the geometry of the system. In general, the flow is so complex that theoretical treatments, which tend to apply to highly idealised situations, have proved to be of little practical utility. Consequently, design methods rely very much on analyses of the behaviour of such systems in practice. While the term 'multiphase flows' embraces the complete spectrum of gas/liquid, liquid/liquid, gas/solid, liquid/solid gas/liquid/solid and gas/liquid/liquid systems, the main concern here is to illustrate the role of non-Newtonian rheology of the liquid phase on the nature of the flow. Attention is concentrated on the simultaneous co-current flow of a gas and a non-Newtonian liquid and the transport of coarse solids in non-Newtonian liquids.

Multi-phase mixtures may be transported horizontally, vertically, or at an inclination to the horizontal in pipes and, in the case of liquid–solid mixtures, in open channels. Although there is some degree of similarity between the hydrodynamic behaviour of the various types of multi-phase flows, the range of physical properties is so wide that each system must be considered separately even when the liquids are Newtonian. Liquids may have densities up to three orders of magnitude greater than gases, but they are virtually incompressible.

The liquids themselves may range from simple Newtonian fluids, such as water, to highly viscous non-Newtonian liquids, such as polymer solutions, fine particle slurries. Indeed, because of the large differences in density and viscosity, the flow of gas–liquid mixtures and liquid–solid (coarse) mixtures must, in practice, be considered separately. For all multi-phase flow systems, it is, however, essential to understand the nature of the interactions between the phases and how these affect the flow patterns, including the way in which the two phases are distributed over the cross-section of the pipe or channel. Notwithstanding the importance of the detailed kinematics of flow, the ensuing discussion is mainly concerned with the overall hydrodynamic behaviour, with particular reference to the following features: flow patterns, average holdup of the individual phases, and the frictional pressure gradient. Flow patterns are strongly influenced by the difference in density between the two phases. In gas–liquid systems, it is always the gas which is the lighter phase and in solid–liquid systems, it is more often than not the liquid. The orientation of the pipe may also play a role. In vertical upward flow, for instance, it will be the lighter phase which will tend to rise more quickly. For liquid–solid mixtures, the slip velocity is of the same order as the terminal settling velocity of the particles but, in a gas–liquid system, it depends upon the flow pattern in a complex manner.

Unlike vertical flow, horizontal flow does not exhibit axial symmetry and the flow pattern is more complex because the gravitational force acts normally to the direction of flow, causing asymmetrical distribution over the pipe cross-section.

In practice, many other considerations affect the design of an installation. For example, wherever particulates are involved, pipe blockage may occur with consequences that can be as serious that it is always necessary to operate under conditions which minimize the probability of its occurring. Abrasive solids may give rise to the undue wear, and high velocities may need to be avoided.

Though the main emphasis in this chapter is on the effects of the non-Newtonian rheology, it is useful to draw analogies with the simpler cases of Newtonian liquids, details of which are much more readily available.

4.2 Two-phase gas–non-Newtonian liquid flow

4.2.1 Introduction

This section deals with the most important characteristics of the flow of a mixture of a gas or vapour and a Newtonian or non-Newtonian liquids in a round pipe. Despite large differences in rheology, two-phase flow of gas–liquid mixtures exhibits many common features whether the liquid is

Newtonian or shows inelastic pseudoplastic behaviour. Applications in the chemical, food and processing industries range from the flow of mixtures of crude oil (which may exhibit non-Newtonian characteristics) and gas from oil well heads to that of vapour–liquid mixtures in boilers and evaporators.

The nature of the flow of gas–liquid mixtures is complex, and the lack of knowledge concerning local velocities of the individual phases makes it difficult to develop any method of predicting the velocity distribution. In many instances, the gas (or vapour) phase may be flowing considerably faster than the liquid and continually accelerating as a result of its expansion, as the pressure falls. Either phase may be in laminar or in turbulent flow, albeit that laminar flow, (especially for a gas) does not have such a clear cut meaning as in the flow of single fluids. In practice, Newtonian liquids will most often be in turbulent flow, whereas the flow of non-Newtonian liquids is more often streamline because of their high apparent viscosities.

Additional complications arise if there is heat transfer from one phase to another such as that encountered in the tubes of a condenser or boiler. Under these conditions, the mass flowrate of each phase is progressively changing as a result of the vapour condensing or the liquid vaporising. However, this phenomenon is of little relevance to the flow of gas and non-Newtonian liquid mixtures.

Consideration will now be given in turn to three particular aspects of gas–liquid flow which are of practical importance (i) flow patterns or regimes (ii) holdup, and (iii) frictional pressure gradient.

4.2.2 Flow patterns

For two-phase cocurrent gas–liquid flow, there is the wide variety of possible flow patterns which are governed principally by the physical properties (density, surface tension, viscosity of gas, rheology of liquid), input fluxes of the two phases and the size and the orientation of the pipe. Since the mechanisms responsible for holdup and momentum transfer (or frictional pressure drop) vary from one flow pattern to another, it is essential to have a method of predicting the conditions under which each flow pattern may occur. Before developing suitable methods for the prediction of flow pattern, it is important briefly to define the flow patterns generally encountered in gas–liquid flows. Horizontal and vertical flows will be discussed separately as there are inherent differences in the two cases.

Horizontal flow

The classification proposed by Alves [1954] encompasses all the major and easily recognisable flow patterns encountered in horizontal pipes. These are sketched in Figure 4.1 and are described below.

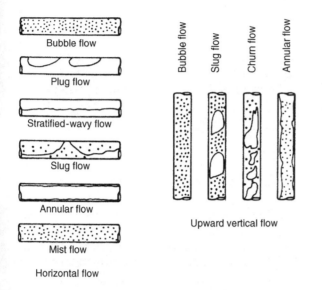

Figure 4.1 *Flow patterns in horizontal and vertical two phase flow*

(i) Bubble flow

This type of flow, sometimes referred to as dispersed bubble flow, is characterised by a train of discrete gas bubbles moving mainly close to the upper wall of the pipe, at almost the same velocity as the liquid. As the liquid flowrate is increased, the bubbles become more evenly distributed over the cross-section of the pipe.

(ii) Plug flow

At increased gas throughput, bubbles interact and coalesce to give rise to large bullet shaped plugs occupying most of the pipe cross-section, except for a thin liquid film at the wall of the pipe which is thicker towards the bottom of the pipe.

(iii) Stratified flow

In this mode of flow, the gravitational forces dominate and the gas phase flows in the upper part of the pipe. At relatively low flowrates, the gas–liquid interface is smooth, but becomes ripply or wavy at higher gas rates thereby giving rise to the so-called 'wavy flow'. As the distinction between the smooth and wavy interface is often ill-defined, it is usual to refer to both flow patterns as stratified-wavy flow.

(iv) Slug flow

In this type of flow, frothy slugs of liquid phase carrying entrained gas bubbles alternate with gas slugs surrounded by thin liquid films. Although plug and slug flow are both well defined, as shown in Figure 4.1, in practice it is often not easy to distinguish between them.

(v) Annular flow

In this type of flow, most of the liquid is carried along the inner wall of the pipe as a thin film, while the gas forms a central core occupying a substantial portion of the pipe cross-section. Some liquid is usually entrained as fine droplets within the gas core. Sometimes the term 'film flow' is also used to describe this flow pattern.

(vi) Mist flow

Mist flow is said to occur when a significant amount of liquid becomes transferred from the annular film to the gas core; at high gas flowrates nearly all of the liquid is entrained in the gas. Thus, some workers regard mist flow to be an extreme case of annular flow.

The above classification of flow patterns is useful in developing models for the flow, but it is important to note that the distinction between any two flow patterns is far from clear cut, especially in the case of bubble, plug and slug flow patterns. Consequently, these latter three flow patterns are often combined and described as 'intermittent flow'. On this basis, the chief flow patterns adopted in this work are dispersed, intermittent, stratified, wavy and annular–mist flow. Detailed descriptions of flow patterns and of the experimental methods for their determination are available in the literature, e.g. see [Hewitt, 1978, 1982; Ferguson and Spedding, 1995]. Although most of the information on flow patterns has been gathered from experiments on gas and Newtonian liquids in co-current flow, the limited experimental work reported to date with inelastic shear-thinning materials suggests that they give rise to qualitatively similar flow patterns [Chhabra and Richardson, 1984] and therefore the same nomenclature will be adopted here.

4.2.3 Prediction of flow patterns

For the flow of gas–Newtonian liquid mixtures, several, mostly empirical, attempts have been made to formulate flow pattern maps [Govier and Aziz, 1982; Hetsroni, 1982; Chisholm, 1983]. The regions over which the different types of flow patterns can occur are conveniently shown as a 'flow pattern map' in which a function of the liquid flowrate is plotted against a function of the gas flowrate and boundary lines are drawn to delineate the various regions. Not only is the distinction between any two flow patterns poorly defined, but the transition from one flow pattern to another may occur over a range of conditions rather than abruptly as suggested in all flow pattern maps. Furthermore, because the flow patterns are usually identified by visual observations of the flow, there is a large element of subjectivity in the assessment of the boundaries.

Most of the data used for constructing such maps have been obtained with the air–water system at or near atmospheric conditions. Although, intuitively, one might expect the physical properties of the two phases to play important roles in determining the transition from one flow pattern to another, it is now generally recognised that they indeed have very little effect [Mandhane *et al.*, 1974; Weisman *et al.*, 1979; Chhabra and Richardson, 1984]. Even shear-thinning behaviour seems to play virtually no role in governing the transition from one flow pattern to another [Chhabra and Richardson, 1984]. Based on these considerations and taking into account the extensive experimental study of Weisman *et al.* [1979], Chhabra and Richardson [1984] modified the widely used flow pattern map of Mandhane *et al.* [1974] as shown in Figure 4.2. On the basis of a critical evaluation of the literature on the flow of mixtures of gas and shear-thinning fluids (aqueous polymer solutions, particulate suspensions of china clay, coal and limestone), this scheme was shown to reproduce about 3700 data points on flow patterns with 70% certainty. The range of experimental conditions which have been used in the compilation of this flow pattern map are: $0.021 \leq V_L \leq 6.1 \, \text{m/s}; 0.01 \leq V_G \leq 55 \, \text{m/s}; 6.35 \leq D \leq 207 \, \text{mm}$ and $0.1 \leq n' \leq 1$. For the systems considered in the preparation of the flow pattern map (Figure 4.2), apparent viscosities of the non-Newtonian liquids at a shear rate of $1 \, \text{s}^{-1}$ varied from $10^{-3} \, \text{Pa·s}$ (water) to $50 \, \text{Pa·s}$. Such few experimental results as are available for visco-elastic polymer solutions are correlated well by this flow pattern map [Chhabra and Richardson, 1984].

Vertical upward flow

In vertical flow, gravity acts in the axial direction giving symmetry across the pipe cross-section. Flow patterns tend to be somewhat more stable, but with

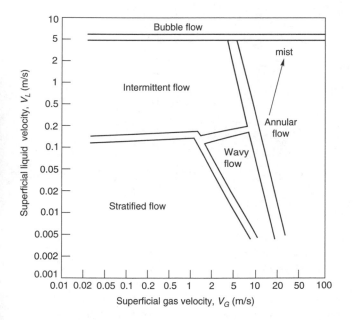

Figure 4.2   *Modified flow pattern map [Chhabra and Richardson, 1984]*

slug flow, oscillations in the flow can occur as a result of sudden changes in pressure as liquid slugs are discharged from the end of the pipe. This effect is also present in horizontal flow.

The flow patterns observed in vertical upward flow of a gas and Newtonian liquid are similar to those shown in Figure 4.1 and are described in detail elsewhere [Barnea and Taitel, 1986]. Taitel *et al.* [1980] have carried out a semi-theoretical study of the fundamental mechanisms responsible for each flow pattern, and have derived quantitative expressions for the transition from one regime to another. This analysis shows a strong dependence on the physical properties of the two phases and on the pipe diameter. Figure 4.3 shows their map for the flow of air–water mixtures in a 38 mm diameter pipe.

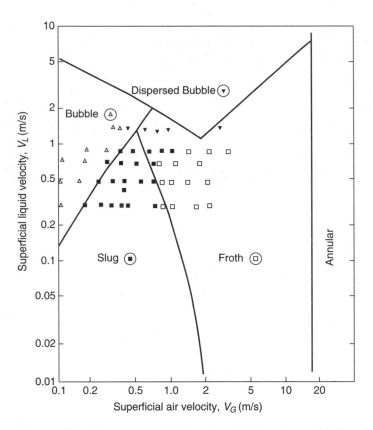

Figure 4.3   *Experimental [Khatib and Richardson, 1984] and predicted [Taitel et al., 1980] flow patterns for upward flow of air and china clay suspensions (D = 38 mm)*

The only study dealing with the vertical two-phase flow in which the liquid is shear-thinning is that of Khatib and Richardson [1984] who worked with suspensions of china clay. Their results compare closely with the predictions of Taitel *et al.* [1980] for air–water mixtures and this suggests that the transition boundaries between the various flow pattern are largely unaffected by the rheology of the liquid and that Figure 4.3 can be used when the liquid, is shear-thinning. However, no such information is available for visco-elastic liquids.

4.2.4 Holdup

Because of the considerable differences in the physical properties (particularly viscosity and density) of gases and liquids, the gas always tends to flow at a higher average velocity than the liquid. Sometimes, this can also occur if the liquid preferentially wets the surface of the pipe and therefore experiences a greater drag. In both cases, the volume fraction (holdup) of liquid at any point in a pipe will be greater than that in the mixture entering or leaving the pipe. Furthermore, if the pressure falls significantly along the pipeline, the holdup of liquid will progressively decrease as a result of the expansion of the gas.

If α_L and α_G are the holdups for liquid and gas respectively, it follows that:

$$\alpha_L + \alpha_G = 1 \tag{4.1}$$

Similarly for the input volume fractions:

$$\lambda_L + \lambda_G = 1 \tag{4.2}$$

λ_L and λ_G may be expressed in terms of the flow rates Q_L and Q_G, at a given point in the pipe, as:

$$\lambda_L = \frac{Q_L}{Q_L + Q_G} = \frac{V_L}{V_L + V_G} \tag{4.3}$$

$$\lambda_G = \frac{Q_G}{Q_L + Q_G} = \frac{V_G}{V_L + V_G} \tag{4.4}$$

where V_L and V_G are the superficial velocities of the two phases. Only under the limiting conditions of no-slip between the two phases and of no significant pressure drop along the pipe will α and λ be equal. Liquid (or gas) holdup along the length of the pipe must be known for the calculation of two-phase pressure drop.

Experimental determination

The experimental techniques available for measuring holdup fall into two categories, namely, direct and indirect methods. The direct method of measurement involves suddenly isolating a section of the pipe by means of quick-acting valves and then determining the quantity of liquid trapped. Good reproducibility may be obtained, as shown by widespread use of this technique for both Newtonian and non-Newtonian liquids [Hewitt *et al.*, 1963; Oliver and Young-Hoon, 1968; Mahalingam and Valle, 1972; Chen and Spedding, 1983]. It yields a volume average value of holdup. Although the method is, in principle, simple, it has two main drawbacks. Firstly, the valves cannot operate either instantaneously or exactly simultaneously. This must lead to inaccuracies and, after each measurement, ample time must be allowed for the flow to reach a steady state. Secondly, it is not practicable to use this method for high temperature and pressure situations and/or when either the gas or the liquid or both is of hazardous nature.

The indirect non-intrusive methods have the advantage of not disturbing the flow. The underlying principle is to measure a physical or electrical property that is strongly dependent upon the composition of the gas–liquid mixture. Typical examples include the measurement of γ-ray or X-ray attenuation [Petrick and Swanson, 1958; Pike *et al*; 1965; Shook and Liebe, 1976], or of change in impendence [Gregory and Mattar, 1973; Shu *et al.*, 1982] or of change in conductivity [Fossa, 1998]. Such methods, however, require calibration and yield values (averaged over the cross-section) at a given position in the pipeline. The γ-ray attenuation method has been used extensively to measure liquid holdup for two-phase flow of mixtures of air and non-Newtonian liquids such as polymer solutions and particulate suspensions in horizontal and vertical pipelines [Heywood and Richardson, 1979; Farooqi and Richardson, 1982; Chhabra *et al.*, 1984; Khatib and Richardson, 1984].

Predictive methods for horizontal flow

Methods available for the prediction of the average value of liquid holdup fall into two categories: those methods which are based on models which utilise information implicit in the flow pattern and those which are entirely empirical. Taitel and Dukler [1980] have developed a semi-theoretical expression for the average liquid holdup and the two-phase pressure gradient for the stratified flow of mixtures of air and Newtonian liquids. Although such analyses attempt to give some physical insight into the flow mechanism, they inevitably entail gross simplifications and empiricism. For instance, Taitel and Dukler [1980] assumed the interface to be smooth and the interfacial friction factor to be the same as that for the gas, but this model tends to underestimate the two-phase pressure drop. This methodology has subsequently been extended to the stratified flow of a gas and power-law liquids [Heywood and Charles, 1979]. Similar idealised models are available for the annular flow of gas and power-law liquids in horizontal pipes [Eissenberg and Weinberger, 1979] but most of them assume the liquid to be in streamline flow and the gas turbulent.

The second category of methods includes purely empirical correlations which disregard the flow patterns and are applicable over stated ranges of the variables. Although such an approach contributes little to our understanding, it does provide the designer with the vital information of a known degree of accuracy and reliability. Numerous empirical expressions are available in the literature for the prediction of the liquid holdup when the liquid is Newtonian and these have been critically evaluated [Mandhane *et al.*, 1975; Govier and Aziz, 1982; Chen and Spedding, 1986]. The simplest and perhaps most widely used correlation is that of Lockhart and Martinelli [1949] which utilises the pressure drop values for single phase flow to define a so-called Lockhart–Martinelli parameter, χ which is:

$$\chi = \left(\frac{-\Delta p_L/L}{-\Delta p_G/L}\right)^{1/2} \tag{4.5}$$

where $(-\Delta p_L/L)$ and $(-\Delta p_G/L)$ are, respectively, the pressure gradients for the flow of liquid and gas alone at the same volumetric flowrates as in the two-phase flow. Although it is based on experimental data for the flow of air–water mixtures in small diameter tubes ($\sim$25 mm) at near atmospheric pressure and temperature, this correlation has proved to be quite successful when applied to other fluids and for tubes of larger diameters. The original correlation, shown in Figure 4.4, consistently over-estimates the value of the liquid holdup (α_L) in horizontal flow of two-phase gas-Newtonian liquid mixtures. This can be

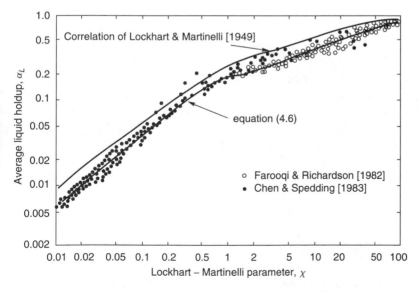

Figure 4.4   *Lockhart–Martinelli correlation for liquid holdup and representative experimental results*

seen in Figure 4.4 which shows the comprehensive data [Chen and Spedding, 1983] for air-water mixtures, of Farooqi [1981] and of Farooqi and Richardson [1982] for the flow of air with aqueous glycerol solutions of various compositions. Taken together, the experimental results shown in Figure 4.4 cover a range of four orders of magnitude of the Lockhart–Martinelli parameter, χ, and average liquid holdups from 0.5% to ~100%. These data cover all the major flow patterns and flow regimes, e.g. nominal streamline and turbulent flow of both gas and liquid. The available experimental results are well represented by the following empirical expressions, as shown in Figure 4.4:

$$\alpha_L = 0.24\chi^{0.8} \qquad 0.01 \leq \chi \leq 0.5 \tag{4.6a}$$

$$\alpha_L = 0.175\chi^{0.32} \qquad 0.5 \leq \chi \leq 5 \tag{4.6b}$$

$$\alpha_L = 0.143\chi^{0.42} \qquad 5 \leq \chi \leq 50 \tag{4.6c}$$

$$\alpha_L = \frac{1}{0.97 + \dfrac{19}{\chi}} \qquad 50 \leq \chi \leq 500 \tag{4.6d}$$

Furthermore, these equations predict values of α_L to within $\pm 1\%$ at the values of χ marking the changeover point between equations. The overall average error is of the order of 7% and the maximum error is about 15%.

Gas–non-Newtonian systems

Because of the widely different types of behaviour exhibited by non-Newtonian fluids, it is convenient to deal with each flow regime separately, depending upon whether the liquid flowing on its own at the same flow rate would be in streamline or turbulent flow. While it is readily conceded that streamline flow does not have as straightforward a meaning in two phase flows as in the flow of single fluids, for the purposes of correlating experimental results, the same criterion is used to delineate the type of flow for non-Newtonian fluids as discussed in Chapter 3 (Section 3.3), and it will be assumed here that the flow will be streamline for $Re_{MR} < 2000$, prior to the introduction of gas.

Streamline flow of liquid

The predictions from equation (4.6) will be compared first with the experimental values of average liquid holdup for cocurrent two-phase flow of a gas and shear-thinning liquids. For a liquid of given rheology (m and n), the pressure gradient ($-\Delta p_L/L$) may be calculated using the methods presented in Chapter 3 but only the power-law model will be used here.

Figure 4.5 shows representative experimental results for average values of liquid holdup α_L, as a function of the parameter χ, together with the predictions of equation (4.6). The curves refer to a series of aqueous china clay suspensions

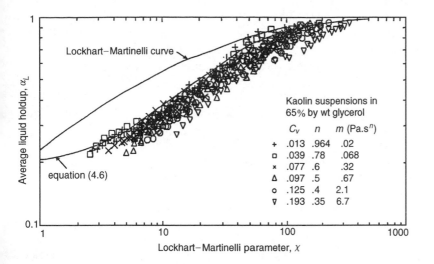

Figure 4.5 *Average liquid holdup data for kaolin suspensions in 65% aqueous glycerol solution in streamline flow (D = 42 mm)*

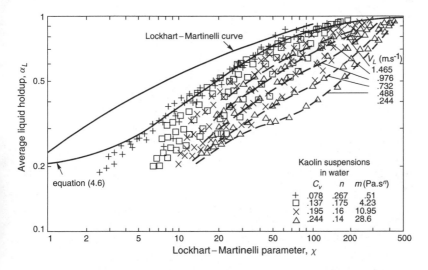

Figure 4.6 *Effect of superficial liquid velocity an average liquid holdup*

in co-current flow with air in a 42 mm diameter horizontal pipe. In addition, Figure 4.6 clearly shows the influence of the liquid superficial velocity (V_L) on the average values of liquid holdup. The results in Figures 4.5 and 4.6 show a similar functional relationship between α_L and χ to that predicted by equation (4.6), but it is seen that this equation over-estimates the value of the average liquid holdup, and distinct curves are obtained for each value of the power-law index (n) and the superficial velocity of the liquid. A detailed examination of the voluminous experimental results reported by different investigators [Oliver and Young-Hoon, 1968; Farooqi and Richardson, 1982; Chhabra and Richardson, 1984] reveals the following features:

(i) For a given value of the power-law index (n), the lower the value of the liquid superficial velocity (V_L), the lower is the average liquid holdup (see Figure 4.6).
(ii) The average liquid holdup decreases as the liquid becomes more shear-thinning (i.e. lower value of n), and the deviation from the Newtonian curve becomes progressively greater.

This suggests that any correction factor which will cause the holdup data for shear-thinning fluids to collapse onto the Newtonian curve, must become progressively smaller as the liquid velocity increases and the flow behaviour index, n, decreases. Based on such intuitive and heuristic considerations, Farooqi and Richardson [1982] proposed a correction factor, J, to be applied to the Lockhart–Martinelli parameter, χ, so that a modified parameter χ_{mod} is defined as:

$$\chi_{mod} = J\chi \tag{4.7}$$

where $$J = \left(\frac{V_L}{V_{L_c}}\right)^{1-n} \tag{4.8}$$

and the average liquid holdup is now given simply by replacing χ with χ_{mod} in equation (4.6), viz.:

$$\alpha_L = 0.24(\chi_{mod})^{0.8} \qquad 0.01 \leq \chi_{mod} \leq 0.5 \tag{4.9a}$$

$$\alpha_L = 0.175(\chi_{mod})^{0.32} \qquad 0.5 \leq \chi_{mod} \leq 5 \tag{4.9b}$$

$$\alpha_L = 0.143(\chi_{mod})^{0.42} \qquad 5 \leq \chi_{mod} \leq 50 \tag{4.9c}$$

$$\alpha_L = \frac{1}{0.97 + \dfrac{19}{\chi_{mod}}} \qquad 50 \leq \chi_{mod} \leq 500 \tag{4.9d}$$

V_{L_c} is the critical velocity for the transition from laminar to turbulent flow. For a given power-law liquid (i.e. known m and n), density and pipe diameter, D, V_{L_c} may be estimated simply by setting the Reynolds number (equation 3.8b) equal to 2000, i.e.

$$Re_{MR} = \frac{\rho V_{L_c}^{2-n} D^n}{8^{n-1} m \left(\dfrac{3n+1}{4n}\right)^n} = 2000 \tag{4.10}$$

For both, $V_L = V_{L_c}$ and/or $n = 1$, the correction factor $J = 1$.

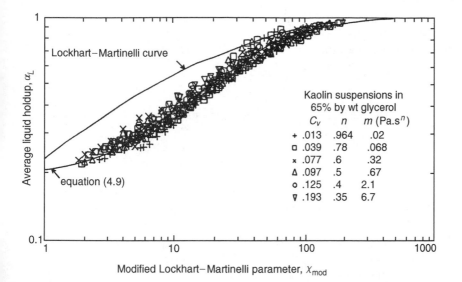

Figure 4.7 *Average liquid holdup as a function of modified parameter* χ_{mod}

In Figure 4.7, the experimentally determined values of average liquid holdup, α_L, are plotted against the modified parameter χ_{mod} for suspensions of kaolin in aqueous glycerol (same data as shown in Figure 4.5) and it will be seen that they are now well correlated by equation (4.9).

Equally good correlations are obtained for the experimental data for two-phase flow of air and nitrogen with aqueous and non-aqueous suspensions of coal [Farooqi *et al.*, 1980] and china clay particles and aqueous solutions of a wide variety of chemically different polymers [Chhabra *et al.*, 1984]. A wide range has been covered ($0.14 \leq n \leq 1$; $0.1 \leq \chi_{mod} \leq \sim 200$) but most data have been obtained in relatively small diameter (3 mm to 50 mm) pipes [Chhabra and Richardson, 1986].

Little is known about the influence of visco-elastic properties of the liquid phase on liquid holdup [Chhabra and Richardson, 1986]. However, Chhabra *et al.* [1984] used aqueous solutions of polyacrylamide (Separan AP-30) as model visco-elastic liquids and a preliminary analysis of these results indicated that equation (4.9) consistently underestimated the value of liquid holdup. Infact the experimental results for visco-elastic liquid and air lie between those predicted by equation (4.9) and equation (4.6). It is thus necessary to introduce an additional parameter to account for visco-elastic effects. For this purpose, a Deborah number was defined as:

$$\text{De} = \frac{\lambda_f V_M}{D} \tag{4.11}$$

where λ_f, the fluid characteristic time, is deduced from the measurement of the primary normal stress difference N_1. Like viscosity, it is generally possible to approximate the variation of N_1 with the shear rate over a limited range by a power m_1 law, eg. equation (1.27) i.e.

$$N_1 = m_1(\dot{\gamma})^{p_1} \tag{4.12}$$

which, in turn, allows the fluid characteristic time λ_f to be defined by equation (1.26):

$$\lambda_f = \left(\frac{m_1}{2m}\right)^{1/(p_1-n)} \tag{4.13}$$

Although the use of the no-slip mixture velocity, V_m, in equation (4.11) is quite arbitrary, it does account for the enhanced shearing of the liquid brought about by the introduction of gas into the pipeline. Over the range of conditions ($0.3 \le$ De ≤ 200 and $2 \le \chi_{\text{mod}} \le 160$), the following simple expression provides a reasonably satisfactory correlation of the available data for α_{LV}, the average value of liquid holdup for visco-elastic liquids:

$$\alpha_{LV} = \alpha_L \left(1 + 0.56 \frac{\text{De}^{0.05}}{\chi_{\text{mod}}^{0.5}}\right) \tag{4.14}$$

α_L is the value of holdup from equation (4.9) in the absence of visco-elastic effects. Although, equation (4.14) does reduce to the limit of $\alpha_{LV} = \alpha_L$ as De $\to 0$, extrapolation outside the range quoted above must be carried out with caution.

Transitional and turbulent flow of liquids (Re$_{MR} > 2000$)

When the non-Newtonian liquid is no longer in streamline flow (prior to the addition of gas), i.e. Re$_{MR} > 2000$, the experimental results for average liquid holdup agree well with those predicted by equation (4.6) and the original Lockhart–Martinelli parameter χ may therefore be used. This is confirmed by the data shown in Figure 4.8 for a variety of shear-thinning liquids including polymer solutions, chalk–water slurries, china clay and coal suspensions [Raut and Rao, 1975; Farooqi et al., 1980; Farooqi and Richardson, 1982; Chhabra et al., 1984]. Scant results available in the literature suggest that equation (4.6) underpredicts the value of holdup for visco-elastic liquids in turbulent flow [Rao, 1997].

Predictive methods for upward vertical flow

The previous discussion on holdup related only to horizontal flow of gas – non-Newtonian liquid mixtures. Very few experimental results are available for holdup in vertical upward flow with shear-thinning liquids [Khatib and Richardson, 1984]. These authors used a γ-ray attenuation method

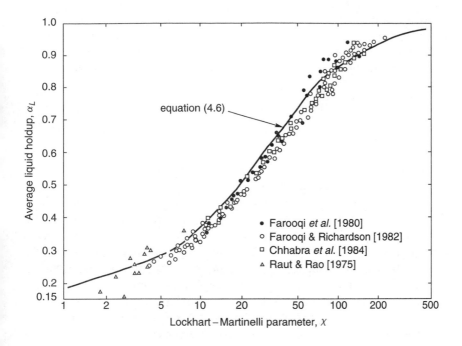

Figure 4.8 *Experimental and predicted (equation 4.6) values of liquid holdup under turbulent conditions for liquid*

to measure the average as well as instantaneous values of liquid holdup for shear-thinning suspensions of china clay and air flowing upwards in a 38 mm diameter pipe. The average values of liquid holdup in streamline flow are in line with the predictions from equation (4.9).

Thus, in summary, average liquid holdup can be estimated using equation (4.6) for Newtonian liquids under all flow conditions, and for non-Newtonian liquids in transitional and turbulent regimes ($Re_{MR} > 2000$).

For the streamline flow of shear-thinning fluids ($Re_{MR} < 2000$), it is necessary to use equation (4.9). A further correction must be introduced (equation 4.14) for visco-elastic liquids. Though most of the correlations are based on horizontal flow, preliminary results indicate that they can also be applied to the vertical upward flow of mixtures of gas and non-Newtonian liquids.

4.2.5 Frictional pressure drop

Generally, methods for determining the frictional pressure drop begin by using a physical model of the two-phase system, and then applying an approach similar to that for single phase flow. Thus, in the so-called separated flow model, the two phases are first considered to be flowing separately and allowance is then made for the effect of interfacial interactions. Irrespective of the type of flow and the rheology of the liquid phase, the total pressure gradient $(-\Delta p_{TP}/L)$ in horizontal flow consists of two components which represent the frictional and acceleration contribution respectively, i.e.

$$\left(-\frac{\Delta p_{TP}}{L}\right) = \left(-\frac{\Delta p_f}{L}\right) + \left(-\frac{\Delta p_a}{L}\right) \qquad (4.15)$$

Both a momentum balance and an energy balance for two-phase flow through a horizontal pipe may be written as expanded forms of those for single phase flow. The difficulty of proceeding in this manner is that local values of important variables such as in-situ velocities and holdups of the individual phases are not known and cannot readily be predicted. Some simplification is possible if it is assumed that each phase flows separately in the channel and occupies a fixed fraction of the pipe, but there are additional complications stemming from the difficulty of specifying interfacial conditions and the effect of gas expansion along the pipe length. As in the case of single phase flow of a compressible medium, the shear stress is no longer simply linked to the pressure gradient because the expansion of the gas results in the acceleration of the liquid phase. However, as a first approximation, it may be assumed that the total pressure drop can be expressed simply as the sum of a frictional and acceleration components:

$$(-\Delta p_{TP}) = (-\Delta p_f) + (-\Delta p_a) \qquad (4.16)$$

For upward flow of gas–liquid mixtures, an additional term $(-\Delta p_g)$ attributable to the hydrostatic pressure, must be included on the right hand side of equation (4.16), and this depends on the liquid holdup which therefore must be estimated.

Thus, complete analytical solutions for the equations of motion are not possible (even for Newtonian liquids) because of the difficulty of defining the flow pattern and of quantifying the precise nature of the interactions between the phases. Furthermore, rapid fluctuations in flow frequently occur and these cannot be readily incorporated into analysis. Consequently, most developments in this field are based on dimensional considerations aided by data obtained from experimental measurements. Great care must be exercised, however, when using these methods outside the limits of the experimental work.

Good accounts of idealised theoretical developments in this field are available for mixtures of gas and Newtonian liquids [Govier and Aziz, 1982; Hetsroni, 1982; Chisholm, 1983] and the limited literature on mixtures of gas and non-Newtonian liquids has also been reviewed elsewhere [Mahalingam, 1980; Chhabra and Richardson, 1986; Bishop and Deshpande, 1986].

Practical methods for estimating pressure loss

Over the years, several empirical correlations have been developed for the estimation of the two-phase pressure drop for the flow of gas–liquid mixtures, with and without heat transfer. Most of these, however, relate to Newtonian liquids, though some have been extended to include shear-thinning liquid behaviour. As the pertinent literature for Newtonian fluids and for non-Newtonian fluids has been reviewed extensively [see references above], attention here will be confined to the methods which have proved to be most reliable and have therefore gained wide acceptance.

Gas–Newtonian liquid systems

The most widely used method for estimating the pressure drop due to friction is that proposed by Lockhart and Martinelli [1949] and subsequently improved by Chisholm [1967]. It is based on a physical model of separated flow in which each phase is considered separately and then the interaction effect is introduced. In this method, the two phase pressure drop due to friction $(-\Delta p_{TP})$, is expressed in terms of dimensionless drag ratios, ϕ_L^2 or ϕ_G^2 defined by the following equations:

$$\phi_L^2 = \frac{-\Delta p_{TP}/L}{-\Delta p_L/L} \tag{4.17}$$

$$\phi_G^2 = \frac{-\Delta p_{TP}/L}{-\Delta p_G/L} \tag{4.18}$$

These equations, in turn, are expressed as functions of the Lockhart–Martinelli parameter χ, defined earlier in equation (4.5). Obviously, the drag ratios are inter-related since $\phi_G^2 = \chi^2\phi_L^2$. Furthermore, Lockhart and Martinelli [1949] used a flow classification scheme depending upon whether the gas/liquid flow is nominally in the laminar–laminar, the laminar–turbulent, the turbulent–laminar or the turbulent–turbulent regime. Notwithstanding the inherently fluctuating nature of two-phase flows and the dubious validity of such a flow classification, the regime is ascertained by calculating the value of the Reynolds number (based on superficial velocity) for each phase. The flow is said to be laminar if this Reynolds number is smaller than 1000 and turbulent if it is greater than 2000, with mixed type of flow in the intermediate zone. Figure 4.9 shows the original correlation of Lockhart and Martinelli [1949] who suggested that the $\phi_L - \chi$ curve should be used for $\chi \geq 1$ and the $\phi_G - \chi$ curve for $\chi \leq 1$. Even though this correlation is based on data for the air–water system in relatively small diameter pipes, it has proved to be of value for the flow of other gas–liquid systems in pipes of diameters up to 600 mm. The predictions are well within ±30%, but in some cases, errors upto 100% have also been reported. It is paradoxical that this method has been

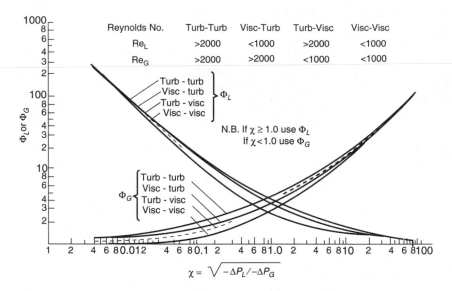

Figure 4.9   *Two-phase pressure drop correlation of Lockhart and Martinelli [1949]*

found to perform poorly for the simplest geometric system, namely, stratified flow! The main virtue of this method lies, however, in its simplicity and in the fact that no prior knowledge of the flow pattern is needed. This is in contrast to the theoretical models which invariably tend to be flow-pattern dependent.

Chisholm [1967] has developed an algebraic form of relation between ϕ_L^2 and χ:

$$\phi_L^2 = 1 + \frac{C_0}{\chi} + \frac{1}{\chi^2} \tag{4.19}$$

where, for air-water mixtures, the values of C_0 are as follows:

Gas	Liquid	C_0
Laminar	Laminar	5
Laminar	Turbulent	10
Turbulent	Laminar	12
Turbulent	Turbulent	20

Further correction is needed if the densities of the two phases are appreciably different from those of air and water [Chisholm, 1967]. Extensive comparisons between the predictions of equation (4.19) and experimental values embracing all the four regimes show satisfactory agreement [Chhabra and Richardson, 1986].

Gas–non-Newtonian liquid systems

As remarked earlier, analytical treatments of two-phase flow are of limited value and this applies equally for non-Newtonian liquids. The relatively simple flow patterns, of annular and stratified flow, for power-law liquids have received some attention in the literature. For annular flow, some workers have assumed that the thin liquid film at the wall behaves like a laminar film flowing between two parallel plates [Mahalingam and Valle, 1972] while others [Oliver and Young Hoon, 1968; Eissenberg and Weinberger, 1979] have approximated the flow area to be an annulus, with no inner wall. Likewise, Heywood and Charles [1979] and Bishop and Deshpande [1986] have modified the Taitel and Dukler's [1980] idealised model for stratified flow to include power-law fluids. In most cases, the interface is assumed to be smooth (free from ripples) and the interfacial friction factor has been approximated by that for a gas flowing over a solid surface. These and other simplifications account for the fact that values of pressure drop may deviate from the experimental data by a factor of up to four. On the other hand, experimental work in this field has yielded results which can be used to predict pressure drops over a wide range of conditions. The results obtained with the liquid in laminar flow ($\text{Re}_{MR} < 2000$) or in turbulent flow ($\text{Re}_{MR} > 2000$) prior to the introduction of gas will now be treated separately. Shear-thinning fluids are found to exhibit completely different behaviour from Newtonian liquids in streamline conditions whereas in turbulent flow, the non-Newtonian properties appear to be of little consequence.

Laminar conditions

When a gas is introduced into a shear-thinning fluid in laminar flow, the frictional pressure drop may, in some circumstances, actually be reduced below the value for the liquid flowing alone at the same volumetric rate. As the gas flowrate is increased, the two-phase pressure drop decreases, then passes through a minimum (maximum drag reduction) and finally increases again and eventually exceeds for the flow of liquid alone. This effect which has been observed with flocculated suspensions of fine kaolin and anthracite coal and with shear-thinning polymer solutions occurs only where the flow of liquid on its own would be laminar. A typical plot of drag ratio (ϕ_L^2) as a function of superficial air velocity is shown in Figure 4.10, for a range of values of the liquid superficial velocity, and the corresponding values of its Reynolds number Re_{MR} are given. The liquid is a 24.4% (by volume) kaolin suspension in water which exhibits power-law rheology. An analysis of a large number of experimental results identifies the following salient features:

(i) For a liquid with known values of m, n, ρ, the value of the minimum drag ratio (ϕ_L^2)$_{\text{min}}$ decreases as the liquid velocity is lowered.

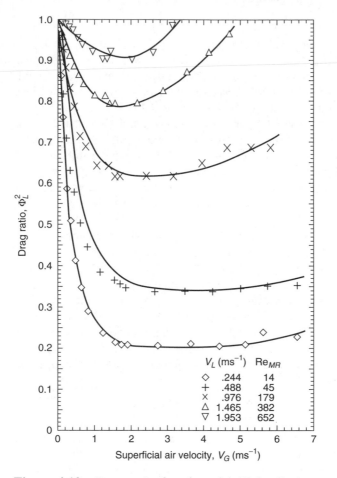

Figure 4.10   *Drag ratio data for a 24.4% kaolin-in-water suspension in a 42 mm diameter pipe*

(ii) The superficial gas (air) velocity needed to achieve maximum drag reduction increases as the liquid velocity decreases.

(iii) The higher the degree of shear-thinning behaviour (i.e. the smaller the value of n), the greater is the extent of the drag reduction obtainable.

If, for a given liquid (i.e. fixed values of m and n for a power-law fluid), the drag ratio is plotted against the no-slip mixture velocity, $V_M(= V_L + V_G)$, as opposed to the superficial gas velocity, it is found that the minima in ϕ_L^2 all occur at approximately the same mixture velocity, irrespective of the liquid flowrate (see Figure 4.11). Furthermore, this minimum always occurs when Re_{MR} (based on V_M rather than V_L) is approximately 2000, corresponding to

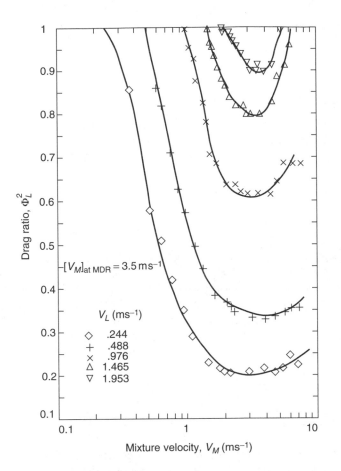

Figure 4.11 *Drag ratio data for a 24.4% (by volume) kaolin-in-water suspension as a function of mixture velocity*

the upper limit of streamline flow. This suggests that the value of ϕ_L^2 continues to fall progressively until the liquid is no longer in streamline flow. Thus, at low flowrates of liquid, more air can be injected before this point is reached.

At first sight, it seems rather anomalous that on increasing the total volumetric throughput by injection of air, the frictional pressure drop can actually be lower than that for the flow of liquid alone. Also, the magnitude of the effect can be very large, with values of ϕ_L^2 as low as 0.2 (obtained with highly shear-thinning china clay suspensions), i.e. the two-phase pressure drop can be reduced by a factor of 5 by air injection. The mechanism by which this can occur may be illustrated using a highly idealised model. Suppose that the gas and liquid form a series of separate plugs, as depicted schematically in Figure 4.12.

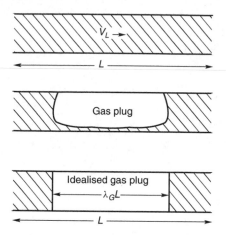

Figure 4.12   *Idealised plug flow model*

For the two-phase flow, the total pressure drop will be approximately equal to the sum of the pressure drops across the individual liquid slugs, the pressure drop across the gas slugs being negligible in comparison with that for the liquid, slugs. For a power-law fluid in laminar flow at a velocity of V_L in a pipe of length L, the pressure drop $(-\Delta p_L)$ is given by (equation 3.6, Chapter 3):

$$-\Delta p_L = A_1 L V_L^n \tag{4.20}$$

where A_1 is a constant for a given pipe (D) and fluid $(m$ and $n)$.

The addition of gas has two effects: the length of pipe in contact with liquid is reduced, and the velocity of the liquid plug is increased. If λ_L (defined by equation 4.3) is the input volume fraction of liquid, then in the absence of slip, the wetted length of pipe is reduced to $L\lambda_L$ and the velocity of the liquid plug is increased to V_L/λ_L. The two phase pressure drop is:

$$-\Delta p_{TP} = A_1(L\lambda_L)(V_L/\lambda_L)^n \tag{4.21}$$

The drag ratio, ϕ_L^2, is obtained as

$$\phi_L^2 = \frac{-\Delta p_{TP}}{-\Delta p_L} = (\lambda_L)^{1-n} \tag{4.22}$$

For a shear-thinning fluid $n < 1$ and $\lambda_L < 1$, the drag ratio ϕ_L^2 must be less than unity, and hence a reduction in pressure drop occurs as a result of the presence of the air. The lower the value of n and the larger the value of λ_L, the greater will the effects be, and this situation is qualitatively consistent with experimental observations. It should be noted that any effects due to the expansion of the gas in the pipeline have not been considered here.

This simple model is likely to under-estimate the magnitude of $(-\Delta p_{TP})$ and ϕ_L^2 because the liquid and gas will not form idealised plugs and there will be some slip between the two phases. It has been found experimentally that equation (4.22) does apply at low air velocities ($<\sim 1$ m/s), but at higher gas flowrates the model holds progressively less well. In the limiting case of a Newtonian liquid ($n = 1$), equation (4.22) yields $\phi_L^2 = 1$, for all values of λ_L and the two phase pressure drop would be unaffected by air injection. In practice, because gas will always disturb the flow, there will be additional pressure losses, and the two-phase pressure drop will always increase, with the introduction of a gas.

Drag reduction can also occur with a fluid exhibiting an apparent yield stress [Farooqi *et al.*, 1980].

Maximum drag reduction (MDR)

As noted earlier, for a given liquid and pipe, the minimum value of ϕ_L^2 occurs at a constant value of the no-slip mixture velocity which corresponds approximately to $Re_{MR} \sim 1000-2000$. This implies that ϕ_L^2 attains a minimum value $(\phi_L^2)_{\min}$ when the flow in the liquid plug no longer remains streamline. Values of $(\phi_L^2)_{\min}$ have been correlated against the correction factor $J = (V_L/V_{L_c})^{1-n}$ introduced earlier in connection with the prediction of liquid holdup. Thus:

$$(\phi_L^2)_{\min} = J^{0.205} \qquad\qquad 0.6 \le J \le 1 \qquad\qquad (4.23a)$$

$$(\phi_L^2)_{\min} = 1 - 0.0315J^{-2.25} \quad 0.35 \le J \le 0.6 \qquad\qquad (4.23b)$$

$$(\phi_L^2)_{\min} = 1.9J \qquad\qquad 0.05 \le J \le 0.35 \qquad\qquad (4.23c)$$

Figure 4.13 compares the predictions from equation (4.23) with representative experimental results for both aqueous polymer solutions and particulate suspensions in pipes of diameters up to 200 mm. It will be noted that equation (4.23) is particularly useful in estimating, a priori, the minimum achievable drag ratio as it requires a knowledge only of the properties of liquid (ρ, m, n) and the operating conditions (D, V_L); the corresponding gas velocity is calculated as ($V_{L_c} - V_L$).

General method for estimation of two phase pressure loss

The discussion so far has related to the drag reduction occurring when a gas is introduced into a shear-thinning fluid initially in streamline flow. A more general method is required for the estimation of the two phase pressure drop for mixtures of gas and non-Newtonian liquids. The well-known Lockhart–Martinelli [1949] method will now be extended to encompass shear-thinning liquids, first by using the modified Lockhart–Martinelli parameter, χ_{mod} (equation 4.8). Figure 4.14 shows a comparison between

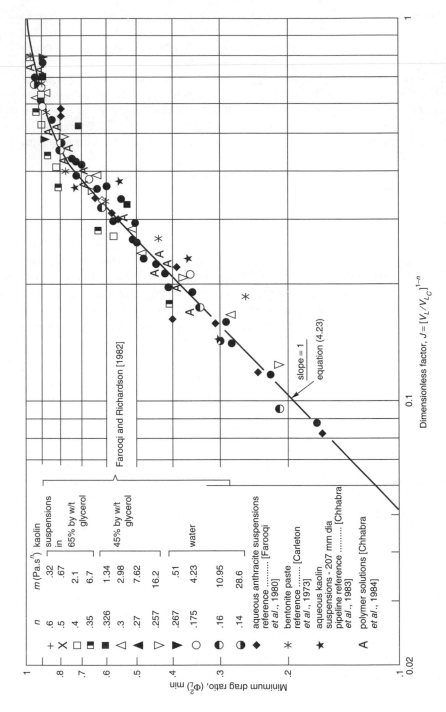

Figure 4.13   *Correlation for minimum drag ratio*

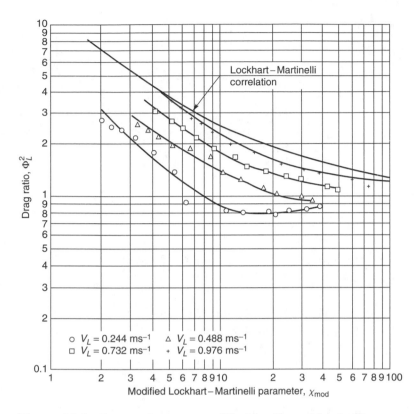

Figure 4.14 *Drag ratio versus modified Lockhart–Martinelli parameter for cocurrent flow of air and a china clay suspension in a 42 mm diameter pipe*

the Lockhart–Martinelli correlation and typical experimental measurements of two phase pressure drop for air and a china clay suspension ($m = 0.67 \text{ Pa·s}^n$, $n = 0.50$) flowing cocurrently in a 42 mm diameter horizontal pipe. The conditions are such that the gas would be in turbulent flow and the kaolin suspension in streamline flow if each phase were flowing on its own. The value of J ranges from 0.16 to 0.65 for these conditions. Evidently, as the liquid velocity increases, the experimental values of the drag ratio move towards the correlation of Lockhart and Martinelli [1949], approximated here by equation (4.19) with χ replaced by χ_{mod}. Indeed, when a large amount of data culled from various sources in the literature is analysed in this fashion, the deviations from the predictions of equation (4.19) range from $+60\%$ to -800%, the experimental values being generally overestimated. Based on these observations, Dziubinski and Chhabra [1989] empirically modified the drag ratio to give

$$(\phi_L^2)_{\text{mod}} = \phi_L^2 / J \qquad (4.24)$$

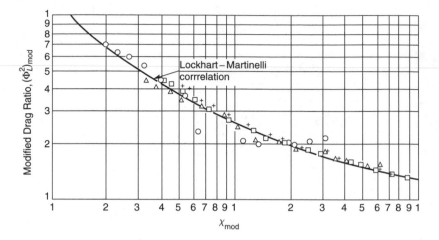

Figure 4.15 *Modified drag ratio versus modified Lockhart–Martinelli parameter (same data as shown in Figure 4.14)*

The results shown in Figure 4.14 are re-plotted in Figure 4.15 using the modified drag ratio, $(\phi_L^2)_{\text{mod}}$ and the modified Lockhart–Martinelli parameter, χ_{mod}; data points are now seen to straddle the original correlation of Lockhart and Martinelli [1949], i.e. equation (4.19). Indeed, this approach reconciles nearly 1500 data points relating to the streamline flow of liquid with an error of ±40% which is comparable with the uncertainty associated with the original correlation for Newtonian liquids. The validity of this approach has been tested over the following ranges of conditions as: $0.10 \leq n \leq 0.96$; $2.9 \leq D \leq 207$ mm; $0.17 \leq V_L \leq 2$ m/s; $0.11 \leq V_G \leq 23$ m/s.

It is emphasised, however, that because the adaptation of the correlation for non-Newtonian fluids is entirely empirical and that the same factor J appears in both abscissa and ordinate, great caution must be exercised in using this method outside the limits of the variables employed in its formulation.

Turbulent flow

For both Newtonian and non-Newtonian liquids in turbulent flow, the addition of gas always results in an increase in the pressure drop and gives values of drag ratio, ϕ_L^2, in excess of unity. Using χ, both the graphical correlation of Lockhart and Martinelli in Figure 4.9 and equation (4.19) satisfactorily represent the data, as illustrated in Figure 4.16 for turbulent flow of both gas and liquid, as also argued recently by Rao [1997].

In a recent study, Dziubinski [1995] has put forward an alternative formulation for the prediction of the two-phase pressure drop for a gas and shear-thinning liquid mixture in the intermittent flow regime. By analogy with

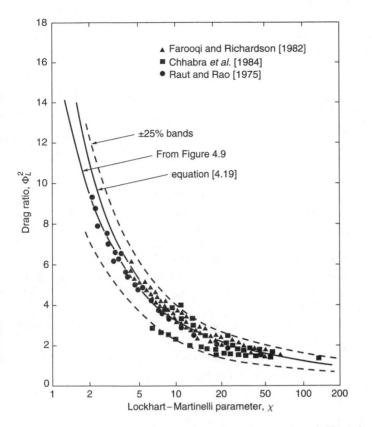

Figure 4.16 *Experimental and predicted (equation 4.19) values of drag ratio for turbulent flow of polymer solutions and suspensions*

the flow of single phase fluids, he introduced a loss coefficient Λ defined as:

$$\Lambda = \frac{\tau_w \rho D^2}{\mu^2} \qquad (4.25)$$

For power-law fluids in streamline flow, Dziubinski's expression for the drag ratio is:

$$\phi_L^2 = \frac{1 + 1.036 \times 10^{-4} (\text{Re}_{TP})^{1.235}}{1 + 1.036 \times 10^{-4} (\text{Re}_L)^{1.235}} \lambda_L^{1-n} \qquad (4.26)$$

where λ_L is the input liquid fraction, equation (4.3), and the two Reynolds numbers in equation (4.26) are defined as:

$$\text{Re}_{TP} = \frac{\rho V_M^{2-n} D^n}{8^{n-1} m \left(\dfrac{3n+1}{4n} \right)^n} \qquad (4.27)$$

and Re_L is based on the superficial velocity of the liquid. Similarly, for the turbulent flow of gas/pseudoplastic liquid mixtures ($Re_{TP} > 2000$) his expression in terms of the loss coefficient Λ_{TP} is:

$$\Lambda_{TP} = 0.0131\lambda_L\Delta^{-5}\exp(1.745\Delta - 0.634\lambda_L)(Re^*_{TP})^{1.75} \quad (4.28)$$

where $\quad \Delta = \dfrac{3n+1}{4n},\quad$ and

$$Re^*_{TP} = \Delta^2 Re_{TP}$$

Attention is drawn to the fact that the values of m and n for use in turbulent region are deduced from the data in the laminar range at the values of $(8V_L/D)$ which is only the *nominal* shear rate at the tube wall for streamline flow, and thus this aspect of the procedure is completely empirical. Dziubinski [1995] stated that equation (4.26) reproduced the same experimental data as those referred to earlier with an average error of $\pm 15\%$, while equation (4.28) correlated the turbulent flow data with an error of $\pm 25\%$. Notwithstanding the marginal improvement over the method of Dziubinski and Chhabra [1989], it is reiterated here that both methods are of an entirely empirical nature and therefore the extrapolation beyond the range of experimental conditions must be treated with reserve.

More recently, based on the notion of the fractional pipe surface in contact with the liquid, Kaminsky [1998] has developed a new method for the prediction of the two-phase pressure drop for the flow of a mixture of gas and a power-law fluid. This method is implicit in pressure gradient (and therefore requires an interative solution) and also necessitates additional information about the fraction of the pipe surface in contact with liquid which is not always available.

Vertical (upward) flow

The interpretation of results for vertical flow is more complicated since they are strongly dependent on the in-situ liquid holdup which , in turn, determines the hydrostatic component of the pressure gradient. Khatib and Richardson [1984] reported measured values of the two-phase pressure drop and liquid holdup for the vertical upward co-current flow of air and aqueous china clay suspensions in a 38 mm diameter pipe. Representative results are shown in Figures 4.17 and 4.18 for air–water and air–china clay suspensions, respectively. In all cases, as the air flow rate is increased, the total pressure gradient decreases, passes through a minimum and then rises again. Although the minimum pressure gradient occurs at about the same value of the no-slip mixture velocity as in horizontal flow, it has no connection with the laminar–turbulent transition. There is a minimum in the curve because, as the gas flow rate is increased,

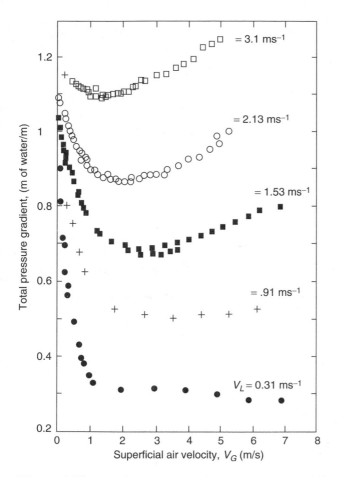

Figure 4.17   *Total pressure gradient for the upward flow of air–water mixtures in a 38 mm diameter pipe*

the holdup of liquid decreases, but the frictional pressure gradient increases due to the higher liquid velocities.

The frictional pressure drop may be estimated by subtracting the hydrostatic component (calculated from the holdup) from the total pressure gradient, as shown in Figure 4.19. It will be seen that under certain conditions, particularly at low liquid flow rates, the frictional component appears to approach zero. For the flow of air–water mixtures, 'negative friction losses' are well documented in the literature. This anomaly arises because not all of the liquid present in the pipe contributes to the hydrostatic pressure, because some liquid may form a film at the pipe wall. This liquid is sometimes flowing downwards and most of its weight is supported by an upward shear force at the wall. The drag exerted by the gas on the liquid complements the frictional force at the pipe wall.

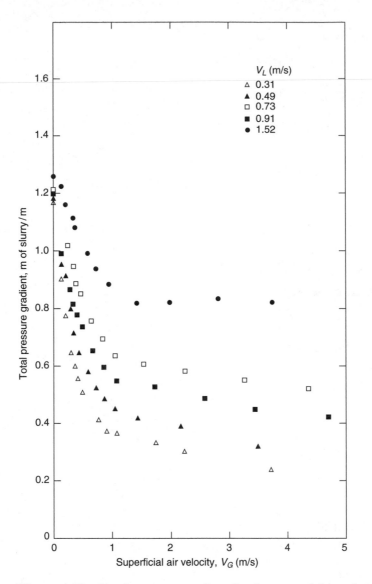

Figure 4.18   *Total pressure gradient for the upward flow of air and 18.9% (by volume) aqueous kaolin suspension*

It is, in principle, possible to split the liquid holdup, measured at a given position, into two components; one associated predominantly with liquid slugs and the other with liquid films at the walls which do not contribute to the hydrostatic pressure. Then, using the effective hydrostatic pressure, a more realistic value for the frictional pressure drop can be obtained. Because it

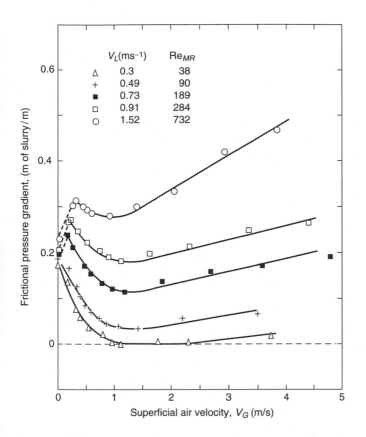

Figure 4.19 *Frictional component of pressure gradient for the results shown in Figure 4.18*

is not possible to make sufficiently accurate measurements of the separate components of the liquid holdup, there will be large errors in the amended friction terms calculated in this fashion.

4.2.6 Practical applications and optimum gas flowrate for maximum power saving

Drag reduction offers the possibility of lowering both the pressure drop and the power requirements in slurry pipelines. Air injection can be used, in practice, in two ways:

(i) To reduce the pressure drop, and hence the upstream pressure in a pipeline, for a given flow rate of shear-thinning liquid.
(ii) To increase the throughput of liquid for a given value of pressure drop.

Air injection can also be beneficial because it may be easier to re-start pumping after a shutdown as the pipe will not be completely full of slurry. On the other hand, if the pipeline follows an undulating topography, difficulties can arise from air collecting at the high points.

Air injection may sometimes be an alternative to deflocculation before pumping. In general, less power is required to pump deflocculated slurries but highly consolidated sediments may form on shutdown and these may be difficult to resuspend when pumping is resumed. Furthermore, deflocculating agents are expensive and may be undesirable contaminants of the product.

However, additional energy will be required to compress the air to a pressure in excess of the upstream pressure. Thus, the circumstances under which there will be a net saving of power for pumping will be strongly dependent on the relative efficiences of the slurry pump and the gas compressor, and on the specific plant layout. Dziubinski and Richardson [1985] have addressed this problem and the salient features of their study are summarised here. They introduced a power saving coefficient, ψ, defined as

$$\psi = \frac{N_L - N_{TP} - N_G}{N_L} \tag{4.29}$$

where N_L is the power needed for pumping the slurry on its own; N_{TP} is the power for pumping the two-phase mixture and N_G is the power for compressing the gas prior to injection into the liquid.

It can be readily seen that the power saving coefficient can be expressed as a function of the efficiences of the pump and the compressor, the superficial velocities of the gas and the slurry and the two-phase pressure drop. Furthermore, Dziubinski and Richardson [1985] noted that the conditions for the maximum drag reduction coincide approximately with the range of conditions over which the simple plug model applies. Thus, they expressed the two phase pressure drop and the $(-\Delta p_L)$ term appearing in the expression for ψ in terms of V_L and V_G. Finally, for a given slurry (ρ, m, n), pump and compressor efficiences, pipe dimensions (D, L) and the slurry flow rate (V_L), they obtained the optimum value of V_G by setting $d\psi/dV_G = 0$. Based on extensive computations, they concluded that, unless the efficiency of the compressor exceeds that of the pump, there will be no net power saving. Also, the liquid flow rate must be well into laminar regime and it must be moderately shear-thinning ($n < \sim0.5$). Finally, the maximum power saving occurs at a much lower gas velocity than that needed for the maximum drag reduction. Reference must be made to their original paper for a detailed treatment of this aspect of two-phase flow.

Example 4.1

Air is injected into a 50 m long horizontal pipeline (of 42 mm diameter) carrying a china clay slurry of density 1452 kg/m^3. The rheological behaviour of the slurry follows the power-law model, with $m = 5.55\,\text{Pa·s}^n$ and $n = 0.35$. The volumetric flowrates of air and liquid are $7.48\,\text{m}^3/\text{h}$ and $1.75\,\text{m}^3/\text{h}$ respectively. The air is introduced into the pipeline at 20°C and at a pressure of 1.2 bar. Ascertain the flow pattern occurring in the pipeline. Estimate (a) the average liquid holdup at the midpoint (b) the pressure gradient for the two-phase flow (c) the maximum achievable drag reduction and the air velocity to accomplish it.

Solution

$$\text{Cross-sectional area of pipe} = \frac{\pi}{4}D^2$$

$$= \frac{\pi}{4}(42 \times 10^{-3})^2 = 1.38 \times 10^{-3}\,\text{m}^2$$

$$\text{Superficial liquid velocity, } V_L = \frac{1.75}{3600} \times \frac{1}{1.38 \times 10^{-3}}$$

$$= 0.35\,\text{m/s}$$

$$\text{Superficial gas velocity, } V_G = \frac{7.48}{3600} \times \frac{1}{1.38 \times 10^{-3}}$$

$$= 1.5\,\text{m/s}$$

From Figure 4.2, it can be seen that the flow pattern is likely to be of the intermittent type under these conditions.

For liquid flowing on its own, the power-law Reynolds number,

$$\text{Re}_{MR} = \frac{\rho V_L^{2-n} D^n}{8^{n-1} m \left(\dfrac{3n + 1}{4n} \right)^n} \qquad \text{(eq. (3.8b))}$$

$$= \frac{1452 \times (0.35)^{2-0.35} (42 \times 10^{-3})^{0.35}}{8^{0.35-1}(5.55) \left(\dfrac{3 \times 0.35 + 1}{4 \times 0.35} \right)^{0.35}} = 51.6$$

$\therefore$ the flow is streamline and the Fanning friction factor is given by equation (3.8a),

$$f = \frac{16}{\text{Re}_{MR}} = \frac{16}{51.6} = 0.31$$

$$-\frac{\Delta p_L}{L} = \frac{2f\rho V_L^2}{D} = \frac{2 \times 0.31 \times 1452 \times (0.35)^2}{0.042} = 2630\,\text{Pa/m}$$

For air alone, one can follow the same method to estimate $(-\Delta p_G/L)$. The viscosity of air at 20°C is $1.8 \times 10^{-5}\,\text{Pa·s}$ and the density is estimated by assuming it to be an ideal gas, at mean pressure of 1.1 bar:

$$\rho_G = \frac{pM}{RT} = \frac{1.1 \times 1.013 \times 10^5 \times 29}{8314 \times 293} = 1.33\,\text{kg/m}^3$$

$$\therefore \text{ Reynolds number, } Re_G = \frac{\rho_G V_G D}{\mu_G} = \frac{1.33 \times 1.5 \times 42 \times 10^{-3}}{1.8 \times 10^{-5}}$$

$$= 4460$$

and the friction factor is calculated using the Blasius formula

$$f = 0.079 Re_G^{-0.25} = 0.079(4460)^{-0.25} = 0.0096$$

$$\therefore -\frac{\Delta p_G}{L} = \frac{2f \rho_G V_G^2}{D} = \frac{2 \times 0.0096 \times 1.33 \times 1.5^2}{0.042} = 1.35 \, \text{Pa/m}$$

The Lockhart–Martinelli parameter, χ, is evaluated as:

$$\chi = \sqrt{\frac{-\Delta p_L/L}{-\Delta p_G/L}} = \sqrt{\frac{2630}{1.35}} = 44.2$$

Since the liquid is in laminar flow, the correction factor J must be calculated. The critical liquid velocity corresponding to $Re_{MR} = 2000$ is estimated from the relation

$$Re_{MR} = \frac{\rho(V_{L_c})^{2-n} D^n}{8^{n-1} m \left(\frac{3n+1}{4n}\right)^n} = 2000$$

Substituting for ρ, D, m and n, solving for V_{L_c}

$$V_{L_c} = \left[\frac{(2000)(8)^{0.35-1}(5.55)\left(\frac{3 \times 0.35 + 1}{4 \times 0.35}\right)^{0.35}}{1452 \times (0.042)^{0.35}}\right]^{1/(2-0.35)}$$

$$= 3.21 \, \text{m/s}$$

$$\therefore J = (V_L/V_{L_c})^{1-n} = \left(\frac{0.35}{3.21}\right)^{1-0.35} = 0.237$$

$$\therefore \chi_{\text{mod}} = \chi J = 44.2 \times 0.237 = 10.46$$

The average liquid holdup is given by equation (4.9c):

$$\alpha_L = 0.143 \chi_{\text{mod}}^{0.42} = 0.143 \times 10.46^{0.42} = 0.38$$

i.e. on average 38% of the pipe cross-section is filled with liquid. This fraction will, however, continually change along the pipe length as the pressure falls.

The two-phase pressure gradient is estimated using equation (4.19) in terms of χ_{mod} and $\phi_{L_{\text{mod}}}^2$. Since the liquid is in streamline flow and the gas is in turbulent regime, $C = 12$ and

$$\phi_{L_{\text{mod}}}^2 = 1 + \frac{12}{\chi_{\text{mod}}} + \frac{1}{(\chi_{\text{mod}})^2} = 1 + \frac{12}{10.46} + \frac{1}{(10.46)^2} = 2.16$$

or $$\phi_L^2 = J\phi_{L_{\text{mod}}}^2 = 0.237 \times 2.16 = 0.51$$

Hence, $-\Delta p_{TP}/L = 0.51 \times (-\Delta p_L/L) = 0.51 \times 2630 = 1340 \, \text{Pa/m}$.

∴ Total pressure drop over 50 m pipe length $= 1340 \times 50 = 67\,\text{kPa}$.

It is also of interest to contrast this value with the prediction of equation (4.26). Here, $V_M = V_L + V_G = 0.35 + 1.50 = 1.85\,\text{m/s}$.

$$\therefore \text{Re}_{TP} = \frac{1452 \times 1.85^{2-0.35} \times 0.042^{0.35}}{8^{0.35-1} \times 5.55 \times \left(\dfrac{3 \times 0.35 + 1}{4 \times 0.35}\right)^{0.35}} = 805$$

and $\qquad \lambda_L = \dfrac{V_L}{V_G + V_L} = \dfrac{0.35}{1.5 + 0.35} = 0.189$

From equation (4.26),

$$\phi_L^2 = \left\{ \frac{1 + 1.036 \times 10^{-4} \times 805^{1.235}}{1 + 1.036 \times 10^{-4} \times (51.6)^{1.235}} \right\} (0.189)^{1-0.35} = 0.47$$

$\therefore (-\Delta p_{TP})/L = 0.47 \times 2630 = 1230\,\text{Pa/m}$ which is about 10% lower than the value calculated above. However, both these values of 1340 Pa/m and 1230 Pa/m compare well with the corresponding experimental value of 1470 Pa/m.

The maximum achievable drag reduction is calculated by using equation (4.23c):

$$\phi_{L_{\min}}^2 = 1.9\,J = 1.9 \times 0.237 = 0.45$$

The corresponding air velocity is obtained simply by subtracting the value of V_L from V_{L_c}, i.e. $3.21 - 0.35 = 2.86\,\text{m/s}$.

However, for the relative efficiencies of the slurry pump and the compressor in the ratio of 1 to 2, the approach of Dziubinski and Richardson [1985] yields the optimum gas velocity to be 0.2 m/s, which is much smaller than the value of 2.86 m/s for maximum drag reduction.□

4.3 Two-phase liquid–solid flow (hydraulic transport)

Hydraulic transport is the conveyance of particulate matter in liquids. Although most of the earlier applications of the technique used water as the carrier medium (and hence the term hydraulic), there are now many industrial plants, particularly in the minerals, mining and power generation industries, where particles are transported in a variety of liquids which may exhibit either Newtonian or non-Newtonian flow behaviour. Transport may be in vertical or horizontal or inclined pipes, but in the case of long pipelines, it may follow the undulations of the land over which the pipeline is installed. The diameter and length of the pipeline and its inclination, the properties of the solids (size, shape and density) and of the liquid (density, viscosity, Newtonian or non-Newtonian), and the flow rates all influence the nature of the flow and the pressure gradient. Design methods are, in general, not very reliable, especially for the transportation of coarse particles; Therefore, the estimated values of

Table 4.1   *Important variables in slurry pipelines*

Component	Parameters
Solids:	shape, size and size distribution, density, strength, abrasiveness
Liquid:	type of liquid (Newtonian, pseudoplastic, viscoplastic), corrosive nature, density, rheological properties and their temperature dependence, stability
Pipeline:	diameter and length, its orientation, fittings, valves, material of construction
Operating conditions:	flow rates of liquid and solids, concentration, type of pump, etc.

the pressure drop and power should be treated with caution. In practice, it is more desirable and important to ensure that the system operates reliably, and without the risk of blockage and without excessive erosion, than to achieve optimal operating conditions in relation to power requirements.

Table 4.1 lists the most important variables which must be considered in designing the facility and in estimating pressure drop and power consumption.

It is customary to divide suspensions into two broad categories – fine particle suspensions in which the particles are reasonably uniformly distributed in the liquid with little separation; and coarse suspensions in which particles, if denser than the liquid, tend to separate out and to travel predominantly in the lower part of a horizontal pipe (at a lower velocity than the liquid); in a vertical pipe the solids may have an appreciably lower velocity than the liquid. Although, this is obviously not a very clear cut classification and is influenced by the flow rate and concentration of solids, it does nonetheless provide a convenient initial basis for classifying the flow behaviour of liquid–solid mixtures.

Fine particles usually form fairly homogeneous suspensions which do not separate to any significant extent during flow. In high concentration suspensions, settling velocities of the particles are small in comparison with the liquid velocities (under normal operating conditions) and in turbulent flow, the eddies in the liquid phase keep particles suspended. In practice, turbulent conditions will prevail, except when the liquid has a very high viscosity (such as in coal–oil slurries) or exhibits non-Newtonian behaviour. In addition, concentrated flocculated suspensions are frequently conveyed in streamline flow when they behave essentially as single phase shear-thinning liquids (e.g. flocculated kaolin and coal suspensions).

Depending upon its state, the suspension may exhibit Newtonian or non-Newtonian behaviour. It is often a good approximation to treat it as a pseudo-homogeneous single phase systems by ascribing to it an effective density and viscosity. Thus, one can use the methods outlined in Chapter 3 to estimate the pressure gradient in terms of the flow rate. Attention will now be focussed on the transportation of coarse particles in non-Newtonian carrier media which offer two advantages when the flow is streamline: firstly, the effective or apparent viscosity of a shear-thinning fluid is a maximum at the centre of the pipe and this facilitates the suspension of the particles (though some of this effect may be offset by the propensity for migration across streamlines and the enhanced settling velocities in sheared fluids); secondly, the apparent viscosity will be minimum at the pipe wall, as a result of which the frictional pressure gradient will be low and will increase only relatively slowly as the liquid velocity is raised. Furthermore, if the fluid exhibits a yield stress, suspension of coarse particles in the central part of the pipeline will be further assisted. In practice with the transport of particulate matter of wide size distribution (e.g. coal dust to large lumps), the fine colloidal particles tend to form a pseudo-homogeneous shear-thinning medium of enhanced apparent viscosity and density in which the coarse particles are conveyed. On the other hand, the heavy medium may consist of fine particles of a different solid, particularly one of higher density such as in the transport of cuttings in drilling muds in drilling applications. In such a case, it is necessary to separate and re-cycle the heavy medium. The use of heavy carrier media can be advantageous when the coarse particles are transported in suspension rather than as a sliding bed and may enable operation to be carried out under streamline flow conditions.

In spite of these potential benefits, only a few studies dealing with the transport of coarse particles in heavy media have been reported. Charles and Charles [1971] investigated the feasibility of transporting 216 µm sand particles in highly shear-thinning clay suspensions ($0.24 \leq n \leq 0.35$) and they concluded that energy requirements could be reduced by a factor of six when using heavy media as opposed to water. Similarly, Ghosh and Shook [1990] reported slight reduction in head loss for the transport of 600 µm sand particles in a 52 mm diameter pipe in a moderately shear-thinning carboxymethyl cellulose solution; however, no reduction in head loss was observed for 2.7 mm pea gravel particles, presumably because these large particles were conveyed in the form of a sliding bed. Indeed Duckworth *et al.* [1983, 1986] successfully conveyed coal particles (up to 19 mm) in a pipe of 250 mm diameter in a slurry of fine coal which behaved as an ideal Bingham plastic fluid. However, in none of these studies has an attempt been made to develop a general method for the prediction of pressure gradient in such applications.

In an extensive experimental study, Chhabra and Richardson [1985] transported coarse gravel particles (3.5, 5.7, 8.1 mm) in a 42 mm diameter horizontal pipe in a variety of carrier fluids, including Newtonian liquids of high viscosity, pseudoplastic china clay suspensions and polymer solutions. However, a majority of the coarse particles were seen to be transported in the form of a sliding bed along the bottom of the pipe while the liquid flow could be either streamline or turbulent. In this mode of hydraulic transport, the resistance to motion of bed of particles due to friction between the solids and the pipe wall is balanced by the force due to the hydraulic pressure gradient and, following the procedure of Newitt *et al.* [1955], a force balance gives:

$$k_1 \gamma C Q (\rho_S - \rho_L) g = i_S Q \rho_L g \qquad (4.30)$$

where γ is the friction coefficient between solids and pipe wall; Q is the total volume of suspension of concentration C in the control volume; k_1 is a system constant and i_S is the hydraulic gradient attributable to the presence of solids. Upon rearrangement,

$$i_S = k_1 \gamma C \left(\frac{\rho_S - \rho_L}{\rho_L} \right) = k_1 \gamma C (s - 1) \qquad (4.31)$$

where $s = (\rho_S / \rho_L)$ and $\quad i_S = i - i_L$ $\qquad (4.32)$

where i and i_L are the hydraulic pressure gradients, respectively for the flow of mixture (total) and of liquid (heavy medium) alone at the same volumetric rate. Substituting $i_L = 2 f_L V^2 / gD$ and eliminating i_S, yields:

$$\frac{i - i_L}{i_L} = \frac{k_1 \gamma C (s - 1)}{2 f_L V^2 / gD}$$

$$\text{or} \quad f_L \frac{(i - i_L)}{i_L} = k_2 \frac{gDC(s - 1)}{V^2} \qquad (4.33)$$

where D is the pipe diameter and C is the volume fraction of coarse solids, and k_2 is a constant to be determined from experimental results. Although, in equation (4.30), C should be the in-situ concentration but in view of the fact that such measurements are generally not available it is customary to replace it with the concentration of solids in the discharged mixture.

From equation (4.33), it would be expected that in the moving bed regime, $f_L (i - i_L) / i_L$ would vary linearly with the concentration C in the discharged mixture. Chhabra and Richardson [1985] found that the following modified correlation represented their own data and those of Kenchington [1978] (who transported 750 μm sand particles in 13 and 25 mm pipes in a kaolin suspension) somewhat better than equation (4.33) as seen in Figure 4.20:

$$\left(\frac{i - i_L}{C i_L} \right) f_L = 0.55 \left[\frac{gD(s - 1)}{V^2} \right]^{1.25} \qquad (4.34)$$

There are insufficient reliable results in the literature for expressions to be given for the pressure gradients in other flow regimes.

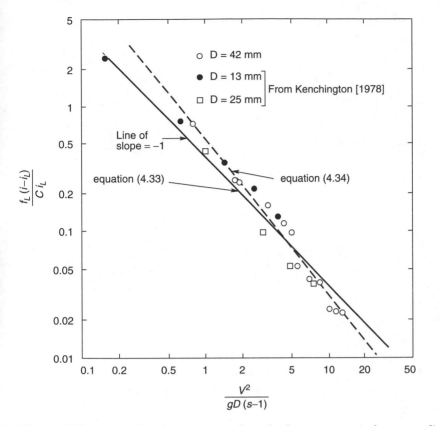

Figure 4.20   *Overall representation of results for transport in heavy media*

Example 4.2

A china clay slurry ($m = 9.06\,\text{Pa·s}^n$; $n = 0.19$, $\rho_L = 1210\,\text{kg/m}^3$) is used to transport 5 mm gravel particles (nearly spherical) of density $2700\,\text{kg/m}^3$ at a mean mixture velocity of 1.25 m/s in a horizontal pipe of 50 mm diameter. If transport is in the sliding-bed regime and the discharged mixture contains 22% (by volume) gravel particles. Estimate the pressure gradient for the mixture flow.

Solution

First, the value of the friction factor f_L for the flow of liquid alone at the same average velocity is estimated. The Reynolds number is given by eq. (3.8b):

$$\text{Re}_{MR} = \frac{\rho_L V^{2-n} D^n}{8^{n-1} m \left(\dfrac{3n+1}{4n} \right)^n} = \frac{1210 \times (1.25)^{2-0.19} (50 \times 10^{-3})^{0.19}}{(8^{0.19-1})(9.06) \left(\dfrac{3 \times 0.19 + 1}{4 \times 0.19} \right)^{0.19}} = 532$$

As the flow of the china clay slurry is laminar,

$$f = \frac{16}{\text{Re}_{MR}} = \frac{16}{532} = 0.03$$

$$i_L = \frac{2f_L V^2}{gD} = \frac{2 \times 0.03 \times 1.25^2}{9.81 \times 50 \times 10^{-3}}$$

$$= 0.192 \text{ m of china clay slurry per m of pipe}$$

Substituting values in equation (4.34),

$$i = i_L + 0.55 \frac{C i_L}{f_L} \left[\frac{gD(s-1)}{V^2} \right]^{1.25}$$

$$= 0.192 + \frac{0.55 \times 0.22 \times 0.192}{0.03}$$

$$\times \left[\frac{9.81 \times 50 \times 10^{-3} \left(\frac{2700}{1210} - 1 \right)}{1.25^2} \right]^{1.25}$$

$$= 0.192 + 0.236 = 0.428 \text{ m of china clay slurry/m}$$

or $\quad \left(-\dfrac{\Delta p}{L} \right) = i \times \rho_L \times g = 0.428 \times 1210 \times 9.81 = 5080 \, \text{Pa/m}$

Note that about (0.236/0.428), i.e. 55% of the total pressure drop is attributable to the presence of coarse gravel particles which in this flow regime is independent of the gravel size.□

4.4 Further reading

Brown, N.P. and Heywood, N.I., (Eds) *Slurry Handling: Design of Solid–Liquid Systems.* Elsevier, London (1991).
Chhabra, R.P., in *Civil Engineering Practice* (edited by Cheremisinoff, P.N. Cheremisinoff, N.P. and Cheng, S.L. Technomic, Lancaster, PA) **2** (1988) 251.
Chhabra, R.P. and Richardson, J.F., in *Encyclopedia of Fluid Mechanics* (edited by Cheremisinoff, N.P. Gulf, Houston)˙**3** (1986) 563.
Govier, G.W. and Aziz, K., *The Flow of Complex Mixtures in Pipes.* Krieger, Malabar, FL (1982).
Mahalingam, R., *Adv. Transport Process.* **1** (1980) 58.
McKetta, J.J. (Ed.) *Piping Design Handbook.* Marcel–Dekker, New York (1992).
Shook, C.A. and Roco, M.C., *Slurry Flow: Principles and Practice.* Butterworth-Heinemann, Stoneham, MA (1991).

4.5 References

Alves, G.E., *Chem. Eng. Prog.* **50** (1954) 449.
Barnea, D. and Taitel, Y., in *Encyclopedia of Fluid Mechanics.* (edited by Cheremisinoff, N.P. Gulf, Houston) **3** (1986) Chapter 17.

Bishop, A.A. and Deshpande, S.D., *Int. J. Multiphase Flow.* **12** (1986) 977.

Carleton, A.J., Cheng, D.C.-H. and French, R.J., *Pneumotransport 2. (BHRA Fluid Eng.*, Bedford, UK) paper F-2 (1973).

Charles, M.E. and Charles, R.A., in *Advances in Solid–Liquid Flow and its Applications.* (edited by Zandi, I. Pergamon, Oxford) (1971) 187.

Chen, J.J. and Spedding, P.L., *Int. J. Multiphase Flow.* **9** (1983) 147.

Chhabra, R.P., Farooqi, S.I. and Richardson, J.F., *Chem. Eng. Res. Des.* **62** (1984) 22.

Chhabra, R.P., Farooqi, S.I. Richardson, J.F. and Wardle, A.P., *Chem. Eng. Res. Des.* **61** (1983) 56.

Chhabra, R.P. and Richardson, J.F., *Can. J. Chem. Eng.* **62** (1984) 449.

Chhabra, R.P. and Richardson, J.F., *Chem. Eng. Res. Des.* **63** (1985) 390.

Chhabra, R.P. and Richardson, J.F., in *Encyclopedia of Fluid Mechanics.* (edited by Cheremisinoff, N.P. Gulf, Houston) **3** (1986) 563.

Chisholm, D., *Int. J. Heat Mass Transf.* **10** (1967) 1767.

Chisholm, D., *Two Phase Flow in Pipelines and Heat Exchangers.*, George Goodwin, London (1983).

Duckworth, R.A., Pullum, L and Lockyear, C.F., *J. Pipelines.* **3** (1983) 251.

Duckworth, R.A., Pullum, L, Addie, G.R. and Lockyear, C.F., *Hydrotransport 10, BHR Group.* Bedford (1986), Paper # C2.

Dziubinski, M., *Chem. Eng. Res. Des.* **73** (1995) 528.

Dziubinski, M. and Chhabra, R.P., *Int. J. Eng. Fluid Mech.* **2** (1989) 63.

Dziubinski, M. and Richardson, J.F., *J. Pipelines.* **5** (1985) 107.

Eissenberg, F.G. and Weinberger, C.B., *AICEJ.* **25** (1979) 240.

Farooqi, S.I., *PhD Thesis*, University of Wales (1981).

Farooqi, S.I., Heywood, N.I. and Richardson, J.F., *Trans. Inst. Chem. Engrs.* **58** (1980) 16.

Farooqi, S.I. and Richardson, J.F., *Trans. Inst. Chem. Engrs.* **60** (1982) 292 & 323.

Ferguson, M.E.G. and Spedding, P.L., *J. Chem. Tech. Biotechnol.* **62** (1995) 262.

Fossa, M., *Flow Meas. & Instrumentation* **9** (1998) 103.

Ghosh, T. and Shook, C.A., in *Freight Pipelines.* (edited by Liu, H. and Round, G.F. Hemisphere, New York) p. 281 (1990).

Govier, G.W. and Aziz, K., *The Flow of Complex Mixtures in Pipes.* Krieger, R.E. Malabar, FL (1982).

Gregory, G.A. and Mattar, L., *J. Can. Pet. Tech.* 12 No. **2** (1973) 48.

Hetsroni, G., (Ed.) *Handbook of Multiphase Systems*, McGraw Hill, New York (1982).

Hewitt, G.F., *Measurement of Two-Phase Flow Parameters.* Academic, New York (1978).

Hewitt, G.F., in *Handbook of Multiphase Systems.* (edited by Hetsroni, G., McGraw Hill, New York) (1982) p. 2–25.

Hewitt, G.F., King, I. and Lovegrove, P.C., *Brit. Chem. Eng.* **8** (1963) 311.

Heywood, N.I. and Charles, M.E., *Int. J. Multiphase Flow.* **5** (1979) 341.

Heywood, N.I. and Richardson, J.F., *Chem, Eng, Sci,* **34** (1979) 17.

Kaminsky, R.D., *J. Energy Res. Technol. (ASME)* **120** (1998) 2.

Kenchington, J.M., *Hydrotransport 5, BHR Group.* Bedford, U.K., (1978) Paper # D7.

Khatib, Z. and Richardson, J.F., *Chem. Eng. Res. Des.* **62** (1984) 139.

Lockhart, R.W. and Martinelli, R.C., *Chem. Eng. Prog.* **45** (1949) 39.

Mahalingam, R., *Adv. Transport Process.* **1** (1980) 58.

Mahalingam, R. and Valle, M.A., *Ind. Eng. Chem. Fundam.* **11** (1972) 470.

Mandhane, J.M., Gregory, G.A. and Aziz, K., *Int. J. Multiphase Flow.* **1** (1974) 537.

Mandhane, J.M., Gregory, G.A. and Aziz, K., *J. Pet. Technol.* **27** (1975) 1017.

Newitt, D.M., Richardson, J.F. Abbott, M. and Turtle, R.B., *Trans. Inst. Chem. Engrs.* **33** (1955) 93.

Oliver, D.R. and Young–Hoon, A., *Trans. Inst. Chem. Engrs.* **46** (1968) 106.

Petrick, P. and Swanson, B.S., *Rev. Sci. Instrum.* **29** (1958) 1079.

Pike, R.W., Wilkins, B. and Ward, H.C., *AlChEJ.* **11** (1965) 794.

Rao, B.K., *Int. J. Heat Fluid Flow* **18** (1997) 559.

Raut, D.V. and Rao, M.N., *Indian J. Technol.* **13** (1975) 254.

Shook, C.A. and Liebe, J.O., *Can. J. Chem. Eng.* **54** (1976) 118.

Shu, M.T., Weinberger, C.B. and Lee, Y.H., *Ind. Eng. Chem. Fundam.* **21** (1982) 175.

Spedding, P.L. and Chen, J.J., in *Encyclopedia of Fluid Mechanics.* (edited by Cheremisinoff, N.P. Gulf, Houston) **3** (1986) Chapter 18.

Taitel, Y., Barnea, D. and Dukler, A.E., *AlChEJ.* **26** (1980) 345.

Weisman, J., Duncan, D. Gibson, J. and Crawford, T., *Int. J. Multiphase Flow.* **5** (1979) 437.

4.6 Nomenclature

		Dimensions in **M, L, T**
C	Volumetric concentration of solids in discharged mixture (–)	$M^0L^0T^0$
C_o	Constant, equation (4.19) (–)	$M^0L^0T^0$
C_v	Volume fraction of solid in suspension (–)	$M^0L^0T^0$
D	Pipe diameter (m)	L
De	Deborah number (–)	$M^0L^0T^0$
f	Fanning friction factor (–)	$M^0L^0T^0$
g	acceleration due to gravity (m/s^2)	LT^{-2}
i	pressure gradient (m of liquid/m of pipe length)	$M^0L^0T^0$
J	correction factor (–)	$M^0L^0T^0$
L	length of pipe (m)	L
m	power-law consistency coefficient (Pa·s^n)	$ML^{-1}T^{n-2}$
m_1	power-law coefficient for first normal stress difference (Pa.s^{p_1})	$ML^{-1}T^{p'-2}$
m'	apparent power-law consistency coefficient (Pa.s$^{n'}$)	$ML^{-1}T^{p'-2}$
n	power-law flow behaviour index (–)	$M^0L^0T^0$
n'	apparent power-law index (–)	$M^0L^0T^0$
N	power (W)	ML^2T^{-3}
N_1	first normal stress difference (Pa)	$ML^{-1}T^{-2}$
p_1	index in first normal stress difference, equation (4.12) (–)	$M^0L^0T^0$
$-\Delta p/L$	pressure gradient (Pa/m)	$ML^{-2}T^{-2}$
Q	volumetric flow rate (m^3/s)	L^3T^{-1}
Re_{MR}	Metzner–Reed Reynolds number (–)	$M^0L^0T^0$
s	specific gravity of solids (–)	$M^0L^0T^0$
V	superficial velocity (m/s)	LT^{-1}

Greek letters

α	average holdup (–)	$M^0L^0T^0$
γ	coefficient of friction (–)	$M^0L^0T^0$
$\dot{\gamma}$	shear rate(s^{-1})	T^{-1}
λ	input fraction (–)	$M^0L^0T^0$
λ_f	fluid characteristic time, equation (4.13) (s)	T
Λ	loss coefficient (–)	$M^0L^0T^0$
ρ	density (kg/m^3)	ML^{-3}

		Dimensions in **M, L, T**
μ	viscosity (Pa·s)	$\mathbf{ML^{-1}T^{-1}}$
ϕ^2	drag ratio (−)	$\mathbf{M^0L^0T^0}$
χ	Lockhart−Martinelli parameter (−)	$\mathbf{M^0L^0T^0}$
ψ	coefficient, equation (4.29) (−)	$\mathbf{M^0L^0T^0}$
τ_w	wall shear stress (Pa)	$\mathbf{ML^{-1}T^{-2}}$

Subscripts

a	accelerational contribution
f	frictional contribution
G	gas
g	gravitational
L	liquid
L_c	corresponding to liquid Reynolds number of 2000
Lv	visco-elastic liquid
M	mixture
MDR	maximum drag reduction
min	minimum
mod	modified
s	Solid
TP	two-phase

Chapter 5
Particulate systems

5.1 Introduction

In many practical applications, we need to know the force required to move a solid object through a surrounding fluid, or conversely, the force that a moving fluid exerts on a solid as the fluid flows past it. Many processes for the separations of particles of various sizes, shapes and materials depend on their behaviour when subjected to the action of a moving fluid. Frequently, the liquid phase may exhibit complex non-Newtonian behaviour whose characteristics may be measured using falling-ball viscometry. Furthermore, it is often necessary to calculate the fluid dynamic drag on solid particles in process equipment, for example for slurry pipelines, fixed and fluidised beds. Similarly, in the degassing of polymer melts prior to processing, bubbles rise through a still mass of molten polymer. Likewise, the movement of oil droplets and polymer solutions in narrow pores (albeit strongly influenced by capillary forces) occurs in enhanced oil recovery operations. The settling behaviour of a particle is also strongly influenced by the presence of other neighbouring particles as in concentrated suspensions. Furthermore, it is often desirable to keep the active component uniformly suspended, as in many pharmaceutical products, paints, detergents, agro-chemicals, emulsions and foams.

Frequently, the particles are in the form of clusters (such as in fixed and some fluidised beds) and ensembles as in foams, dispersions and emulsions. However, experience with Newtonian fluids has shown that the hydrodynamics of systems consisting of single particles, drops and bubbles serves as a useful starting point for understanding the mechanics of the more complex multi-particle systems which are not amenable to rigorous analysis. This chapter aims to provide an overview of the developments in the field of non-Newtonian fluid–particle systems. In particular, consideration is given to the drag force, wall effects and settling velocity of single spherical and non-spherical particles, bubbles and drops in various types of non-Newtonian fluids (particularly based on power-law and Bingham plastic models). Flow in packed and fluidized beds and hindered settling are then considered. Detailed accounts and extensive bibliographies on this subject are available elsewhere [Chhabra, 1986, 1993a,b; Ghosh *et al.*, 1994].

Unlike the flows considered in Chapter 3 which were essentially unidirectional, the fluid flows in particulate systems are either two- or three-dimensional and hence are inherently more difficult to analyse theoretically, even in the creeping (small Reynolds number) flow regime. Secondly, the results are often dependent on the rheological model appropriate to the fluid and a more generalised treatment is not possible. For instance, there is no 'standard non-Newtonian drag curve' for spheres, and the relevant dimensionless groups depend on the fluid model which is used. Most of the information in this chapter relates to time-independent fluids, with occasional reference to visco-elastic fluids.

5.2 Drag force on a sphere

All bodies immersed in a fluid are subject to a buoyancy force. In a flowing fluid (or in the situation of relative velocity between the fluid and the object), there is an additional force which is made up of two components: the skin friction (or viscous drag) and the form drag (due to the pressure distribution) as shown schematically in Figure 5.1. At low velocities, no separation of the boundary layer takes place, but as the velocity is increased, separation occurs and the skin friction forms a gradually decreasing proportion of the total fluid dynamic drag on the immersed object. The conditions of flow over a spherical particle are characterised by the Reynolds number; the exact form of the latter depends upon the rheological model of the fluid.

The total drag force is obtained by integrating the components of the forces attributable to skin friction and form drag in the direction of fluid motion (z-coordinate) that act on an elemental area of the surface of the sphere, i.e.

$$F_D \quad = \quad F_t \quad + \quad F_n \tag{5.1}$$

(total drag   (skin friction   (form drag
   force)   force)   force)

The total drag force, $F_D = F_t + F_n$, is often expressed using a dimensionless drag coefficient C_D as

$$C_D = \frac{F_D}{\left(\frac{1}{2}\rho V^2\right)(\pi R^2)} \tag{5.2}$$

The shear stress and pressure distributions necessary for the evaluation of the integrals implicit in equation (5.1) can, in principle, be obtained by solving the continuity and momentum equations. In practice, however, numerical solutions are often necessary even at low Reynolds numbers. Since detailed discussions of this subject are available elsewhere, [Chhabra, 1993a] only the significant results are presented here for power-law and viscoplastic fluid models.

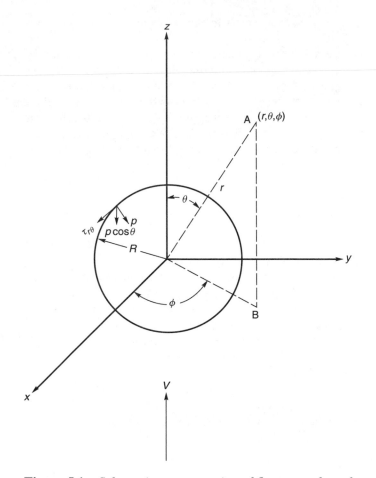

Figure 5.1   *Schematic representation of flow around a sphere*

5.2.1 Drag on a sphere in a power-law fluid

A simple dimensional analysis (see example 5.1) of this flow situation shows that the drag coefficient can be expressed in terms of the Reynolds number and the power-law index, i.e.

$$C_D = f(\text{Re}, n) \tag{5.3}$$

Often for the creeping flow region ($\text{Re} \ll 1$), the numerical results may be expressed as a deviation factor, $X(n)$, in the relation between drag coefficient and Reynolds number obtained from Stokes law

$$C_D = \frac{24}{\text{Re}} X(n) \tag{5.4}$$

where $Re = \rho V^{2-n} d^n / m$, d being the sphere diameter. The numerical values of $X(n)$ for both shear-thinning and shear-thickening fluid behaviour are listed in Table 5.1 [Gu and Tanner, 1985; Tripathi *et al.*, 1994; Tripathi and Chhabra, 1995]. Evidently, shear-thinning causes drag increase (X > 1), and drag reduction (X < 1) occurs in shear-thickening ($n > 1$) fluids.

Table 5.1 *Values of X(n) for a sphere*

n	X(n)
1.8	0.261
1.6	0.390
1.4	0.569
1.2	0.827
1.0	1.002
0.9	1.14
0.8	1.24
0.7	1.32
0.6	1.382
0.5	1.42
0.4	1.442
0.3	1.458
0.2	1.413
0.1	1.354

Based on detailed analysis of experimental results and numerical simulations, the creeping flow occurs in shear-thinning fluids up to about $Re \sim 1$, even though a visible wake appears only when Re for the sphere reaches the value of 20 [Tripathi *et al.*, 1994].

In the case of creeping sphere motion in Newtonian fluids, the skin and form friction contributions are in the proportion of 2 to 1. This ratio continually decreases with increasing degree of pseudoplasticity, the two components becoming nearly equal at $n \sim 0.4$; these relative proportions, on the other hand, increase with increasing degree of shear-thickening behaviour.

Numerical predictions of drag on a sphere moving in a power-law fluid are available for the sphere Reynolds number up to 130 [Tripathi *et al.*, 1994; Graham and Jones, 1995] and the values of drag coefficient are best represented by the following expressions with a maximum error of 10% for shear-thinning fluids [Graham and Jones, 1995]:

$$C_D = \frac{35.2(2)^n}{Re^{1.03}} + n \left\{ 1 - \frac{20.9(2)^n}{Re^{1.11}} \right\} \quad 0.2(2)^n \leq Re \leq 24(2)^n \quad (5.5a)$$

$$C_D = \frac{37(2)^n}{\text{Re}^{1.1}} + 0.25 + 0.36\,n \qquad\qquad 24(2)^n \leq \text{Re} \leq 100(2)^n \qquad (5.5b)$$

Experimental results for the drag on spheres are now available for Reynolds numbers up to about 1000 and for the power-law index in the range $0.38 \leq n \leq 1$ [Chhabra, 1990]. Extensive comparisons between numerical simulations and experimental data show rather poor agreement in the creeping flow regime (Re < 1) but this improves somewhat as the Reynolds number increases. While the exact reasons for these discrepancies are not known, they have often been attributed to possible visco-elastic behaviour and the choice of inappropriate values of the power-law constants. On the other hand, the standard drag curve for Newtonian fluids correlates the drag results in power-law liquids in the region $1 \leq \text{Re} \leq 1000$ within $\pm 30\%$, as seen in Figure 5.2; this is comparable with the agreement between predictions and data in the range $1 \leq \text{Re} \leq 100$, and confirms that the effect of non-Newtonian viscosity generally diminishes with increasing Reynolds number.

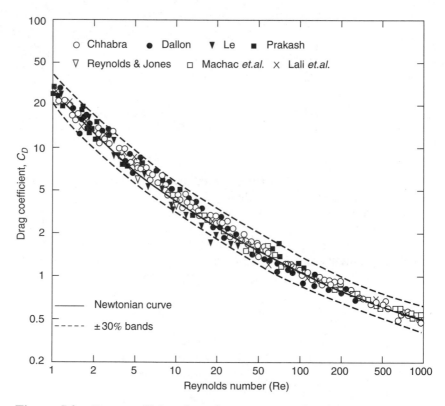

Figure 5.2   *Drag coefficient for spheres in power-law liquids (see Chhabra [1990] for original sources of data)*

Example 5.1

A sphere of diameter d moving at a constant velocity V through a power-law fluid (density ρ, flow index, n and consistency coefficient, m) experiences a drag force, F_D. Obtain the pertinent dimensionless groups of variables.

Solution

There are six variables and three fundamental dimensions (**M, L, T**), and therefore there will be three dimensionless groups. Thus, one can write

$$F_D = \mathrm{f}(\rho, d, m, n, V)$$

Writing dimensions of each of these variables:

$F_D \equiv \mathbf{MLT}^{-2}$ $m \equiv \mathbf{ML}^{-1}\mathbf{T}^{n-2}$

$\rho \equiv \mathbf{ML}^{-3}$ $n \equiv \mathbf{M}^0\mathbf{L}^0\mathbf{T}^0$

$d \equiv \mathbf{L}$ $V \equiv \mathbf{LT}^{-1}$

Choosing ρ, d, V as the recurring set, the fundamental dimensions **M, L** and **T** can be expressed as:

$$\mathbf{L} \equiv d; \quad \mathbf{M} \equiv \rho d^3; \quad \mathbf{T} \equiv d/V$$

and the three π-groups can be formed as:

$F_D/(\rho d^3)(d)(V/d)^2$,   i.e. $F_D/\rho V^2 d^2$

$m/(\rho d^3)(d^{-1})(d/V)^{n-2}$,   i.e. $\dfrac{\rho V^{2-n} d^n}{m}$

and n. Therefore,

$$\frac{F_D}{\rho V^2 d^2} = \mathrm{f}\left(\frac{\rho V^{2-n} d^n}{m}, n\right)$$

By inspection, $(F_D/\rho V^2 d^2) \propto C_D$
∴ $C_D = \mathrm{f}(\mathrm{Re}, n)$ which is the same relationship as given by equation (5.3).□

5.2.2 Drag on a sphere in viscoplastic fluids

By virtue of its yield stress, a viscoplastic material in an unsheared state will support an immersed particle for an indefinite period of time. In recent years, this property has been successfully exploited in the design of slurry pipelines, as briefly discussed in section 4.3. Before undertaking an examination of the drag force on a spherical particle in a viscoplastic medium, the question of static equilibrium will be discussed and a criterion will be developed to delineate the conditions under which a sphere will either settle or be held stationary in a liquid exhibiting a yield stress.

(i) Static equilibrium

The question of whether or not a sphere will settle in an unsheared viscoplastic material has received considerable attention in the literature [Chhabra and Uhlherr, 1988; Chhabra, 1993a]. For the usual case where the sphere is acted upon by gravity, it is convenient to introduce a dimensionless group, Y, which denotes the ratio of the forces due to the yield stress and due to gravity. Neglecting numerical constants, the simplest definition of Y is

$$Y = \frac{\tau_0}{gd(\rho_s - \rho)} \tag{5.6}$$

Thus, small values of Y will favour motion of a sphere. The critical values of Y reported by various investigators [Chhabra and Uhlherr, 1988] fall in two categories. One group, with the value of Y in the range 0.06 ± 0.02, includes the numerical predictions [Beris *et al.* 1985], observations on the motion/no motion of spheres under free fall conditions [Ansley and Smith, 1967] and the residual force upon the cessation of flow [Brookes and Whitmore, 1968]. The second group, with $Y \sim 0.2$, relies on the intuitive consideration that the buoyant weight of the sphere is supported by the vertical component of the force due to the yield stress, and on measurements on a fixed sphere held in an unsheared viscoplastic material [Uhlherr, 1986]. The large discrepancy between the two sets of values suggests that there is a fundamental difference in the underlying mechanisms inherent in these two approaches. Additional complications arise from the fact that the values of yield stress (τ_0) obtained using different methods differ widely [Nguyen and Boger, 1992]. Thus, it is perhaps best to establish the upper and lower bounds on the size and/or density of a sphere that will settle in particular circumstances.

(ii) Flow field

As in the case of the solid plug-like motion of viscoplastic materials in pipes and slits (discussed in Chapter 3), there again exists a bounded zone of flow associated with a sphere moving in a viscoplastic medium, and beyond this zone, the fluid experiences elastic deformation, similar to that in elastic solids [Volarovich and Gutkin, 1953; Tyabin, 1953]. Indeed the difficulty in delineating the interface between the flow and no flow zones has been the main impediment to obtaining numerical solutions to this problem. Furthermore, even within the cavity of shear deformation, there is unsheared material adhering to parts of the sphere surface and this suggests that the yield stress may not act over the entire surface. The existence of such unsheared material attached to a moving sphere has been observed experimentally, but it is rather difficult to estimate its exact shape and size [Valentik and Whitmore, 1965; Atapattu *et al.*, 1995].

Using the laser speckle photographic method, Atapattu *et al.* [1995] measured point velocities in the fluid near a sphere moving at a constant speed

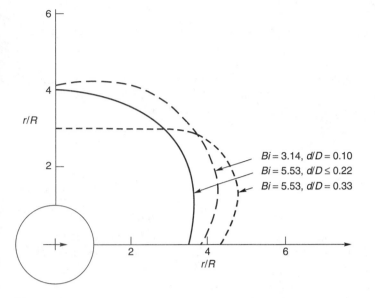

Figure 5.3   *Size of sheared cavity around a sphere moving in a viscoplastic (aqueous carbopol) solution*

on the axis of a cylindrical tube containing viscoplastic carbopol solutions. Notwithstanding the additional effects arising from the walls of the tube, Figure 5.3 shows the typical size and shape of deformation cavity for a range of values of the sphere to tube diameter ratio and the Bingham number, Bi ($= \tau_0^B d / V\mu_B$). The slight difference between the size of cavity in the radial and axial directions should be noted, especially for large values of sphere to tube diameter ratio (d/D), but the deformation envelope rarely extends beyond 4–5 sphere radii. Nor has it been possible to identify small caps of solid regions near the front and rear stagnation points.

Example 5.2

A china clay suspension has a density of 1050 kg/m³ and a yield stress of 13 Pa. Determine the diameter of the smallest steel ball (density 7750 kg/m³) which will settle under its own weight in this suspension.

Solution

Here $\rho = 1050 \, \text{kg/m}^3$; $\rho_s = 7750 \, \text{kg/m}^3$

$\tau_0 = 13 \, \text{Pa}$

From equation (5.6), the sphere will settle only if $Y < \sim 0.04 - 0.05$

Substituting values, $\dfrac{\tau_0}{gd(\rho_s - \rho)} = \dfrac{13}{9.81 \times d \times (7750 - 1050)} \le 0.04$

or $d = 4.9\,\text{mm}.$

For a less conservative estimation, $Y = 0.212$ may be used. The use of this criterion gives $d = 0.93\,\text{mm}$. Thus, a 5 mm sphere will definitely settle in this suspension, but there is an element of uncertainty about the 1 mm steel ball.□

(iii) Drag force

The main difficulty in making theoretical estimates of the drag force on a sphere moving in a viscoplastic medium has been the lack of quantitative information about the shape of the sheared cavity. Both Beris *et al.* [1985] and Blackery and Mitsoulis [1997] have used the finite element method to evaluate the total drag on a sphere moving slowly (creeping regime) in a Bingham plastic medium and have reported their predictions in terms of the correction factor, X, $(= C_D \text{Re}_B/24)$ which now becomes a function of the Bingham number, Bi $(= \tau_0^B d/V\mu_B)$ as:

$$X = 1 + a(\text{Bi})^b \tag{5.7}$$

While Beris *et al.* [1985] evaluated the drag in the absence of walls (i.e. $d/D = 0$), Blackery and Mitsoulis [1997] have numerically computed the value of X for a range of diameter ratios $0 \le d/D \le 0.5$ and up to Bi $= 1000$. For the case of $(d/D) = 0$ (i.e. no wall effects), $a = 2.93$ and $b = 0.83$. In the range $0 \le (d/D) \le 0.5$, the values of a and b vary monotonically in the ranges $1.63 \le a \le 2.93$ and $0.83 \le b \le 0.95$, respectively. As the Bingham number progressively becomes smaller, X would be expected to approach unity. The higher drag $(X > 1)$ in a viscoplastic medium is attributable to the additive effects of viscosity and yield stress.

In addition, many workers have reported experimental correlations of their drag data for spheres falling freely or being towed in viscoplastic media [Chhabra and Uhlherr, 1988; Chhabra, 1993a; Atapattu *et al.*, 1995]; most correlations are based on the use of the Bingham model, though some have found the three parameter Herschel–Bulkley fluid model (equation 1.17) to correlate their data somewhat better [Sen, 1984; Atapattu *et al.*, 1995; Beaulne and Mitsoulis, 1997]. At the outset, it is important to establish the criterion for creeping flow in viscoplastic fluids. For a sphere falling in a Newtonian fluid $(\tau_0 = 0)$, the creeping flow is assumed to occur up to about Re ~ 1. One of the characteristics of creeping flow in a Newtonian fluid is the reciprocal relationship between the Reynolds number and drag coefficient, i.e. $C_D \text{Re} = 24$. For Bingham plastic fluids, intuitively this product must be a function of the Bingham number, as can be seen in equation (5.7). Applying this criterion to the available data, the maximum value of the Reynolds number,

Re $(= \rho V d / \mu_B)$, for creeping flow is given as [Chhabra and Uhlherr, 1988]:

$$\text{Re}_{max} \simeq 100 \, \text{Bi}^{0.4} \tag{5.8}$$

Thus, the greater the Bingham number, the higher is the Reynolds number up to which the creeping flow conditions apply for spheres moving in Bingham plastic fluids.

As mentioned previously, the three parameter Herschel–Bulkley fluid model gives a somewhat better fit of the fluid rheology than the Bingham model. Atapattu *et al.* [1995] put forward the following semi-empirical correlation for drag on spheres in Herschel–Bulkley model liquids:

$$C_D = \frac{24}{\text{Re}}(1 + \text{Bi}^*) \tag{5.9}$$

where the Reynolds number, $\text{Re} = \rho V^{2-n} d^n / m$ and the modified Bingham number, $\text{Bi}^* = \tau_0^H / m (V/d)^n$. Equation (5.9) covers the ranges: $10^{-5} \leq \text{Re} \leq 0.36$; $0.25 \leq \text{Bi}^* \leq 280$; and $0.43 \leq n \leq 0.84$; and it also correlates the scant literature data available in the creeping flow region [Sen, 1984; Hariharaputhiran *et al.*, 1998]. These results are also in line with the numerical predictions for Herschel–Bulkley fluids [Beaulne and Mitsoulis, 1997].

In the intermediate Reynolds number region, though some predictive expressions have been developed, e.g. see Chhabra [1993a] but most of these data are equally well in line with the standard drag curve for Newtonian liquids [Machac *et al.*, 1995].

Thus, in summary, the non-Newtonian characteristics seem to be much more important at low Reynolds numbers and their role progressively diminishes as the inertial effects become significant with the increasing Reynolds number. Therefore, in creeping flow region, equations (5.4), (5.7) and (5.9), respectively, should be used to estimate drag forces on spheres moving in power-law and Bingham model or Herschel–Bulkley fluids. On the other hand, at high Reynolds number, the application of the standard drag curve for Newtonian fluids yields values of drag on spheres which are about as accurate as the empirical correlations available in the literature. The Reynolds number defined as $\rho V^{2-n} d^n / m$ for power-law fluids, as $\rho V d / \mu_B$ for Bingham plastics and as $(\rho V^{2-n} d^n / m)/(1 + \text{Bi}^*)$ for Herschel–Bulkley model fluids must be used in the standard Newtonian drag curve.

5.2.3 Drag in visco-elastic fluids

From a theoretical standpoint, the creeping-flow steady translation motion of a sphere in a visco-elastic medium has been selected as one of the benchmark problems for the validation of procedures for numerical solutions [Walters and Tanner, 1992; Chhabra, 1993a]. Unfortunately, the picture which emerges is not only incoherent but also inconclusive. Most simulation studies are based on the creeping flow assumption (zero Reynolds number) and take into account the influence of fluid visco-elasticity on the drag of a sphere in the absence of shear-thinning behaviour. Early studies suggested a slight reduction ($\sim$5–10%) in drag below the Stokes value, with the amount of drag reduction showing a weak dependence on Deborah or Weissenberg number (defined as $\lambda_f V/d$). However, more recent simulations [Degand and Walters, 1995] suggest that after an initial period of reduction, the drag on a sphere in a visco-elastic medium can exceed that in a Newtonian medium at high values of Deborah number; the latter enhancement is attributed to extensional effects of the fluid. Both drag reduction (up to 25%) and enhancements (up to 200%) compared with the Newtonian value have been observed experimentally [Chhabra, 1993a]. However there is very little quantitative agreement among various workers between the results of numerical simulations and experimental studies. The former seem to be strongly dependent on the details of the numerical procedure, mesh size, etc, while the experimental results appear to be very sensitive to the chemical nature, water purity, etc. of the polymer solutions used. It is not yet possible to interpret and/or correlate experimental results of drag in visco-elastic fluids in terms of measureable rheological properties. Aside from these uncertainties, other time-dependent effects have also been observed. For instance, unlike the monotonic approach to the terminal velocity in Newtonian and power-law type fluids [Bagchi and Chhabra, 1991; Chhabra *et al.*, 1998], a sphere released in a visco-elastic liquid could attain a transitory velocity almost twice that of its ultimate falling velocity [Walters and Tanner, 1992].

On the other hand, the effects of shear-thinning viscosity completely overshadow those of visco-elasticity, at least in the creeping flow region. Indeed, a correlation based on a viscosity model, with zero shear viscosity and/or a characteristic time constant, provides satisfactory representation of drag data when the liquid exhibits both shear-thinning properties and visco-elasticity [Chhabra, 1993a].

5.2.4 Terminal falling velocities

In many process design calculations, it is necessary to know the terminal velocity of a sphere settling in a fluid under the influence of the gravitational field. When a spherical particle at rest is introduced into a liquid, it accelerates until the buoyant weight is exactly balanced by the fluid dynamic drag. Although the so-called terminal velocity is approached asymptotically, the effective transition period is generally of short duration for Newtonian and power-law fluids [Chhabra *et al.*, 1998]. For instance, in the creeping flow regime, the terminal velocity is attained after the particle has traversed a path of length equal to only a few diameters.

For gravity settling of a sphere at its terminal velocity the drag force on it, F_D, is equal to the buoyant weight, i.e.

$$F_D = \frac{\pi d^3}{6}(\rho_s - \rho)g \tag{5.10}$$

Combining equations (5.4) and (5.10), the terminal velocity of a sphere in a power-law fluid (Re < 1):

$$V = \left[\frac{gd^{n+1}(\rho_s - \rho)}{18mX}\right]^{(1/n)} \tag{5.11}$$

In shear-thinning power-law fluids, therefore, the terminal falling velocity shows a stronger dependence on sphere diameter and density difference than in a Newtonian fluid.

This method of calculation is satisfactory provided it is known a priori that the Reynolds number is small (<1). As the unknown velocity appears in both the Reynolds number and the drag coefficient, it is more satisfactory to work in terms of a new dimensionless group, Ar, the so-called Archimedes number defined by:

$$\text{Ar} = C_D\text{Re}^{2/(2-n)} = \frac{4}{3}gd^{(2+n)/(2-n)}(\rho_s - \rho)\rho^{n/(2-n)}m^{2/(n-2)} \tag{5.12}$$

For any given sphere and power-law liquid combination, the value of the Archimedes number can be evaluated using equation (5.12). The sphere Reynolds number can then be expressed in terms of Ar and n as follows:

$$\text{Re} = a\text{Ar}^b \tag{5.13}$$

$$a = 0.1\left[\exp\left(\frac{0.51}{n}\right) - 0.73n\right] \tag{5.14}$$

$$b = \frac{0.954}{n} - 0.16 \tag{5.15}$$

The values calculated from equations (5.13) to (5.15) represent about 400 data points in visco-inelastic fluids ($0.4 \leq n < 1$; $1 \leq \text{Re} \leq 1000$; $10 \leq \text{Ar} \leq 10^6$) with an average error of 14% and a maximum error of 21%. Finally, in view of the fact that non-Newtonian characteristics exert little influence on the drag, the use of predictive correlations for terminal falling velocities in Newtonian media yields only marginally larger errors for power-law fluids. Finally, attention is drawn to the fact that the estimation of terminal velocity in viscoplastic liquids requires an iterative solution, as illustrated in example 5.4.

Example 5.3

For spheres of equal terminal falling velocities, obtain the relationship between diameter and density difference between particle and fluid for creeping flow in power law fluids.

Solution

From equation (5.11), the terminal settling velocity of a sphere increases with both its density and size. For two spheres of different diameters d_A, d_B and densities, ρ_{SA} and ρ_{SB}, settling in the same fluid, the factor X is a function of n only (see Table 5.1) and

$$\frac{V_A}{V_B} = \left[\frac{d_A^{n+1}(\rho_{SA} - \rho)}{d_B^{n+1}(\rho_{SB} - \rho)}\right]^{(1/n)}$$

Thus for pseudoplastic fluids ($n < 1$), the terminal velocity is more sensitive to both sphere diameter and density difference than in a Newtonian fluid and it should, in principle, be easier to separate closely sized particles. For equal settling velocities,

$$\frac{d_B}{d_A} = \left(\frac{\rho_{SA} - \rho}{\rho_{SB} - \rho}\right)^{1/(n+1)}$$

For $n = 1$, this expression reduces to its Newtonian counterpart.□

Example 5.4

Estimate the terminal settling velocity of a 3.18 mm steel sphere (density = 7780 kg/m³) is a viscoplastic polymer solution of density 1000 kg/m³. The flow curve for the polymer solution is approximated by the three parameter Herschel–Bulkley model as:

$$\tau = 3.3 + 3.69(\dot{\gamma})^{0.53}$$

The settling may be assumed to occur in creeping flow region.

Solution

In the creeping flow region, the drag coefficient is given by equation (5.9), i.e.

$$C_D = \frac{24}{\text{Re}}(1 + \text{Bi}^*) \tag{5.9}$$

The other dimensionless groups are:

$$C_D = \frac{4}{3}\frac{gd}{V^2}\left(\frac{\rho_s - \rho}{\rho}\right)$$

$$\text{Re} = \frac{\rho V^{2-n} d^n}{m}$$

$$\text{Bi}^* = \frac{\tau_0}{m(V/d)^n}$$

Trial and error solution is needed as the unknown velocity appears in all of these groups. The other values (in S.I. units) are:

$$\tau_0^H = 3.3\,\text{Pa}; \quad m = 3.69\,\text{Pa·s}^{0.53}; \quad n = 0.53; \quad d = 3.18 \times 10^{-3}\,\text{m}$$

$$\rho_s = 7780\,\text{kg/m}^3; \quad \rho = 1000\,\text{kg/m}^3; \quad g = 9.81\,\text{m/s}^2$$

Substituting these values:

$$C_D = \frac{0.2813}{V^2} \quad \text{or} \quad V = \sqrt{\frac{0.2813}{C_D}}$$

$$\text{Re} = \frac{(1000)(3.18 \times 10^{-3})^{0.53}V^{2-0.53}}{3.69} = 12.86V^{1.47}$$

$$\text{Bi}^* = \frac{\tau_0^H}{m(V/d)^n} = \frac{3.3 \times (3.18 \times 10^{-3})^{0.53}}{3.69 \times V^{0.53}} = 0.0424V^{-0.53}$$

Assume a value of $V = 15\,\text{mm/s} = 15 \times 10^{-3}\,\text{m/s}$.

$$\therefore \text{Bi}^* = 0.393; \quad \text{Re} = 12.86 \times (15 \times 10^{-3})^{1.47} = 0.0268$$

Now from equation (5.9), the value of C_D:

$$C_D = \frac{24}{\text{Re}}(1 + \text{Bi}^*) = \frac{24}{0.0268}(1 + 0.393)$$

$$= 1248$$

$$\therefore \text{ the velocity, } V = \sqrt{\frac{0.2813}{C_D}} = \sqrt{\frac{0.2813}{1248}} = 0.015\,\text{m/s} = 15\,\text{mm/s}$$

which matches with the assumed value. Also, in view of the small value of the Reynolds number ($\text{Re}_{\text{max}} \sim 70$, from equation 5.8), the assumption of the creeping flow is justified.□

5.2.5 Effect of container boundaries

The problem discussed so far relates to the motion of a single spherical particle in an unbounded, or effectively infinite, expanse of fluid. If other particles are present in the neighbourhood of the sphere, its settling velocity will be influenced and the effect will become progressively more marked as the concentration of particles increases. There are three contributory factors. First, as the particles settle, the displaced liquid flows upwards. Secondly, the particle experiences increased buoyancy force owing to the higher density of the suspension. Finally, the flow pattern of the liquid relative to the particles will be altered thereby affecting the velocity gradients. The sedimentation of concentrated suspensions in non-Newtonian fluids is discussed in section 5.2.6 while the effect of the vessel walls is discussed here.

The walls of the vessel containing the liquid exert an extra retarding effect on the terminal falling velocity of the particle. The upward flow of the displaced liquid, not only influences the relative velocity, but also sets up a velocity profile in the confined geometry of the tube. This effect may be quantified by introducing a wall factor, f, which is defined as the ratio of the terminal falling velocity of a sphere in a tube, V_m, to that in an unconfined liquid, V, viz.,

$$f = \frac{V_m}{V} \tag{5.16}$$

The experimental determination of the setling velocity in an infinite medium requires the terminal falling velocity of a sphere to be measured in tubes of different diameters and then extrapolating these results to $(d/D) = 0$, as shown in Figure 5.4 for a series of plastic spheres falling in a 0.5% Methocel solution. When the settling occurs in the creeping flow region (Re < 1), the measured falling velocity shows a linear dependence on the diameter ratio and can readily be extrapolated to $(d/D) = 0$. The available experimental results in Newtonian and power-law liquids indicate that the wall factor, f, is independent of the

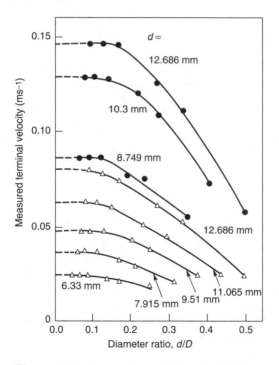

Figure 5.4   *Dependence of terminal falling velocity of spheres in a 0.5% aqueous hydroxyethyl cellulose solution* ($Re_m > 1$) • *PVC spheres* △ *Perspex spheres*

sphere Reynolds number (based on the measured velocity, V_m) both at small ($<\sim 1$) and large ($>\sim 1000$) values of the Reynolds number [Chhabra and Uhlherr, 1980; Uhlherr and Chhabra, 1995; Chhabra *et al.*, 1996]. Based on an extensive experimental study in the range of conditions $0.5 \leq n \leq 1$; $0.01 \leq \text{Re}_m \leq 1000$ and $(d/D) \leq 0.5$, the wall factor can be empirically correlated with the diameter ratio and the sphere Reynolds number as [Chhabra and Uhlherr, 1980]:

$$\frac{(1/f) - (1/f_\infty)}{(1/f_0) - (1/f_\infty)} = [1 + 1.3\text{Re}_m^2]^{-1/3} \tag{5.17}$$

where f_0 and f_∞, the values of the wall factor in the low and high Reynolds number regions respectively are given by:

$$f_0 = 1 - 1.6\frac{d}{D} \tag{5.18}$$

and

$$f_\infty = 1 - 3\left(\frac{d}{D}\right)^{3.5} \tag{5.19}$$

While it is readily recognised that the creeping flow occurs up to about $\text{Re}_m \sim 1$, the critical value of the Reynolds number corresponding to the upper asymptotic value, f_∞, is strongly dependent upon the value of (d/D), e.g. ranging from $\text{Re}_m \sim 30$–40 for $(d/D) = 0.1$ to $\text{Re}_m \sim 1000$ for $(d/D) = 0.5$. Qualitatively, the additional retardation caused by the walls of the vessel is less severe is power-law fluids than that in Newtonian fluids under otherwise identical conditions; the effect becoming progressively less important with the increasing Reynolds number and/or decreasing diameter ratio. The wall effect is even smaller in visco-elastic liquids [Chhabra, 1993a].

Sedimenting particles are also subject to an additional retardation as they approach the bottom of the containing vessel because of the influence of the lower boundary on the flow pattern. No results are available on this effect for non-Newtonian fluids and therefore the corresponding expressions for Newtonian fluids offer the best guide [Clift *et al.*, 1978], at least for inelastic shear-thinning fluids. For instance, in the creeping flow regime and $d/D < 0.1$, the effect is usually expressed as

$$\frac{V_m}{V} = \frac{1}{1 + 1.65(d/L)} \tag{5.20}$$

where L is the distance from the bottom of the vessel.

5.2.6 Hindered settling

As mentioned earlier, the terminal falling velocity of a sphere is also influenced by the presence of neighbouring particles.

In concentrated suspensions, the settling velocity of a sphere is less than the terminal falling velocity of a single particle. For coarse (non-colloidal) particles in mildly shear-thinning liquids ($1 > n \geq 0.8$) [Chhabra *et al.*, 1992], the expression proposed by Richardson and Zaki [1954] for Newtonian fluids applies at values of $\mathrm{Re}(= \rho V^{2-n} d^n / m)$ up to about 2:

$$\frac{V_0}{V} = (1 - C)^Z \tag{5.21}$$

where V_0 is the hindered settling velocity of a suspension of uniform size spheres at a volume fraction C; V is the terminal falling velocity of a single sphere in the same liquid, Z is a constant which is a function of the Archimedes number and (d/D) and is given as [Coulson and Richardson, 1991]:

$$\frac{4.8 - Z}{Z - 2.4} = 0.0365 \, \mathrm{Ar}^{0.57} [1 - 2.4(d/D)^{0.27}] \tag{5.22}$$

where for power-law liquids, the Archimedes number, Ar, is defined by equation (5.12).

In visco-elastic fluids, some internal clusters of particles form during hindered settling and the interface tends to be diffuse [Allen and Uhlherr, 1989, Bobbroff and Phillips, 1998]. For $100-200\,\mu\mathrm{m}$ glass spheres in visco-elastic polyacrylamide solutions, significant deviations from equation (5.22) have been observed.

Example 5.5

Estimate the hindered settling velocity of a 25% (by volume) suspension of $200\,\mu\mathrm{m}$ glass beads in an inelastic carboxymethyl cellulose solution ($n = 0.8$ and $m = 2.5\,\mathrm{Pa \cdot s}^n$) in a 25 mm diameter tube. The density of glass beads and of the polymer solution are $2500\,\mathrm{kg/m}^3$ and $1020\,\mathrm{kg/m}^3$ respectively.

Solution

The velocity of a single glass bead is calculated first. In view of the small size and rather high consistency coefficient of the solution, the particle Reynolds number will be low, equation (5.11) can be used. From Table 5.1, $\mathrm{X}(n) = 1.24$ corresponding to $n = 0.8$.

Substituting values in equation (5.11):

$$V = \left[\frac{gd^{n+1}(\rho_s - \rho)}{18mX} \right]^{(1/n)}$$

$$= \left[\frac{9.81 \times (200 \times 10^{-6})^{0.8+1}(2500 - 1020)}{18 \times 2.5 \times 1.24} \right]^{1/(0.8)}$$

$$= 4.97 \times 10^{-6}\,\mathrm{m/s} \text{ or } 4.97\,\mu\mathrm{m/s}.$$

Check the value of Reynolds number, Re:

$$\text{Re} = \frac{\rho V^{2-n} d^n}{m} = \frac{1020 \times (4.97 \times 10^{-6})^{2-0.8}(200 \times 10^{-6})^{0.8}}{2.5}$$

$$= 1.93 \times 10^{-7} \ll 1$$

Therefore, the settling occurs in the creeping flow region and the equation (5.11) is valid.

The Archimedes number is given by equation (5.12) as:

$$\text{Ar} = \frac{4}{3} g d^{(2+n)/(2-n)} (\rho_s - \rho) \rho^{n/(2-n)} m^{2/(n-2)}$$

$$= \left(\frac{4}{3}\right) \times 9.81 \times (200 \times 10^{-6})^{(2+0.8)/(2-0.8)}$$

$$\times (2500 - 1020)(1020)^{(0.8)/(2-0.8)}(2.5)^{2/(0.8-2)}$$

$$= 0.000\,996\,4$$

The value of Z is evaluated from equation (5.22):

$$\frac{4.8 - Z}{Z - 2.4} = 0.0365 \times (0.000\,996\,4)^{0.57} \left[1 - 2.4 \left(\frac{200 \times 10^{-6}}{25 \times 10^{-3}}\right)^{0.27}\right]$$

$$= 0.000\,247$$

and

$$Z \simeq 4.8$$

$\therefore$

$$\frac{V_0}{V} = (1 - C)^Z$$

or

$$V_0 = 4.97 \times 10^{-6} \times (1 - 0.25)^{4.8} = 1.25 \times 10^{-6}\,\text{m/s}$$

This velocity is about a quarter of the value for a single particle.$\square$

5.3 Effect of particle shape on terminal falling velocity and drag force

A spherical particle is unique in that it presents the same projected area to the oncoming fluid irrespective of its orientation. For non-spherical particles, on the other hand, the orientation must be specified before the drag force can be calculated. The drag force on spheroidal (oblates and prolates) particles moving in shear-thinning and shear-thickening power-law fluids ($0.4 \leq n \leq 1.8$) have been evaluated for Reynolds numbers up to 100 [6, 7]. The values of drag coefficient are given in the original papers [Tripathi *et al.*, 1994; Tripathi and Chhabra, 1995] and the main trends are summarised here. For pseudo-plastic fluids ($n < 1$), creeping flow occurs for Re up to about 1 (based on equal volume sphere diameter) and for dilatant fluids ($n > 1$) up to about 0.2–0.5. For a given Reynolds number and aspect ratio (minor/major axis), the drag on oblates is less than that on a sphere of equal volume whereas for prolate particles, it is higher. The drag force in the creeping flow region is higher for shear-thinning fluids than for Newtonian fluids; this is consistent with the behaviour observed for a sphere. The influence of power-law index, however, diminishes with increasing particle Reynolds number. The opposite effect is observed with shear-thickening fluids, i.e. the drag is lower than that in a Newtonian fluid.

Many workers have measured drag coefficients for particles, including cylinders, rectangular prisms, discs, cones settling at their terminal velocities in power-law fluids. Work in this area has recently been reviewed [Chhabra, 1996], but no generalised correlation has yet been proposed. A simple equation which reconciles the bulk of the results for drag on cones, cubes, parallelepipeds, and cylinders (falling axially) settling at terminal condition in power-law fluids (Re < 150; $0.77 \leq n \leq 1$; $0.35 \leq \psi \leq 0.7$) is [Venu Madhav and Chhabra, 1994]:

$$C_D = \frac{32.5}{\text{Re}}(1 + 2.5\text{Re}^{0.2}) \qquad (5.23)$$

where both C_D and Re $\left(\dfrac{\rho V^{2-n} d^n}{m} \right)$ are based on the diameter of the sphere of the same volume; corrections were made for wall effects. The predictions deteriorate progressively as the particle departs from spherical shape, i.e. as sphericity, ψ, decreases.

The scant experimental and theoretical results available for viscoplastic and visco-elastic fluids have been reviewed elsewhere [Chhabra, 1996].

5.4 Motion of bubbles and drops

The drag force acting on a gas bubble or liquid droplet will not, in general, be the same as that acting on a rigid particle of the same shape and size because circulating patterns are set up inside bubbles and drops. While the radial velocity at the interface is zero, the angular velocity, shear, and normal stresses are continuous across the interface for fluid particles and the velocity gradient in the continuous phase (hence shear stress and drag force) is therefore less than that for a rigid particle. In the absence of surface tension and inertial effects (Reynolds number $\ll 1$), the terminal velocity of a fluid sphere falling in an incompressible Newtonian fluid, as calculated from the Stokes' law (equation 5.11 with $n = 1$, $X = 1$) is increased by a factor Q_1 which accounts for the internal circulation:

$$Q_1 = \frac{3}{2}\left(\frac{1 + \delta}{1 + 1.5\delta} \right) \qquad (5.24)$$

where δ is the ratio of the viscosity of the fluid in the sphere to that of the ambient liquid. Clearly, in the limit of $\delta \to \infty$, i.e. for a solid sphere, $Q_1 = 1$ and as $\delta \to 0$, i.e. for a bubble, $Q_1 \to 3/2$. In the intermediate range of viscosity ratios, the internal circulation effects generally decrease with increasing value of δ. With very small droplets, the surface tension forces nullify the tendency for circulation and the droplet falls at a velocity close to that of a solid sphere of the same size. Likewise, even small amounts of surface active agents tend to immobilise the free surface of small fluid particles, thereby inhibiting internal circulation, and these particles again fall at velocities similar to those of solid spheres.

Drops and bubbles, in addition, undergo deformation because of the differences in the pressures acting on various parts of the surface. Thus, when a drop is falling in a quiescent medium, both the hydrostatic and impact pressures will be greater on the forward than on the rear face; this will tend to flatten the drop. Conversely, the viscous drag will tend to elongate it.

The deformation of the drop is resisted by surface tension forces and very small drops (large surface area) therefore remain spherical, whereas large drops (small surface tension forces) are appreciably deformed and the drag is increased. For droplets above a certain size, the deformation is so great that the drag force increases in almost direct proportion to volume and the terminal velocity is almost independent of size. The literature on the motion of fluid particles in Newtonian fluids has been thoroughly reviewed by Clift *et al.* [1978], the corresponding developments in non-Newtonian fluids are briefly summarised here.

The drag coefficient for freely falling spherical droplets (or rising gas bubbles) of Newtonian fluids in power-law liquids at low Reynolds number has been approximately evaluated and, in the absence of surface tension effects, it is given by equation (5.4), i.e.

$$C_D = \frac{24}{\mathrm{Re}} \, X(n, \delta) \qquad\qquad (5.4)$$

where the Reynolds number is defined as $\rho_c V^{2-n} d^n / m_c$, and the correction factor $X(n, \delta)$ now depends on both the power-law index and the ratio of the viscosities of the dispersed and continuous phases. However, in the case of gas bubbles ($\delta = 0$), the correction factor X is a function of the power-law index alone [Hirose and Moo-Young, 1969]:

$$X = 2^n 3^{(n-3)/2} \left\{ \frac{13 + 4n - 8n^2}{(2n+1)(n+2)} \right\} \qquad\qquad (5.25)$$

This equation is valid for only mildly shear-thinning behaviour ($n \geq 0.6$ or so). For a Newtonian fluid ($n = 1$), equation (5.25) gives $X = 1/Q_1 = 2/3$. This simple expression does not, however, account for either surface tension effects or for the shape and changing size (due to expansion) of a freely rising bubble in a stationary power-law medium. The agreement between the predictions of equation (5.25) and the scant data for approximately spherical gas bubbles $(1 \geq n \geq 0.5)$ is reasonable, as seen in Figure 5.5.

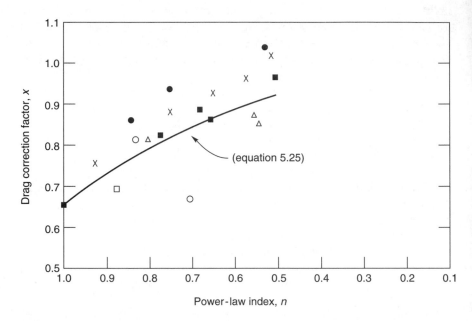

Figure 5.5 *Drag on spherical gas bubbles in power-law fluids in creeping flow region*

No such simple relation exists between the value of the correction factor X and the values of n and δ for droplets, and reference must be made to original or review papers [Nakano and Tien, 1970; DeKee and Chhabra, 1992; Chhabra, 1993a; DeKee et al., 1996]. Droplets are subject to greater drag in power-law fluids than in a Newtonian fluid in the creeping flow region and the effect of shear-thinning becomes progressively less marked with increasing Reynolds number [Nakano and Tien, 1970].

The usefulness of these studies is severely limited by the various simplifying assumptions which have been made concerning shape and surface tension effects. Indeed, a variety of shapes of drops and bubbles has been observed under free fall conditions in non-Newtonian fluids and these differ significantly from those in Newtonian fluids [Clift et al., 1978; Chhabra, 1993a].

In addition, many workers have reported experimental data on various aspects (such as shapes, coalescence, terminal falling velocity) of bubble and drop motion in a non-Newtonian continuous phase.

The bulk of the data on drag on freely falling drops relate to conditions where the viscosity ratio parameter, δ, rarely exceeds 0.001, thereby suggesting that behaviour is really tantamount to that of gaseous spheres. In spite of these difficulties, many correlating expressions for drag on droplets and bubbles in non-Newtonian liquids are available in the literature, but these are too tentative and restrictive to be included here. Thus, as a first approximation, the terminal velocity (or drag) for a droplet may be bounded by treating it first as a rigid sphere and then as a gas bubble of the same size and density as the liquid droplet.

A striking feature is the way the so-called discontinuity in the free rise velocity of a bubble varies with its size. Indeed, Astarita and Appuzzo [1965] noted a six to ten fold increase in the rise velocity at a critical bubble size, as seen in Figure 5.6 for air bubbles in two polymer solutions. Subsequently, similar, though less dramatic results, summarised in a recent review [DeKee *et al.*, 1996], have been reported. The critical bubble radius appears to hover around 2.6 mm, irrespective of the type or the degree of non-Newtonian properties of the continuous phase, and this value is well predicted by the following

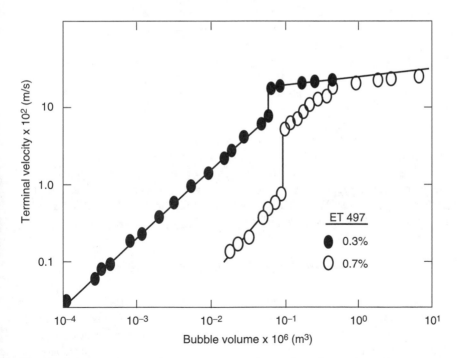

Figure 5.6 *Free rise velocity–bubble volume data showing an abrupt increase in velocity*

equation due to Bond and Newton [1928]:

$$r = \sqrt{\frac{\sigma}{g(\rho_l - \rho_g)}} \qquad (5.26)$$

where σ is the surface tension of the liquid. Indeed, the critical bubble sizes observed experimentally in a variety of inelastic and visco-elastic liquids, and those calculated using equation (5.26) seldom differ by more than 5%. As mentioned earlier, small gas bubbles behave like rigid spheres (no-slip boundary condition at the interface) whereas large ones display a shear-free surface, and this changeover in the type of particle behaviour gives rise to a jump of 50% in terminal velocity for Newtonian fluids. For creeping bubble motion in power-law fluids, the ratio of the terminal falling velocity of a bubble to that of a rigid sphere of the same size and density may be expressed as:

$$\frac{V_b}{V_s} = \left(\frac{X_s}{X_b}\right)^{(1/n)} \qquad (5.27)$$

where the subscripts 'b' and 's' refer to the bubble and sphere respectively. The values of X_s are listed in Table 5.1 and for gas bubbles, they can be estimated from equation (5.25). For $n = 1$, equation (5.27) gives the expected result, namely, $V_b = 1.5V_s$. For a power-law liquid with $n = 0.6$, $X_s = 1.42$ (Table 5.1) and $X_b = 0.93$, and $V_b \simeq 2V_s$. Thus, this mechanism does explain the effect qualitatively, but it does not predict as large an increase in rise velocity as that observed experimentally. Furthermore, visco-elasticity possibly contributes to the abrupt jump in rise velocity. In view of this, equation (5.25) may be applied only when the bubbles display a shear-free interface, tentatively when they are larger than 2.5 mm.

Little is known about the effect of visco-elasticity on the motion of bubbles and drops in non-Newtonian fluids, though a preliminary study suggests that spherical bubbles are subject to a larger drag in a visco-elastic than in an inelastic liquid. Recent surveys clearly reveal the paucity of reliable experimental data on the behaviour of fluid particles in non-Newtonian liquids [Chhabra, 1993a; DeKee *et al.*, 1996].

5.5 Flow of a liquid through beds of particles

The problems discussed so far relate to the motion of single particles and assemblies of particles in stationary non-Newtonian media. Consideration will now be given to the flow of non-Newtonian liquids through a bed of particles, as encountered in a variety of processing applications. For instance, the filtration of polymer melts, slurries and sewage sludges using packed beds, or the leaching of uranium from a dilute slurry of ore in a fluidised bed. Further examples are found in enhanced oil recovery by polymer flooding in which non-Newtonian polymer solutions are forced to flow through a porous rock. It is important therefore to be able to predict pressure drop across such beds.

With downward flow of a liquid, no relative movement occurs between the particles except for that arising from their unstable initial packing. In the streamline region, the pressure drop across the bed is directly proportional to the rate of flow for Newtonian liquids and to the rate of flow raised to a lower (less than unity) exponent for a shear-thinning and to a higher (>1) exponent for a shear-thickening fluid. At rates of flow high enough for turbulence to develop in some of the flow channels, the pressure drop will rise more steeply ($\propto Q^{1.8-2}$).

For upward flow through a bed which is not constrained at the top, the frictional pressure drop will be the same as for the downward flow as long as the structure of the bed is not disturbed by the flow. When the upward flow rate has been increased to the point where the frictional drag on the particles becomes equal to their buoyant weight, rearrangement will occur within the bed which then expands in order to offer less resistance to flow. Once the packing of the bed has reached its loosest stable form, any further increase in flow rate causes the individual particles to separate from one another and become freely supported in the liquid stream and the bed is then said to be fluidised. With further increase in flow rate, the particles move further apart and the bed voidage increases while the pressure difference remains approximately equal to the buoyant weight per unit area of the bed, as shown in Figure 5.7. Up to this stage, the system behaves in a similar manner whether the fluid is a gas or a liquid. Liquid fluidised beds continue to expand in a uniform manner, with the degree of agitation of particles increasing progressively. Fluidisation is then said to be particulate. A qualitatively similar fluidisation behaviour is

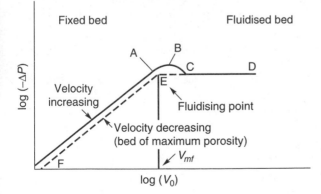

Figure 5.7 *Qualitative pressure drop–flow rate behaviour in fixed and fluidised beds*

obtained with non-Newtonian liquids. Since the main thrust here is on the role of liquid rheology, the behaviour of gas fluidised beds (aggregative type) is not covered in this chapter, but attention is drawn to several excellent books on this subject [Davidson *et al.*, 1985; Coulson and Richardson, 1991].

The estimation of the pressure gradient for the flow of a non-Newtonian liquid through a fixed (or packed) bed is addressed in the next section, fluidised beds are considered in section 5.7.

5.6 Flow through packed beds of particles (porous media)

The flow of non-Newtonian liquids through beds of particles is treated in an analogous way to that adopted in Chapter 3 for the flow through ducts of regular cross-section. No complete analytical solution is, however, possible and a degree of empiricism complemented by the use of experimental results is often necessary. Firstly, however, the basic nature and structure of porous media (or beds of particles) will be briefly discussed.

5.6.1 Porous media

The simplest way of regarding a porous medium is as a solid structure with passages through which fluids can flow. Most naturally occurring minerals (sand, limestones) are consolidated having been subjected to compressive forces for long times. Packed beds of glass beads, catalyst particles, Raschig rings, berl saddles, etc. as used in process equipment are unconsolidated. Unconsolidated media generally have a higher permeability and offer less resistance to flow. Packing may be ordered or random according to whether or not there is a discernable degree of order of the particles, though completely random packing hardly ever occurs as 'order' tends to become apparent as the domain of examination is progressively reduced in size. Cakes and breads are good examples of random media!

Porous media may be characterized at two distinct levels: microscopic and macroscopic. At the microscopic level, the structure is expressed in terms of a statistical description of the pore size distribution, degree of inter-connection and orientation of the pores, fraction of dead pores, etc. In the macroscopic approach, bulk parameters are employed which have been averaged over scales much larger than the size of pores. These two approaches are complementary and are used extensively depending upon the objective. Clearly, the microscopic description is necessary for understanding surface phenomena such as adsorption of macromolecules from polymer solutions and the blockage of pores, etc., whereas the macroscopic approach is often quite adequate for process design where fluid flow, heat and mass transfer are of greatest interest, and the molecular dimensions are much smaller than the pore size. Detailed accounts of micro- and macro-level characterization methods frequently used for porous media are available in the literature [Greenkorn, 1983; Dullien, 1992].

Of the numerous macroscopic parameters used to quantify porous media, those gaining widest acceptance in the literature for describing the flow of single phase fluids are voidage, specific surface, permeability and tortuosity. Their values can often be inferred from experiments on the streamline flow of single phase Newtonian fluids.

(i) Voidage

Voidage, ε, is defined as the fraction of the total volume which is free space available for the flow of fluids, and thus the fractional volume of the bed occupied by solid material is $(1 - \varepsilon)$. Depending upon the nature of the porous medium, the voidage may range from near zero to almost unity. For instance, certain rocks, sandstones etc. have values of the order of 0.15–0.20 whereas fibrous beds and ring packings may have high values of voidage up to 0.95. Obviously, the higher the value of voidage, the lower is the resistance to flow of a fluid.

(ii) Specific surface

In addition, the specific surface, S_B, of the bed affects both its general structure and the resistance it offers to flow. It is defined as the surface area per unit volume of the bed, i.e. m^2/m^3. Hence, S_B can be expressed in terms of the voidage ε and the specific area S of the particles

$$S_B = S(1 - \varepsilon) \tag{5.28}$$

where S is the specific area per unit volume of a particle. Thus for a sphere of diameter d,

$$S = \frac{\pi d^2}{\pi d^3/6} = \frac{6}{d} \tag{5.29}$$

For a given shape, S is inversely proportional to the particle size. Highly porous fibre glasses have specific surface areas in the range $5–7 \times 10^4\,m^2/m^3$ while compact limestones ($\varepsilon \sim 0.04–0.10$) have specific surface areas in the range $\sim 0.2–2 \times 10^6\,m^2/m^3$.

(iii) Permeability

The permeability of a porous medium may be defined by means of the well-known Darcy's law for streamline flow of an incompressible Newtonian fluid

$$\frac{Q}{A} = V_c = \left(\frac{k}{\mu}\right)\left(-\frac{\Delta p}{L}\right) \tag{5.30}$$

where Q is the volume rate of flow of a fluid of viscosity, μ, through a porous medium of area A (normal to flow) under the influence of the pressure gradient $(-\Delta p/L)$, and k is called the permeability of the porous medium.

A porous material is said to have a permeability of 1 darcy if a pressure gradient of 1 atm/cm results in a flow of $1 \, cm^3/s$ of a fluid having viscosity of 1 cP through an area of $1 \, cm^2$. In S.I. units, it is expressed as m^2 and 1 darcy $\simeq 10^{-12} \, m^2$. Evidently, the lower the permeability, the greater is the resistance to flow. Typical values of permeability range from $10^{-11} \, m^2$ for fibre glass to $10^{-14} \, m^2$ for silica powder and limestone.

(iv) Tortuosity

Tortuosity is a measure of the extent to which the path traversed by fluid elements deviates from a straight-line in the direction of overall flow and may be defined as the ratio of the average length of the flow paths to the distance travelled in the direction of flow. Though the tortuosity depends on voidage and approaches unity as the voidage approaches unity, it is also affected by particle size, shape and orientation in relation to the direction of flow. For instance, for plate like particles, the tortuosity is greater when they are oriented normal to the flow than when they are packed parallel to flow. However, the tortuosity factor is not an intrinsic characteristics of a porous medium and must be related to whatever one-dimensional flow model is used to characterise the flow.

5.6.2 Prediction of pressure gradient for flow through packed beds

Many attempts have been made to obtain general relations between pressure drop and mean velocity of flow through porous media or packings, in terms of the bed voidage which is either known or can easily be measured. The following discussion is limited primarily to the so-called capillary tube bundle approach while the other approaches of treating the flow of both Newtonian and non-Newtonian fluids are described in the literature [Happel and Brenner, 1965; Greenkorn, 1983; Dullien, 1992; Chhabra, 1993a].

(i) Streamline flow

The interstitial void space in a porous matrix or bed of particles may be envisaged as consisting of tortuous conduits of complex cross-section but having a constant average area for flow. Thus, flow in a porous medium is equivalent to that in a non-circular conduit offering the same resistance to flow. However, the flow passages in a bed of particles will be oriented and inter-connected in an irregular fashion and the elementary capillary models do not account for these complexities. Despite these difficulties, the analogy between flow through a circular tube and through the channels in a bed of particles, shown schematically in Figure 5.8, provides a useful basis for deriving a general flow rate – pressure drop expression. As seen in Chapter 3, such expressions vary according to the flow model chosen for the fluid. Following Kemblowski *et al.*

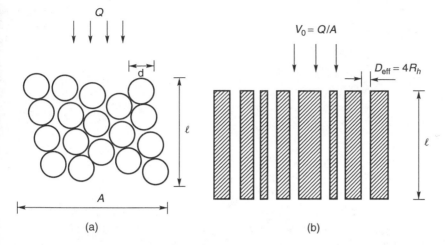

Figure 5.8   *Schematic representation of flow through a bed of uniform spheres (a) and the capillary model idealization (b)*

[1987], we begin by re-writing the well-known Hagen–Poiseuille equation, for the streamline flow of an incompressible Newtonian fluid through circular tubes, as follows:

$$V = \frac{D^2}{32\mu} \left(\frac{-\Delta p}{L} \right) \tag{5.31}$$

Then, for flow through a non-circular duct, this may be re-arranged as

$$V = \frac{D_h^2}{16K_0\mu} \left(\frac{-\Delta p}{L} \right) \tag{5.32}$$

where D_h is the hydraulic mean diameter (4 × Pore volume/surface area of particles) and K_0 is a constant which depends only on the shape of the cross-section. For a circular tube of diameter D, for instance $D_h = D$ and $K_0 = 2$. Likewise, for flow in between two plates separated by a distance $2b$, $D_h = 4b$ and $K_0 = 3$. Kemblowski *et al.* [1987] have further re-arranged equation (5.32) as:

$$\frac{D_h}{4} \left(\frac{-\Delta p}{L} \right) = \mu \left(\frac{4K_0 V}{D_h} \right) \tag{5.33}$$

For an incompressible Newtonian fluid, $(D_h/4) \left(\frac{-\Delta p}{L} \right)$ is the average shear stress at the wall of the flow passage and $(4K_0 V/D_h)$, the shear rate at the wall, which may be regarded as the nominal shear rate for time-independent non-Newtonian fluids. Thus,

$$\langle \tau_w \rangle = \frac{D_h}{4} \left(\frac{-\Delta p}{L} \right) \tag{5.34}$$

and

$$\langle \dot{\gamma}_w \rangle_n = \frac{4 K_0 V}{D_h} \tag{5.35}$$

where $\langle \rangle$ denotes the values averaged over the perimeter of the conduit.

Equations (5.34) and (5.35) can be used to obtain expressions for streamline flow of time-independent fluids through beds of particles, in which case V must be replaced by the mean velocity in the pores or interstices, i.e. V_i and the length L is replaced by the average length of the tortuous path, L_e, traversed by the fluid elements.

For a bed whose structure is independent of its depth, then L_e and L will be linearly related, i.e.

$$L_e = TL \tag{5.36}$$

where T is the tortuosity factor.

The interstitial velocity, V_i, is related to the superficial velocity V_0 by the Dupuit relation which is based on the following considerations.

In a cube of side, l, the volume of the voids is εl^3 and the mean cross-sectional area is the free volume divided by the height, i.e. εl^2. The volumetric flow rate through the cube is given by $V_0 l^2$, so that the average interstitial velocity V_i is given by

$$V_i = \frac{V_0 l^2}{\varepsilon l^2} = \frac{V_0}{\varepsilon} \tag{5.37}$$

Although equation (5.37) is a good approximation for random packings, it does not apply to all regular packings. For instance, for a bed of uniform spheres arranged in cubic packing, $\varepsilon = 0.476$, but the fractional area varies continuously from 0.215 in a plane across the diameters to unity between successive layers. Furthermore, equation (5.37) implicitly assumes that an element of fluid moving at a velocity V_i covers a distance l in the same time as a fluid element of superficial velocity V_0 in an empty tube. This implies that the actual interstitial velocity is likely to be somewhat greater than the value given by equation (5.37). Because an element of fluid in a bed actually travels a distance greater than L, in the Kozeny–Carman capillary model, a correction is made for this effect as:

$$V_i = \frac{V_0}{\varepsilon} T \tag{5.38}$$

Finally, the hydraulic mean diameter D_h must be expressed in terms of the packing characteristics. Thus, for a bed of uniform spheres of diameter d, the hydraulic mean diameter D_h can be estimated as follows:

$$D_h = \frac{4 \times \text{Flow area}}{\text{Wetted perimeter}} = \frac{4 \times \text{volume of flow channels}}{\text{surface area of packing}}$$

$$= \frac{4 \times \dfrac{\dfrac{\text{Volume of flow channels}}{\text{volume of bed}}}{\dfrac{\text{surface area of packing}}{\text{volume of bed}}}}{} = \frac{4\varepsilon}{S_B} \tag{5.39}$$

Substitution from equations (5.28) and (5.29) gives:

$$D_h = \left(\frac{2}{3}\right)\frac{d\varepsilon}{(1-\varepsilon)} \tag{5.40}$$

Note that the wetted surface of the column walls has been neglected, which is justified under most conditions of interest.

Although the early versions of capillary models, namely, the Blake and the Blake–Kozeny models, are based on the use of $V_i = V_0/\varepsilon$ and $L_e = LT$, it is now generally accepted [Dullien, 1992] that the Kozeny–Carman model, using $V_i = V_0 T/\varepsilon$ provides a more satisfactory representation of flow in beds of particles. Using equations (5.36), (5.38) and (5.40), the average shear stress and the nominal shear rate at the wall of the flow passage (equations 5.34 and 5.35) may now be expressed as:

$$\langle \tau_w \rangle = \frac{d\varepsilon}{6(1-\varepsilon)T}\left(\frac{-\Delta p}{L}\right) \tag{5.41}$$

and

$$\langle \dot{\gamma}_w \rangle_n = 6K_0 T\left(\frac{1-\varepsilon}{\varepsilon^2}\right)\frac{V_0}{d} \tag{5.42}$$

For generalised non-Newtonian fluids, Kemblowski *et al.* [1987] postulated that the shear stress at the wall of a pore or 'capillary' is related to the corresponding nominal shear rate at the wall by a power-law type relation:

$$\langle \tau_w \rangle = m'(\langle \dot{\gamma}_w \rangle_n)^{n'} \tag{5.43}$$

where m' and n' are the apparent consistency coefficient and flow behaviour index, respectively, inferred from pressure drop/flow rate data obtained in a packed bed. By analogy with the generalised procedure for streamline flow in circular tubes outlined in Chapter 3, equation (3.30a), m' and n' can be linked to the actual rheological parameters. For a truly power-law fluid, for instance;

$$n' = n \tag{5.44a}$$

$$m' = m\left(\frac{3n+1}{4n}\right)^n \tag{5.44b}$$

It should be, however, noted that the Rabinowitsch–Mooney factor of $((3n+1)/4n)^n$ is strictly applicable only to cylindrical tubes but the limited results available for non-circular ducts suggest that it is nearly independent of the shape of the conduit cross-section [Miller, 1972; Tiu, 1985]. For instance, the values of this factor are within 2–3% of each other for circular tubes and parallel plates over the range $0.1 \leq n \leq 1$.

The cross-section of the channels formed in a bed of spheres would be expected to lie between that of a circular tube and of a plane slit and Kemblowski *et al.* [1987] therefore suggested the use of a mean value of 2.5 for K_0. Considerable confusion also exists in the literature about the value of the tortuosity factor, T. Thus, Carman [1956] proposed a value of $\sqrt{2}$ based on the assumption that the capillaries deviate on average by 45° from the mean direction of flow $(\cos 45 = 1/\sqrt{2})$. On the other hand, if a fluid element follows the surface round the diameter of a spherical particle, the tortuosity factor should equal $\pi/2$. Indeed, the values ranging from ~ 1 to 1.65 have been used in the literature for Newtonian fluids [Agarwal and O'Neill, 1988]. Because T is a function of the geometry of the bed, it has the same value whatever the liquid rheology, provided that it is time-independent, although there is some evidence that T is weakly dependent on flow rate [Dharamadhikari and Kale, 1985]. This is not surprising because macromolecules have a tendency to adsorb on the walls of the pores and, if the flow rate is high enough, the shearing forces may overcome the surface forces; thus polymer molecules become detached thereby making more space available for flow. Thus, flow passages blocked at low flow rates may open up to flow again at high flow rates. Such unusual effects observed with non-Newtonian fluids in porous media are briefly discussed in a later section in this chapter.

A dimensionless friction factor may be defined as:

$$f = \left(\frac{-\Delta p}{L}\right) \frac{d}{\rho V_0^2} \left(\frac{\varepsilon^3}{1-\varepsilon}\right) \tag{5.45}$$

Noting $K_0 = 2.5$ and $T = \sqrt{2}$, combining equations (5.41) to (5.45):

$$f = \frac{180}{\text{Re}^*} \tag{5.46}$$

$$\text{where} \quad \text{Re}^* = \frac{\rho V_0^{2-n} d^n}{m(1-\varepsilon)^n} \left(\frac{4n}{3n+1}\right)^n \left(\frac{15\sqrt{2}}{\varepsilon^2}\right)^{1-n} \tag{5.47}$$

For a Newtonian fluid, $n = 1$, both equations (5.46) and (5.47) reduce to the well known Kozeny–Carman equation. Equation (5.47) correlates most of the literature data on the flow of power-law fluids through beds of spherical particles up to about $\text{Re}^* \sim 1$, though most work to date has been carried out in beds having voidages in the range $0.35 \leq \varepsilon \leq 0.41$ [Kemblowski *et al.*, 1987; Chhabra, 1993a].

Bingham plastic fluids

The flow of viscoplastic fluids through beds of particles has not been studied as extensively as that of power-law fluids. However, since the expressions for the average shear stress and the nominal shear rate at the wall, equations (5.41) and (5.42), are independent of fluid model, they may be used in conjunction with any time-independent behaviour fluid model, as illustrated here for the streamline flow of Bingham plastic fluids. The mean velocity for a Bingham plastic fluid in a circular tube is given by equation (3.13):

$$
V = \frac{D^2}{32\mu_B} \left(\frac{-\Delta p}{L} \right) \left(1 - \frac{4}{3}\phi + \frac{1}{3}\phi^4 \right) \tag{3.13}
$$

where $\phi = \tau_0^B/\tau_w$. This equation can be re-arranged in terms of the nominal shear rate and shear stress at the wall of the pore as:

$$
(\dot{\gamma}_w)_n = \frac{8V}{D} = \frac{\tau_w}{\mu_B} \left(1 - \frac{4}{3}\phi + \frac{1}{3}\phi^4 \right) \tag{5.48}
$$

As seen in Chapter 3, the quantity $(8V/D)$ is the nominal shear rate at the wall (also see equation 5.35, for a circular tube, $D_h = D$, $K_0 = 2$). Substituting for the nominal shear rate and wall shear stress from equations (5.41) and (5.42) in equation (5.48), slight re-arrangement gives:

$$
f = \frac{180}{\mathrm{Re}_B \mathrm{F}(\phi)} \tag{5.49}
$$

where

$$
\mathrm{Re}_B = \frac{\rho V_0 d}{\mu_B} \tag{5.50}
$$

$$
\mathrm{F}(\phi) = 1 - \frac{4}{3}\phi + \frac{\phi^4}{3} \tag{5.51}
$$

and

$$
\phi = \frac{\tau_0^B}{\langle \tau_w \rangle} \tag{5.52}
$$

It should be noted that $T = \sqrt{2}$ and $K_0 = 2.5$ have been used in deriving equation (5.49). Again for the special case of Newtonian fluids, $\tau_0^B = 0$ or $\phi = 0$, $\mathrm{F}(\phi) = 1$ and equation (5.49) reduces to the Kozeny–Carman equation. The scant experimental data available on pressure drop for the streamline flow for Bingham plastic fluids ($\mathrm{Re}_B \sim 1$) is consistent with equation (5.49).

This section is concluded by noting that similar expressions for the friction factor have been derived for a range of purely inelastic fluid models and these have been critically reviewed elsewhere [Chhabra, 1993a,b].

(ii) Transitional and turbulent flow

Because the apparent viscosity of non-Newtonian systems is usually high, flow conditions rarely extend beyond the streamline flow regime. There is no clear cut value of the Reynolds number marking the end of streamline flow. An examination of the available data indicates that the lower the value of the power-law index, the higher the Reynolds number up to which streamline flow occurs. An important factor is that with a range of pore sizes, some can be in laminar flow and others turbulent. Despite these uncertainities, the value of $Re^* \sim 5\text{--}10$ is a good approximation for engineering design calculations.

The capillary bundle approach has also been extended for correlating data on pressure drop in packed beds of spherical particles in the transitional and turbulent regions. Both Mishra *et al.* [1975] and Brea *et al.* [1976] proposed the following empirical method for estimating the 'effective viscosity', μ_{eff}:

$$\mu_{\text{eff}} = m' \left\{ \frac{12V_0(1-\varepsilon)}{d\varepsilon^2} \right\}^{n-1} \tag{5.53}$$

This is then incorporated in the modified Reynolds number, Re', to give

$$Re' = \frac{\rho V_0 d}{\mu_{\text{eff}}(1-\varepsilon)} \tag{5.54}$$

They assumed the 'viscous' and 'inertial' components of the pressure drop to be additive, and proposed the following relationship between the friction factor and the modified Reynolds number:

$$f = \frac{\alpha}{Re'} + \beta \tag{5.55}$$

Based on their experimental data for the flow of power-law fluids in packed and fluidised beds of spheres ($0.7 \leq n \leq 1$; $0.01 \leq Re' \leq 1000$; $0.37 \leq \varepsilon \leq 0.95$), Mishra *et al.* [1975] obtained $\alpha = 150$ and $\beta = 1.75$. With these values, equation (5.55) coincides with the well-known Ergun equation for Newtonian fluids [Ergun, 1952]. On the other hand, the data of Brea *et al.* [1976] encompass somewhat wider ranges of the power-law index ($0.4 \leq n' \leq 1$) and Reynolds number ($0.01 \leq Re' \leq 1700$) but a limited range of voidage ($0.36 \leq \varepsilon \leq 0.40$) and they proposed $\alpha = 160$ and $\beta = 1.75$. A close scrutiny of equation (5.53) shows that it is tantamount to using $K_0 = 2$ (corresponding to a circular cross-section) and $T = (25/12)$, and thus the lower value of K_0 is compensated for by the higher value of the tortuosity factor, T. Although the original papers give mean deviations of 15–16% between the predictions of equation (5.55) and experimental data, careful inspection of the pertinent graphs reveal maximum deviations of up to 100%.

Based on the re-appraisal of the literature data and new data, the following simplified expression provides a somewhat better representation of the data in packed beds, at least for $\varepsilon \le 0.41$ and $Re^* < 100$ [Chhabra, 1993a]:

$$f = \frac{150}{Re^*} + 1.75 \qquad\qquad\qquad (5.56)$$

Thus, it is suggested that for the flow of shear-thinning fluids in packed beds, equation (5.56) should be used for $Re^* < 100$, and for $\varepsilon > 0.41$ and $Re^* > 100$ equation (5.55) is preferable, with $\alpha = 150$ and $\beta = 1.75$.

Example 5.6

Estimate the frictional pressure gradient for the flow of a polymer solution ($m = 3.7\,Pa\cdot s^n$, $n = 0.5$, density $= 1008\,kg/m^3$) at the rate of $0.001\,m^3/s$ through a 50 mm diameter column packed with 1.5 mm leadshots. The average voidage of the packing is 0.39.

Solution

Superficial velocity of flow, $V_0 = (0.001/((\pi/4)(50 \times 10^{-3})^2)) = 0.51\,m/s$
    The Reynolds number of flow:

$$Re^* = \frac{\rho V_0^{2-n} d^n}{m(1-\varepsilon)^n} \left(\frac{4n}{3n+1}\right)^n \left(\frac{15\sqrt{2}}{\varepsilon^2}\right)^{1-n}$$

$$= \frac{(1008)(0.51)^{2-0.5}(1.5 \times 10^{-3})^{0.5}}{3.7(1-0.39)^{0.5}} \left(\frac{4 \times 0.5}{3 \times 0.5 + 1}\right)^{0.5} \left(\frac{15\sqrt{2}}{0.39^2}\right)^{1-0.5}$$

$$= 52$$

$\therefore$ the flow is in the transitional regime. Equation (5.55) or (5.56) may be used. For equation (5.56),

$$f = \frac{150}{Re^*} + 1.75 = \frac{150}{52} + 1.75 = 4.63$$

The pressure gradient, $\left(\frac{-\Delta p}{L}\right)$, across the bed is calculated using this value of f in equation (5.45),

$$\frac{-\Delta p}{L} = \frac{f \rho V_0^2}{d}\left(\frac{1-\varepsilon}{\varepsilon^3}\right) = \frac{(4.63)(1008)(0.51)^2(1-0.39)}{(1.5 \times 10^{-3})(0.39)^3}$$

$$= 8\,300\,000\,Pa/m \quad \text{or} \quad 8.3\,MPa/m.$$

For the sake of comparison, the value of $\left(\frac{-\Delta p}{L}\right)$ using equation (5.55) will also be calculated here. For a power law fluid,

$$m' = m\left(\frac{3n+1}{4n}\right)^n \quad \text{and} \quad n' = n = 0.5$$

$$\therefore \quad m' = 3.7\left(\frac{3 \times 0.5 + 1}{4 \times 0.5}\right)^{0.5} = 4.14\,\text{Pa·s}^n$$

$$\therefore \quad \mu_{\text{eff}} = m'\left\{\frac{12V_0(1-\varepsilon)}{d\varepsilon^2}\right\}^{n'-1}$$

$$= 4.14\left\{\frac{12 \times 0.51 \times (1-0.39)}{(1.5 \times 10^{-3})(0.39)^2}\right\}^{0.5-1} = 0.0324\,\text{Pa·s}$$

The modified Reynolds number, Re', is evaluated using equation (5.54):

$$Re' = \frac{\rho V_0 d}{\mu_{\text{eff}}(1-\varepsilon)} = \frac{(1008)(0.51)(1.5 \times 10^{-3})}{0.0324(1-0.39)} = 39.1$$

This value also suggests that flow is in the transition regime. The corresponding friction factor is:

$$f = \frac{150}{Re'} + 1.75 = \frac{150}{39.1} + 1.75 = 5.59$$

Again using equation (5.45), $\left(\frac{-\Delta p}{L}\right) = 9.6\,\text{MPa/m}$.

This value is only about 15% higher than that calculated previously.□

The above discussion is limited to the flow of inelastic fluids in unconsolidated beds of particles where the pore size is substantially larger than the characteristic dimensions of the polymer molecules. Interaction effects between the walls of the pore and the polymer molecules are then small. Thus, measuring the relationship between pressure drop and flow rate in a packed bed and in a tube would therefore lead to the prediction of the same rheological properties of the fluid. Visco-elastic effects and other phenomena including blockage of pores, polymer adsorption/retention, etc. observed in beds of low permeability or in consolidated systems will be briefly discussed in Section 5.6.7.

5.6.3 Wall effects

In practice, the influence of the confining walls on the fluid flow will be significant only when the ratio of the particle to container diameters is less than about 30 [Cohen and Metzner, 1981]. Particles will pack less closely near the wall, so that the resistance to flow in a bed of smaller diameter may be less than that in a large container at the same superficial velocity. One would expect the effect to be a characteristic of the packing and the column size and information on wall effects for Newtonian fluids can also be used for non-Newtonian fluids, at least for inelastic fluids. Thus, in streamline flow, Coulson [1949] found that, for the given pressure gradient, the superficial velocity of a Newtonian liquid increases by a factor, f_w, defined as:

$$f_w = \left(1 + \frac{1}{2}\frac{S_c}{S}\right)^2 \tag{5.57}$$

where S_c is the surface of the vessel per unit volume of the bed. Thus, for a cylindrical tube of diameter D packed with spheres of diameter d, equation (5.57) becomes:

$$f_w = \left(1 + \frac{d}{3D}\right)^2 \tag{5.58}$$

clearly as (D/d) increases, the correction factor approaches unity.

In contrast to this simple approach, Cohen and Metzner [1981] have made use of the detailed voidage profiles in the radial direction and have treated wall effects in a rigorous manner for both Newtonian and power-law fluids in streamline flow. From a practical standpoint, this analysis suggests the effect to be negligible in beds with $(D/d) > 30$ for inelastic fluids.

5.6.4 Effect of particle shape

The voidage of a bed of particles is strongly influenced by the particle shape, orientation and size distribution. However, voidage correlates quite well with sphericity, as shown in Figure 5.9; values of sphericity for different shape particles have been compiled by Brown *et al.* [1950] and German [1989] amongst others.

For beds of spheres of mixed sizes, the voidage can be significantly less if the smaller size spheres can fill the voids between the larger ones. In general, the voidage of a bed of mixed particles is a function of the volume fractions f of each particle size and the ratio of the sizes. Hence, for a binary-size particle system, the mean voidage of the bed will be a function of d_1/d_2 and f_1, f_2, and the curves of ε versus (d_1/d_2) and f_1/f_2 will pass through minima. Some predictive expressions for voidage of beds of binary and ternary spherical and non-spherical particle systems are available in the literature [Yu *et al.* 1993; Yu and Standish; 1993].

For the flow of Newtonian fluids in packed beds of mixed size spheres, Leva [1957] suggested the use of the volume-mean diameter, whereas the subsequent limited work with Newtonian and power-law fluids [Jacks and

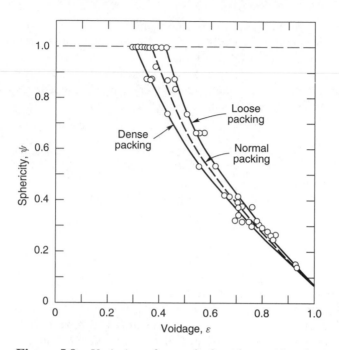

Figure 5.9   *Variation of mean bed voidage with sphericity*

Merrill, 1971; Rao and Chhabra, 1993] shows that it is more satisfactory to use volume/surface mean size:

$$d = \frac{\sum n_i d_i^3}{\sum n_i d_i^2} \tag{5.59}$$

where n_i is the number of spheres of size d_i.

For the flow of power-law fluids through packed beds of cubes, cylinders and gravel chips [Machac and Dolejs, 1981; Chhabra and Srinivas, 1991; Sharma and Chhabra, 1992; Sabiri and Comiti, 1995; Tiu *et al.*, 1997], the few available data for streamline flow correlate well with equation (5.56) if the equal volume sphere diameter, d_s, multiplied by sphericity, ψ, is employed as the effective diameter, $d_{\text{eff}} = d_s \psi$ in the definitions of the Reynolds number and friction factor.

5.6.5 Dispersion in packed beds

Dispersion is the general term which denotes the various types of self-induced mixing processes which can arise during the flow of a fluid. The effects of dispersion are important in packed beds, though they are also present in the simple flow conditions existing in a straight tube. There are two important mechanisms of dispersion – molecular diffusion and mixing arising from the flow pattern within the fluid. An important consequence of dispersion in a packed bed or porous medium is that true plug-flow never occurs (except possibly for viscoplastic materials). Consequently, the performance of oil displacement processes using non-Newtonian polymer solutions, and of packed bed reactors for effecting polymerisation reactions, are adversely affected. For laminar flow in a tube, dispersion arises from random molecular motion and is governed by Fick's law according to which the flux of a component is proportional to the product of its concentration gradient and the molecular diffusivity. Furthermore, the velocity profile for a Newtonian fluid is parabolic with the ratio of the centre-line velocity to the mean velocity equal to 2. For a power-law fluid, this ratio is $(3n + 1)/(n + 1)$ (equation 3.7). Dispersion occurs since elements of fluid take different times to traverse the length of the pipe, depending upon their radial positions and at the exit, elements with a range of residence times mix together. Thus, if a plug of tracer is injected into the entering liquid, it will first appear in the exit stream after the interval of time taken by the fastest moving fluid element at the axis to travel the length of the pipe. Then tracer will appear later in the liquid issuing at progressively greater distances from the axis of the pipe. Because the fluid velocity approaches zero at the wall, some tracer will still appear after a long period.

In turbulent flow, molecular diffusion is augmented by the presence of turbulent eddies and mixing is more intensive though, due to the flatter velocity profiles in tubes, the role of velocity gradient in dispersion diminishes.

In a bed of particles, the effects of dispersion will generally be greater than in a straight tube, partly because of the successive contractions and expansions in the flow passages. Radial mixing readily occurs in the flow passages or cells because a liquid element enters them with high kinetic energy, much of which is converted into rotational motion within the cells. Also, the continually changing velocity promotes dispersion in a bed of particles. Wall effects can be significant because of channelling through the region of high voidage near the wall.

At low flow rates, molecular diffusion dominates and cell mixing (attributable to the development of rotational motion) contributes relatively little to the overall dispersion. At high flow rates, however, mixing in a packed bed may be modelled by considering it to consist of a series of mixing cells, each being of the same size as the packing itself. Irrespective of the actual mechanism, dispersion processes may be characterised by a dispersion coefficient. In packed beds, dispersion is generally anisotropic, except at very low velocities; that is, the dispersion coefficients D_L in the axial (longitudinal) and D_R in the radial directions will generally be different. The process may normally be considered to be linear in so far as the rate of dispersion is proportional to the concentration gradient multiplied by the corresponding dispersion coefficient. However, unlike the molecular diffusion coefficient, D_L and D_R are strongly dependent on the flow regime and the bed geometry. In fact, the dispersion coefficient is analogous to the eddy kinematic viscosity which has been discussed in connection with momentum transfer in turbulent flow (section 3.6.4 in Chapter 3).

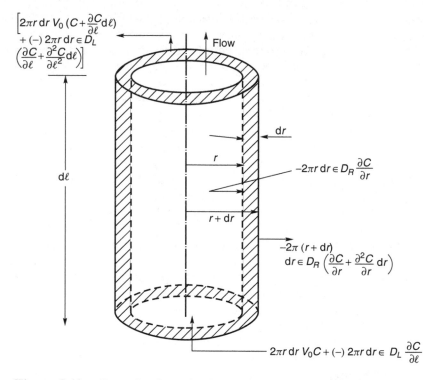

Figure 5.10   *Control volume for the derivation of the differential equation for dispersion*

The differential equation for dispersion in a cylindrical bed of voidage ε can be readily derived by considering the material balance over a differential annular control volume of length dl and of thickness dr, as shown in Figure 5.10. On the basis of a dispersion model, it is seen that the concentration C of a tracer will depend upon the axial position l, radial position r and time, t. A material balance over this element gives [Coulson and Richardson, 1991]:

$$\frac{\partial C}{\partial t} + \frac{V_0}{\varepsilon}\frac{\partial C}{\partial l} = D_L\frac{\partial^2 C}{\partial l^2} + \frac{1}{r}D_R\frac{\partial}{\partial r}\left(r\frac{\partial C}{\partial r}\right) \tag{5.60}$$

Longitudinal dispersion coefficients can be evaluated by injecting a flat pulse of tracer into the bed so that $\partial C/\partial r = 0$. The values of D_L can be estimated by matching the predicted (equation 5.60) and the actual (measured) changes in shape of a pulse of tracer as it passes between two locations (axial direction) in the bed. The results are often expressed in terms of dimensionless groups, namely, Peclet number, Pe, Reynolds number, Re and Schmidt number, Sc.

Wen and Yim [1971] reported a few results on axial dispersion coefficients (D_L) for the flow of two weakly shear-thinning polymer (PEO) solutions ($n = 0.81$ and 0.9) through a bed packed with glass spheres ($d = 4.76$ and 14.3 mm) of voidages 0.4 and 0.5. Over the range ($7 \leq \mathrm{Re_1} \leq 800$), their results did not deviate substantially from the correlation developed by these authors previously for Newtonian fluids:

$$Pe = 0.2 + 0.011(\mathrm{Re_1})^{0.48} \tag{5.61}$$

$$\text{where} \quad Pe = V_0 d/D_L \quad \text{and} \quad \mathrm{Re_1} = \frac{\rho V_0^{2-n} d^n}{m' 8^{n-1}} \tag{5.61a}$$

Subsequent experimental work [Payne and Parker, 1973] was carried out with aqueous polyethylene oxide solutions having values of n in the range $0.53 \leq n < 1$, flowing in a packed bed of $350\,\mu m$ glass beads ($\varepsilon = 0.365$). The Reynolds number was of the order of 10^{-3} thereby rendering the second term on the right hand side of equation (5.61) negligible. However, under these conditions, the Peclet number showed weak dependence on the power-law index, decreasing from $Pe = 0.2$ for $n = 1$ to $Pe \sim 0.1$ for $n = 0.53$, thereby suggesting that dispersion was greater in shear-thinning fluids.

In the only reported study of radial mixing of non-Newtonian rubber solutions in packed beds [Hassell and Bondi, 1965], the quality of mixing was found to deteriorate rapidly with the increasing viscosity.

5.6.6 Mass transfer in packed beds

Little work has been reported on mass transfer between non-Newtonian fluids and particles in a fixed bed. Kumar and Upadhyay [1981] measured the rate of dissolution of benzoic acid spheres and cylindrical pellets in an aqueous carboxymethyl cellulose solution ($n = 0.85$) and, using the plug flow model, proposed the following correlation for mass transfer in terms of the j_m factor:

$$\varepsilon j_m = \frac{0.765}{(\mathrm{Re'})^{0.82}} + \frac{0.365}{(\mathrm{Re'})^{0.39}} \tag{5.62}$$

where ε is the voidage of the bed; $\mathrm{Re'}$ is the modified Reynolds number defined by equations (5.53) and (5.54); and the j_m factor is:

$$j_m = \frac{k_c}{V_0} Sc^{2/3} \tag{5.63}$$

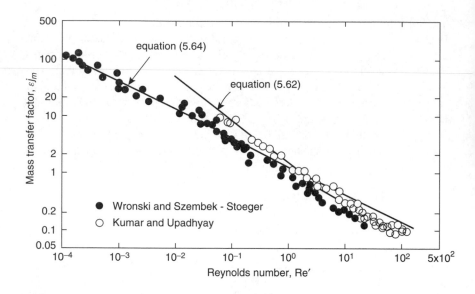

Figure 5.11   *Correlation of liquid–solid mass transfer for power-law fluids in packed beds*

where k_c is the mass transfer coefficient and the Schmidt number, $Sc = \mu_{\text{eff}}/\rho D_{AB}$ with the effective viscosity, μ_{eff}, as defined in equation (5.53). This correlation was found to represent the experimental data with a mean error of $\pm 10\%$ for the range of conditions: $0.1 \leq Re' \leq 40$ and $800 \leq Sc \leq 72\,000$, but for a single value of $n = 0.85$. For cylindrical pellets, the characteristic linear dimension was taken as the diameter of the sphere of equal volume multiplied by the sphericity. The validity of equation (5.62) has subsequently been confirmed independently [Wronski and Szembek-Stoeger, 1988] but it does over-predict the j_m factor at low Reynolds numbers ($Re' \leq 0.1$), as can be seen in Figure 5.11. The following correlation due to Wronski and Szembek–Stoeger [1988] takes account of most of the data in the literature and offers some improvement over equation (5.62), especially in the low Reynolds number region:

$$\varepsilon j_m = [0.097(Re')^{0.30} + 0.75(Re')^{0.61}]^{-1} \tag{5.64}$$

No analogous heat transfer studies have been reported with non-Newtonian fluids.

5.6.7 Visco-elastic and surface effects in packed beds

So far the flow of inelastic non-Newtonian fluids through unconsolidated beds of particles has been considered. The pore size has been substantially larger than the characteristic dimensions of the polymer molecules or of the flocs. In such cases, there is negligible or no interaction between the walls of the pore and the polymer molecules or flocs, and the rheological properties (values of m and n, for instance) may be inferred from pressure drop–flow rate data obtained in packed beds and tubes. Visco-elastic effects and other phenomena such as blockage of pores, polymer adsorption, and apparent slip effects, observed in low permeability and consolidated systems will now be briefly discussed.

(i) Visco-elastic effects

Visco-elastic effects may be quantified in terms of a Deborah or Weissenberg number (De). For flow in packed beds, De is simply defined as:

$$\text{De} = \frac{\lambda_f V_0}{d} = \frac{\lambda_f}{(d/V_0)} \tag{5.65}$$

where λ_f is a characteristic time of the fluid and (d/V_0) is a measure of the time-scale of the flow. The characteristic time, λ_f, is determined from rheological measurements but the time-scale of the process clearly depends upon the kinematic conditions, principally, flow rate and packing size. At low velocities, the time-scale of flow (d/V_0) is large as compared with λ_f and De is therefore small. Thus, a fluid element is able continually to adjust to its changing flow geometry and area, and no effects arise from visco-elasticity. In other-words, the pressure loss through a bed of particles is determined essentially by the effective viscosity of the fluid at low flow rates and Deborah numbers. With gradual increase in flow rate, (d/V_0) decreases progressively (hence the value of De increases) and the fluid elements are no longer able to re-adjust themselves to the rapidly changing flow conditions. The energy expended in squeezing (extensional flow) through 'throats' or 'narrow passages' in the bed is dissipated, as the fluid element has insufficient time ($\lambda_f > (d/V_0)$) to relax or recover its original state; this, in turn, may result in a substantially increased pressure drop. If the loss coefficient $\Lambda (= f\text{Re}^*)$ is plotted against Deborah number, or simply (V_0/d), if λ_f is constant for a fluid. Λ remains constant (at 150 or 180) at low values of (V_0/d) and, beyond a critical value of (V_0/d), it begins to rise gradually, as seen in Figure 5.12 which refers to the flow of aqueous solutions of hydroxy propyl guar (a polymer used as a cracking fluid in drilling and enhanced oil recovery operations) through a packed bed of 200 to 900 μm glass beads [Vorwerk and Brunn, 1994]. As polymer concentration is increased, visco-elastic behaviour becomes more marked, and values of $(\Lambda/180)$ begin to deviate from unity at progressively lower values of (V_0/d), as shown in Figure 5.12.

The occurrence of very high pressure drops is well documented in the literature for a range of chemically different polymer melts and solutions

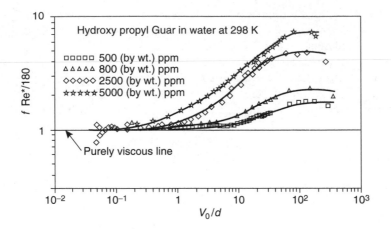

Figure 5.12   *Visco-elastic effects in packed beds*

[Chhabra, 1993a], but there is little evidence concerning the critical value of Deborah number at which visco-elastic effects are significant or the extent to which Λ is increased as a result of visco-elasticity. The nature of the polymer solution and the geometry of the bed interact in a complex manner. For instance, Marshall and Metzner [1967] observed visco-elastic effects at De $\simeq 0.05 - 0.06$ whereas Michele [1977] reported the critical value of De to be as large as 3! The main difficulty stems from the fact that there is no simple method of calculating the value of λ_f in an unambiguous manner; some workers have inferred its value from steady shear data (first normal stress difference and/or shear viscosity) while others have used transient rheological tests to deduce the value of λ_f. However, polymer solutions certainly undergo substantial extensional deformation, in addition to shearing, in the narrow capillaries in beds of particles and in other porous matrices such as screens, mats and dense rocks. No satisfactory correlations for the frictional pressure gradient for visco-elastic fluids in packed beds are therefore available.

(ii) Anomalous surface effects

As mentioned previously, additional complications arise when the size of the pores or flow capillaries in the porous matrix in only slightly larger than the polymer molecules themselves. Polymer molecules may then be retained by both adsorption on to the walls of the pores and by mechanical entrapment, thereby leading to partial or even complete blockage of pores and increased pressure drops. The exact cause of the sharp rise in pressure gradient as shown in Figure 5.12 is therefore difficult to pinpoint. The existence of a slip velocity at the wall of the pore may explain many of the observations. However, at present there is no way of assessing *a priori* whether slip effects would be important or not in any new situation. Extensive reviews of the developments in this field are available in the literature [Cohen, 1988; Agarwal *et al.*, 1994].

5.7 Liquid–solid fluidisation

5.7.1 Effect of liquid velocity on pressure gradient

As shown schematically in Figure 5.7 for the upward flow of a liquid through a bed of particles, a linear relation is obtained between the pressure gradient and the superficial velocity (on logarithmic coordinates) up to the point where the bed is fluidised and where expansion of the bed starts to occur (*A*), but the slope of the curve then gradually diminishes as the bed expands. As the liquid velocity is gradually increased, the pressure drop passes through a maximum value (*B*) and then falls slightly and eventually attains a nearly constant value, independent of liquid velocity (*CD*). If the velocity of flow is reduced again, the bed contracts until it reaches the condition where the particles are just resting in contact with one another (*E*); it then has the maximum stable voidage for a fixed bed of the particles in question. No further change in the voidage of the bed occurs as the velocity is reduced, provided it is not shaken or vibrated. The pressure drop (*EF*) in this re-formed packed bed may then be less than that in the original bed at the same velocity. If the liquid velocity were now to be increased again, the new curve (*EF*) would normally be re-traced and the slope will suddenly drop to zero at the fluidising point *E*; this condition is difficult to produce, however, because the bed tends to become consolidated again as a result of stray vibrations and disturbances.

In an ideal fluidised bed, the pressure drop corresponding to ECD in Figure 5.7 is equal to the buoyant weight of the bed per unit area. In practice, however, deviations from this value may be observed due to channelling and interlocking of particles. Point *B* is situated above *CD* because the frictional forces between the particles must be overcome before the rearrangement of particles can occur.

Figure 5.13 shows representative experimental results of $\log(-\Delta p)$ versus $\log(V_0)$ for the flow of an aqueous carboxymethyl cellulose solution ($n = 0.9$) through a bed of 3.57 mm glass spheres in a 100 mm diameter column [Srinivas and Chhabra, 1991]. In spite of the scatter of the data at the onset of fluidisation, the regions up to *A* (fixed bed) and *CD* (fluidised) can be clearly seen in this case.

The velocity corresponding to the incipient fluidising point can also be calculated from the relation presented in Section 5.6, by equating the pressure drop across the bed to its buoyant weight per unit area. The voidage at the onset of fluidisation should correspond to the maximum value attainable in the fixed bed. Hence, in a fluidised bed, the total frictional force on the particles

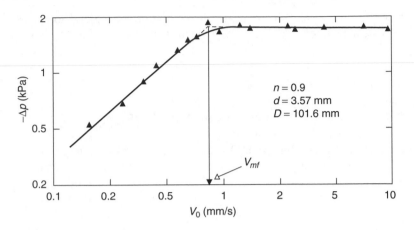

Figure 5.13   *Experimental pressure drop–superficial velocity curve and determination of minimum fluidisation velocity*

must be equal to the buoyant weight of the bed. Thus, in a bed of height L and voidage ε:

$$-\Delta p = (\rho_s - \rho)(1 - \varepsilon)Lg \tag{5.66}$$

where g is the acceleration due to gravity.

There may be some discrepancy between the observed and calculated minimum fluidisation velocities. This difference is attributable to channelling or wall effects, or to the agglomeration of small particles. Equation (5.66) applies from the initial expansion of the bed up to high values (**ca.** 0.95) of voidage.

For streamline flow of a power-law fluid through a fixed bed of spherical particles, the relationship between liquid velocity V_0, bed voidage ε and pressure drop $(-\Delta p)$ is given by equation (5.46). Substituting for $(-\Delta p)$ from equation (5.66), equation (5.46) becomes:

$$V_0 = \left[\frac{(\rho_s - \rho)g\varepsilon^3 d^{n+1}}{180\,m(1 - \varepsilon)^n} \left(\frac{4n}{3n+1} \right)^n \left(\frac{15\sqrt{2}}{\varepsilon^2} \right)^{1-n} \right]^{(1/n)} \tag{5.67}$$

For $n = 1$, equation (5.67) reduces to the well-known Kozeny–Carman equation;

$$V_0 = 0.005\,55 \frac{(\rho_s - \rho)g\varepsilon^3 d^2}{\mu(1 - \varepsilon)} \tag{5.68}$$

where μ is the Newtonian fluid viscosity.

5.7.2 Minimum fluidising velocity

At the point of incipient fluidisation, the bed voidage, ε_{mf}, depends on the shape and size range of the particles, but is approximately equal to 0.4 for isometric particles. The minimum fluidising velocity V_{mf} for a power-law fluid in streamline flow is then obtained by substituting $\varepsilon = \varepsilon_{mf}$ in equation (5.67). Although this equation applies only at low values of the bed Reynolds numbers (<1), this is not usually a limitation at the high apparent viscosities of most non-Newtonian materials.

At high velocities, the flow may no longer be streamline, and a more general equation must be used for the pressure gradient in the bed, such as equation (5.56). Substituting $\varepsilon = \varepsilon_{mf}$ and $V_0 = V_{mf}$, this equation becomes:

$$f_{mf} = \frac{150}{\mathrm{Re}^*_{mf}} + 1.75 \tag{5.69}$$

Replacement of f_{mf} by the Galileo number eliminates the unknown velocity V_{mf} which appears in Re^*_{mf}. Multiplying both sides of equation (5.69) by $(\mathrm{Re}^*_{mf})^{2/(2-n)}$:

$$\mathrm{Ga}_{mf} = 150(\mathrm{Re}^*_{mf})^{n/(2-n)} + 1.75(\mathrm{Re}^*_{mf})^{2/(2-n)} \tag{5.70}$$

where $\mathrm{Ga}_{mf} = f_{mf}(\mathrm{Re}^*_{mf})^{2/(2-n)}$

$$= \frac{(\rho_s - \rho)gd\varepsilon^3_{mf}}{\rho V^2_{mf}} \cdot (\mathrm{Re}^*_{mf})^{2/(2-n)} \tag{5.71}$$

For a given liquid (known m, n, ρ) and particle (ρ_s, d, ε_{mf}) combination, equation (5.70) can be solved for Re^*_{mf} which in turn enables the value of the minimum fluidising velocity to be calculated, as illustrated in example 5.7.

Example 5.7

A bed consists of uniform glass spheres of size 3.57 mm diameter (density = 2500 kg/m³). What will be the minimum fluidising velocity in a polymer solution of density, 1000 kg/m³, with power-law constants: $n = 0.6$ and $m = 0.25$ Pa·s^n? Assume the bed voidage to be 0.4 at the point of incipient fluidisation.

Solution

First calculate the value of the Galielo number using equation (5.71) which after substitution for Re^*_{mf} from equation (5.47) becomes:

$$\mathrm{Ga}_{mf} = \frac{(\rho_s - \rho)gd\varepsilon^3_{mf}}{\rho} \left[\frac{\rho d^n}{m(1-\varepsilon_{mf})^n} \left(\frac{4n}{3n+1}\right)^n \left(\frac{15\sqrt{2}}{\varepsilon^2_{mf}}\right)^{1-n} \right]^{2/(2-n)}$$

Substituting the numerical values:

$$\text{Ga}_{mf} = \frac{(2500 - 1000)(9.81)(3.57 \times 10^{-3})(0.4)^3}{1000}$$

$$\times \left[\frac{1000(3.57 \times 10^{-3})^{0.6}}{0.25(1 - 0.4)^{0.6}} \left(\frac{4 \times 0.6}{3 \times 0.6 + 1} \right)^{0.6} \times \left(\frac{15\sqrt{2}}{(0.4)^2} \right)^{1-0.6} \right]^{2/(2-0.6)}$$

$$= 83.2$$

Substituting this value in equation (5.70):

$$150(\text{Re}_{mf}^*)^{(0.6)/(2-0.6)} + 1.75(\text{Re}_{mf}^*)^{2/(2-0.6)} = 83.2$$

or $\qquad 150(\text{Re}_{mf}^*)^{0.43} + 1.75(\text{Re}_{mf}^*)^{1.43} = 83.2$

By trial and error procedure, $\text{Re}_{mf}^* = 0.25$. Thus,

$$\frac{\rho V_{mf}^{2-n} d^n}{m(1 - \varepsilon_{mf})^n} \left(\frac{4n}{3n + 1} \right)^n \left(\frac{15\sqrt{2}}{\varepsilon_{mf}^2} \right)^{1-n} = 0.25$$

On substituting numerical values and solving, the unknown velocity, $V_{mf} = 0.00236\,\text{m/s} = 2.36\,\text{mm/s}$.

This value compares well with the experimental value of 2.16 mm/s for this system [Srinivas and Chhabra, 1991].□

5.7.3 Bed expansion characteristics

Liquid–solid fluidised systems are generally characterised by the regular expansion of the bed which takes place as the liquid velocity increases from the minimum fluidisation velocity to a value approaching the terminal falling velocity of the particles. The general form of relation between velocity and bed voidage is found to be similar for both Newtonian and inelastic power-law liquids. For fluidisation of uniform spheres by Newtonian liquids, equation (5.21), introduced earlier to represent hindered settling data, is equally applicable:

$$\frac{V_0}{V} = (1 - C)^Z = \varepsilon^Z \qquad\qquad (5.21)$$

where V_0 is now the superficial velocity of the liquid in the fluidised bed and V is the terminal settling velocity of a single sphere in the same liquid; Z is a constant related to the particle Archimedes number and to the particle-to-vessel diameter ratio by equation (5.22).

Figure 5.14 shows a typical bed expansion behaviour for 3.57 mm glass spheres fluidised by aqueous carboxymethyl cellulose solutions ($n = 0.84$ and $n = 0.9$), and the behaviour is seen to conform to the form of equation

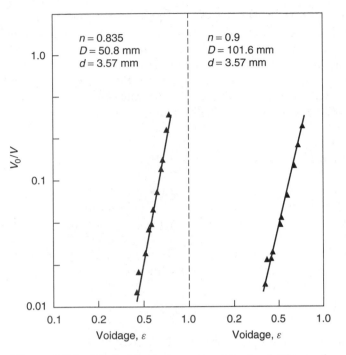

Figure 5.14 *Typical bed expansion data for 3.57 mm glass spheres fluidised by shear-thinning polymer solutions [Srinivas and Chhabra, 1991]*

(5.21). Qualitatively similar results have been reported by many workers, and in a recent review [Chhabra, 1993a,b] it has been shown that equation (5.22) correlates most of the data available for inelastic power-law fluids $(0.6 \leq n \leq 1; (d/D) \leq 0.16)$. Values calculated from equation (5.22) and experimental values of Z differ by less than 10%. This suggests that the modified Archimedes number (equation 5.12) takes account of power-law shear-thinning behaviour. On the other hand, much larger values of Z have been reported for fluidisation with visco-elastic polymer solutions [Briend *et al.*, 1984; Srinivas and Chhabra, 1991], but no systematic study has been made to predict the value of Z for visco-elastic liquids.

5.7.4 Effect of particle shape

Little is known about the influence of particle shape on the minimum fluidising velocity and bed expansion of liquid fluidised beds even for Newtonian liquids [Couderc, 1985; Flemmer *et al.*, 1993]. The available scant data suggests that, if the diameter of a sphere of equal volume is used together with its sphericity factor, satisfactory predictions of the minimum fluidising velocity are obtainable from the expressions for spherical particles. Only one experimental study has been identified [Sharma and Chhabra, 1992] which is of relevance, and in that gravel chips (of sphericity of 0.6) were fluidised by aqueous carboxymethyl cellulose solutions ($0.78 \leq n \leq 0.91$). If an effective particle size (diameter of a sphere of equal volume multiplied by its sphericity) is used in lieu of sphere diameter in equation (5.70), the values of V_{mf} are predicted with a mean error of 18%. Similarly, the values of Z were within $\pm 10\%$ of the predictions of equation (5.22), although the latter correlation consistently underpredicted the value of Z for gravel chips.

5.7.5 Dispersion in fluidised beds

The material presented earlier on dispersion in packed beds (Section 5.6.5) is also relevant to fluidised beds. However, in the only experimental study reported [Wen and Fan, 1973], longitudinal dispersion coefficient D_L was measured by introducing a fluorescein dye into beds of glass and of aluminium spheres fluidised by mildly shear-thinning carboxymethyl cellulose solutions ($n = 0.86$ and $n = 0.89$). The limited results are well correlated by the following modified form of equation (5.61):

$$\text{Pe}\chi = 0.2 + 0.011\text{Re}_1^{0.48} \tag{5.72}$$

where $\chi = 1$ for packed beds and $\chi = (V_0/V_{mf})^{2-n}$ for fluidised systems; the Reynolds number is based on the superficial velocity.

5.7.6 Liquid–solid mass transfer in fluidised beds

Kumar and Upadhyay [1980, 1981] have measured the rates of dissolution of spheres and cylindrical pellets of benzoic acid in a bed fluidised by an aqueous carboxymethyl cellulose solution ($n = 0.85$). Mass transfer coefficients were calculated on the basis of a plug-flow model from the measured rate of weight loss of the particles, and using the log mean value of concentration driving force at the inlet and outlet of the bed. Their data were satisfactorily correlated using equation (5.62) (as can be seen in Figure 5.15), though the results are over-predicted at low values of the Reynolds number. Subsequently, Hwang *et al.* [1993] have measured concentrations within a fluidised bed of cylindrical pellets of benzoic acid for a number of aqueous solutions of carboxymethyl cellulose ($0.63 \leq n \leq 0.92$). The longitudinal concentration profile was adequately represented by a model combining plug-flow with axial dispersion. Using equation (5.72) to calculate D_L, together with the experimentally determined concentration profiles, mass transfer coefficients were evaluated in terms of the particle Reynolds number ($0.01 \leq \text{Re}' \leq 600$) and the Schmidt number. They proposed the following correlation for the j_m factor defined by equation (5.63):

$$\log(\varepsilon j_m) = 0.169 - 0.455 \log \text{Re}' - 0.0661(\log \text{Re}')^2 \tag{5.73}$$

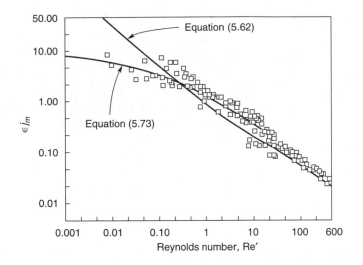

Figure 5.15 *Correlations for liquid–solid mass transfer in beds fluidised by power-law liquids*

The effective viscosity used in the definitions of the Reynolds and Schimdt numbers is estimated from equation (5.53). Figure 5.15 also includes the predictions of equation (5.73) along with the data of Hwang *et al.* [1993]. The scatter is such that, although both equations (5.62) and (5.73) give similar predictions at Reynolds numbers greater than about 1, equation (5.73) seems better in the low Reynolds number region.

5.8 Further reading

Chhabra, R.P., *Bubbles, Drops and Particles in Non-Newtonian Fluids*. CRC Press, Boca Raton, FL (1993).

Clift, R., Grace, J. and Weber, M.E. *Bubbles, Drops and Particles*. Academic, New York (1978).

Coulson, J.M. and Richardson, J.F. *Chemical Engineering*. Vol. II, 4th edn. Butterworth-Heinemann, Oxford (1991).

Davidson, J.F., Clift, R. and Harrison, D. (editors), Fluidisation, 2nd edn. Academic, New York (1985).

Dullien, F.A.L., *Porous Media: Fluid Transport and Pore Structure*. 2nd edn. Academic, New York (1992).

5.9 References

Agarwal, P.K. and O'Neill, B.K., *Chem. Eng. Sci.* **43** (1988) 2487.

Agarwal, U.S., Dutta, A. and Mashelkar, R.A., *Chem. Eng. Sci.* **49** (1994) 1693.

Allen, E. and Uhlherr, P.H.T., *J. Rheol.* **33** (1989) 627.

Ansley, R.W. and Smith, T.N., *AIChEJ.* **13** (1967) 1193.

Astarita, G. and Appuzzo, G., *AIChEJ.* **11** (1965) 815.

Atapattu, D.D., Chhabra, R.P. and Uhlherr, P.H.T., *J. Non-Newt. Fluid Mech.* **59** (1995) 245.
Bagchi, A. and Chhabra, R.P., *Powder Technol.* **68** (1991) 85.
Beauline, M. and Mitsoulis, E., *J. Non-Newt. Fluid Mech.* **72** (1997) 55.
Beris, A.N., Tsamapolous, J., Armstrong, R.C. and Brown, R.A., *J. Fluid Mech.* **158** (1985) 219.
Blackery, J. and Mitsoulis, E., *J. Non-Newt. Fluid Mech.* **70** (1997) 59.
Bobroff, S. and Phillips, R.J., *J. Rheol,* **42** (1998) 1419.
Bond, W.N. and Newton, D.A., *Phil. Mag.* **5** (1928) 794.
Brea, F.M., Edwards, M.F. and Wilkinson, W.L., *Chem. Eng. Sci.* **31** (1976) 329.
Briend, P., Chavarie, C., Tassart, M. and Hlavacek, B., *Can. J. Chem. Eng.* **62** (1984) 26.
Brookes, G.F. and Whitmore, R.L., *Rheol. Acta.* **7** (1968) 188.
Brookes, G.F. and Whitmore, R.L., *Rheol. Acta.* **8** (1969) 472.
Brown, G.G. and associates, Unit Operations, Wiley, New York (1950).
Carman, P.C., *Flow of Gases through Porous Media.* Butterworths, London (1956).
Chhabra, R.P., in *Encyclopedia of Fluid Mechanics.* (edited by Cheremisinoff, N.P., Gulf, Houston), Vol. 1 (1986) 983.
Chhabra, R.P., *Chem. Eng. & Process.* **28** (1990) 89.
Chhabra, R.P., *Bubbles, Drops and Particles in Non-Newtonian Fluids.* CRC Press, Boca Raton, FL (1993a).
Chhabra, R.P., *Adv. Heat Transfer.* **23** (1993b) 178.
Chhabra, R.P., Advances in Transport Processes **9** (1993c) 501.
Chhabra, R.P., in *Handbook of Polymer Processing Technology.* (edited by Cheremisinoff, N.P. Marcel-Dekker, New York), Chapter 1 (1996).
Chhabra, R.P., Soares, A.A and Ferreira, J.M., *Can. J. Chem. Eng.* **76** (1998) 1051.
Chhabra, R.P. and Srinivas, B.K., *Powder Technol.* **67** (1991) 15.
Chhabra, R.P. and Uhlherr, P.H.T., *Chem. Eng. Commun.* **5** (1980) 115.
Chhabra, R.P. and Uhlherr, P.H.T., in *Encyclopedia of Fluid Mechanics.* (edited by Cheremisinoff, N.P., Gulf, Houston) **7** (1988) 253.
Chhabra, R.P., Uhlherr, P.H.T. and Richardson, J.F., *Chem. Eng. Sci.* **51** (1996) 4532.
Chhabra, R.P., Unnikrishnan, A. and Nair, V.R.U., *Can. J. Chem. Eng.* **70** (1992) 716.
Clift, R., Grace, J. and Weber, M.E., *Bubbles, Drops and Particles.* Academic, New York (1978).
Cohen, Y., in *Encyclopedia of Fluid Mechanics.* (edited by Cheremisinoff, N.P., Gulf, Houston) **7** (1988) Chapter 14.
Cohen, Y. and Metzner, A.B., *AIChEJ.* **27** (1981) 705.
Couderc, J.-P., in *Fluidisation.* (edited by Davidson, J.F., Clift, R. and Harrison, D. Academic, New York) 2nd edn. (1985) Chapter 1.
Coulson, J.M., *Trans. Inst. Chem. Engrs.* **27** (1949) 237.
Coulson, J.M. and Richardson, J.F., *Chemical Engineering.* Vol. 2, 4th edn. Butterworth-Heinemann, Oxford (1991).
Davidson, J.F., Clift, R. and Harrison, D. (editors), Fluidisation, 2nd edn. Academic, New York (1985).
Degand, E. and Walters, K., *J. Non-Newt. Fluid Mech.* **57** (1995) 103.
DeKee, D. and Chhabra, R.P., in *Transport Processes in Bubbles, Drops and Particles.* (edited by Chhabra, R.P. and DeKee, D. Hemisphere, New York), (1992) Chapter 2.
DeKee, D., Chhabra, R.P. and Rodrigue, D, in *Handbook of Polymer Processing Technology.* (edited by Cheremisinoff, N.P. Marcel-Dekker, New York) (1996) Chapter 3.
Dharamadhikari, R.V. and Kale, D.D., *Chem. Eng. Sci.* **40** (1985) 427.
Dullien, F.A.L., *Porous Media: Fluid Transport and Pore Structure.* 2nd edn. Academic, New York (1992).
Ergun, S., *Chem. Eng. Prog.* **48** (Feb, 1952) 89.
Flemmer, R.C., Pickett, J. and Clark, N.N., *Powder Technol.* **77** (1993) 123.
German, R.M., *Particle Packing Characteristics*, Metal Powder Industries Federation. Princeton, NJ (1989).

Ghosh, U.K., Upadhyay, S.N. and Chhabra, R.P., *Adv. Heat Transf.* **25** (1994) 251.
Graham, D.I. and Jones, T.E.R., *J. Non-Newt, Fluid Mech.* **54** (1995) 465.
Greenkorn, R.A., *Flow Phenomena in Porous Media.* Marcel-Dekker, New York (1983).
Gu, D. and Tanner, R.I., *J. Non-Newt, Fluid Mech.* **17** (1985) 1.
Happel, J. and Brenner, H., *Low Reynolds Number Hydrodynamics.* Prentice Hall, Englewood Cliffs, NJ (1965).
Hariharaputhiran, M., Shankar Subramanian, R., Campbell, G.A. and Chhabra, R.P., *J. Non-Newt. Fluid Mech.* **79** (1998) 87.
Hassell, H.L. and Bondi, A., *AIChEJ.* **11** (1965) 217.
Hirose, T. and Moo-Young, M., *Can. J. Chem. Eng.* **47** (1969) 265.
Hwang, S.-J., Lu, C.-B. and Lu, W.-J., *Chem. Eng. J.* **52** (1993) 131.
Jacks, J.-P. and Merrill, R.P., *Can. J. Chem. Eng.* **49** (1971) 699.
Kemblowski, Z., Dziubinski, M. and Sek, J., *Advances in Transport Processes.* **5** (1987) 117.
Kumar, S. and Upadhyay, S.N., *Lett. Heat Mass Transf.* **7** (1980) 199.
Kumar, S. and Upadhyay, S.N., *Ind. Eng. Chem. Fundam.* **20** (1981) 186.
Leva, M., *Chem. Eng.* **64** (1957) 245.
Machac, I. and Dolejs, V., *Chem. Eng. Sci.* **36** (1981) 1679.
Machac, I., Ulbrichova, I., Elson, T.P. and Cheesman, D.J., *Chem. Eng. Sci.* **50** (1995) 3323.
Marshall, R.J. and Metzner, A.B., *Ind. Eng. Chem. Fundam.* **6** (1987) 393.
Michele, H., *Rheol. Acta.* **16** (1977) 413.
Miller, C., *Ind. Eng. Chem. Fundam.* **11** (1972) 524.
Mishra, P., Singh, D. and Mishra, I.M., *Chem. Eng. Sci.* **30** (1975) 397.
Nakano, Y. and Tien, C., *AIChEJ.* **16** (1970) 569.
Nguyen, Q.D. and Boger, D.V., *Annu. Rev. Fluid Mech.* **24** (1992) 47.
Payne, L.W. and Parker, H.W., *AIChEJ.* **19** (1973) 202.
Rao, P.T. and Chhabra, R.P., *Powder Technol.* **77** (1993) 171.
Richardson, J.F. and Zaki, W.N., *Trans. Inst. Chem. Engrs.* **32** (1954) 35.
Sabiri, N.E. and Comiti, J., *Chem. Eng. Sci.* **50** (1995) 1193.
Sen, S., *MS Dissertation.* Brigham Young University, Provo, Utah (1984).
Sharma, M.K. and Chhabra, R.P., *Can. J. Chem. Eng.* **70** (1992) 586.
Srinivas, B.K. and Chhabra, R.P., *Chem. Eng. & Process.* **29** (1991) 121.
Tiu, C., in *Developments in Plastics Technology-2.* (edited by Whelan, A and Craft, J.L., Elsevier, Amsterdam) (1985) Chapter 7.
Tiu, C., Zhou, J.Z.Q., Nicolae, G., Fang, T.N. and Chhabra, R.P., *Can. J. Chem. Eng.* **75** (1997) 843.
Tripathi, A., Chhabra, R.P. and Sundararajan, T., *Ind. Eng. Chem. Res.* **33** (1994) 403.
Tripathi, A. and Chhabra, R.P., *AIChEJ.* **42** (1995) 728.
Tyabin, N.V., *Colloid, J., (USSR)* **19** (1953) 325.
Uhlherr, P.H.T., *Proc. 9th National Conf. on Rheology*, Adelaide, Australia. p. 231 (1986).
Uhlherr, P.H.T. and Chhabra, R.P., *Can. J. Chem. Eng.* **73** (1995) 918.
Valentik, L. and Whitmore, R.L., *Brit. J. Appl. Phys.* **16** (1965) 1197.
Venu Madhav, G. and Chhabra, R.P., *Powder Technol.* **78** (1994) 77.
Volarovich, M.P. and Gutkin, A.M., *Colloid J.* (USSR) **15** (1953) 153.
Vorwerk, J. and Brunn, P.O., *J. Non-Newt. Fluid Mech.* **51** (1994) 79.
Walters, K. and Tanner, R.I., in *Transport Processes in Bubbles, Drops and Particles.* (edited by Chhabra, R.P. and DeKee, D., Hemisphere, New York) (1992) Chapter 3.
Wen, C.Y. and Fan, L.-S., *Chem. Eng. Sci.* **28** (1973) 1768.
Wen, C.Y. and Yim, J., *AIChEJ.* **17** (1971) 1503.
Wronski, S. and Szembek-Stoeger, M., *Inzynieria Chem. Proc.* **4** (1988) 627.
Yu, A.-B. and Standish, N., *Powder Technol.* **74** (1993) 205.
Yu, A.-B., Standish, N. and McLean, A., *J. Amer. Ceram. Soc.* **76** (1993) 2813.

5.10 Nomenclature

Dimensions
in **M, L, T**

A	area of flow (m^2)	$\mathbf{L}^2$
Ar	Archimedes number (–)	$\mathbf{M}^0\mathbf{L}^0\mathbf{T}^0$
Bi	Bingham number (–)	$\mathbf{M}^0\mathbf{L}^0\mathbf{T}^0$
Bi*	modified Bingham number (–)	$\mathbf{M}^0\mathbf{L}^0\mathbf{T}^0$
C	volume fraction of solids in a suspension or in a fluidised bed (–)	$\mathbf{M}^0\mathbf{L}^0\mathbf{T}^0$
C_D	drag coefficient (–)	$\mathbf{M}^0\mathbf{L}^0\mathbf{T}^0$
C_{DN}	drag coefficient in a Newtonian fluid (–)	$\mathbf{M}^0\mathbf{L}^0\mathbf{T}^0$
D	tube or vessel diameter (m)	$\mathbf{L}$
D_h	hydraulic mean diameter for a packed bed (m)	$\mathbf{L}$
De	Deborah number (–)	$\mathbf{M}^0\mathbf{L}^0\mathbf{T}^0$
D_L	longitudinal dispersion coefficient (m^2/s)	$\mathbf{L}^2\mathbf{T}^{-1}$
D_R	radial dispersion coefficient (m^2/s)	$\mathbf{L}^2\mathbf{T}^{-1}$
d	sphere diameter (m)	$\mathbf{L}$
F_D	total drag force on a particle (N)	$\mathbf{MLT}^{-2}$
F_n	normal component of drag force due to pressure (N)	$\mathbf{MLT}^{-2}$
F_t	tangential component of drag force due to shear stress (N)	$\mathbf{MLT}^{-2}$
f	friction factor for flow in packed beds (–) or wall correction factor for particle settling in a tube (–)	$\mathbf{M}^0\mathbf{L}^0\mathbf{T}^0$
f_0	value of wall correction factor in streamline region (–)	$\mathbf{M}^0\mathbf{L}^0\mathbf{T}^0$
f_∞	value of wall correction factor in the high Reynolds number region (–)	$\mathbf{M}^0\mathbf{L}^0\mathbf{T}^0$
f_W	wall correction factor for packed beds (–)	$\mathbf{M}^0\mathbf{L}^0\mathbf{T}^0$
g	acceleration due to gravity (m/s^2)	$\mathbf{LT}^{-2}$
Ga	Galileo number (–)	$\mathbf{M}^0\mathbf{L}^0\mathbf{T}^0$
He	Hedström number (–)	$\mathbf{M}^0\mathbf{L}^0\mathbf{T}^0$
j_m	j factor for mass transfer (–)	$\mathbf{M}^0\mathbf{L}^0\mathbf{T}^0$
K_0	constant dependent on shape of cross-section (–)	$\mathbf{M}^0\mathbf{L}^0\mathbf{T}^0$
k	permeability of porous medium (m^2)	$\mathbf{L}^2$
k_c	mass transfer coefficient (m/s)	$\mathbf{LT}^{-1}$
L	height of bed (m)	$\mathbf{L}$
L_e	effective path length traversed by a fluid element (m)	$\mathbf{L}$
m	power law consistency coefficient (Pa·s^n)	$\mathbf{ML}^{-1}\mathbf{T}^{n-2}$
m'	apparent power law consistency coefficient (Pa·s$^{n'}$), equation (5.48)	$\mathbf{ML}^{-1}\mathbf{T}^{n'-2}$
n	flow behaviour index (–)	$\mathbf{M}^0\mathbf{L}^0\mathbf{T}^0$
n'	apparent flow behaviour index (–)	$\mathbf{M}^0\mathbf{L}^0\mathbf{T}^0$
p	pressure (Pa)	$\mathbf{ML}^{-1}\mathbf{T}^{-2}$
$(-\Delta p/L)$	pressure gradient (Pa/m)	$\mathbf{ML}^{-2}\mathbf{T}^{-2}$
Pe	Peclet number (–)	$\mathbf{M}^0\mathbf{L}^0\mathbf{T}^0$
Q	volumetric flow rate (m^3/s)	$\mathbf{L}^3\mathbf{T}^{-1}$
R	sphere radius (m)	$\mathbf{L}$
Re	particle Reynolds number (–)	$\mathbf{M}^0\mathbf{L}^0\mathbf{T}^0$
Re$_B$	particle Reynolds number based on Bingham plastic viscosity (–)	$\mathbf{M}^0\mathbf{L}^0\mathbf{T}^0$
Re$'$, Re*	modified Reynolds number for packed beds (–), equations (5.59) and (5.53)	$\mathbf{M}^0\mathbf{L}^0\mathbf{T}^0$

		Dimensions in $\mathbf{M}$, $\mathbf{L}$, $\mathbf{T}$
Re_1	modified Reynolds number, equation (5.61b) (–)	$\mathbf{M^0L^0T^0}$
r	coordinate (m)	$\mathbf{L}$
S	specific surface of a particle (m^{-1})	$\mathbf{L^{-1}}$
S_B	specific surface of a bed of particles (m^{-1})	$\mathbf{L^{-1}}$
T	tortuosity factor (–)	$\mathbf{M^0L^0T^0}$
V	terminal falling velocity of a sphere (m/s)	$\mathbf{LT^{-1}}$
V_i	interstitial or pore velocity (m/s)	$\mathbf{LT^{-1}}$
V_m	terminal falling velocity of a sphere in a bounded fluid medium (m/s)	$\mathbf{LT^{-1}}$
V_0	superficial velocity (m/s)	$\mathbf{LT^{-1}}$
X	drag correction factor (–)	$\mathbf{M^0L^0T^0}$
Y	yield parameter (–)	$\mathbf{M^0L^0T^0}$
Z	fluidisation/hindered settling index (–)	$\mathbf{M^0L^0T^0}$

Greek letters

$\dot{\gamma}_w$	wall shear rate (s^{-1})	$\mathbf{T^{-1}}$
δ	ratio of viscosity of the fluid sphere and that of the continuous medium (–)	$\mathbf{M^0L^0T^0}$
ε	voidage (–)	$\mathbf{M^0L^0T^0}$
θ	coordinate (–)	$\mathbf{M^0L^0T^0}$
λ_f	fluid relaxation time (s)	$\mathbf{T}$
μ	viscosity (Pa·s)	$\mathbf{ML^{-1}T^{-1}}$
ρ	density (kg/m^3)	$\mathbf{ML^{-3}}$
τ_0	yield stress, (Pa)	$\mathbf{ML^{-1}T^{-2}}$
$\tau_{r\theta}$	$r\theta$-component of stress, (Pa)	$\mathbf{ML^{-1}T^{-2}}$
ϕ	τ_0^B/τ_w	$\mathbf{M^0L^0T^0}$

Subscripts/superscripts

b	bubble
s	solid
B	Bingham model parameter
eff	effective.

Chapter 6

Heat transfer characteristics of non-Newtonian fluids in pipes

6.1 Introduction

In many chemical and processing applications, fluids need to be heated or cooled and a wide range of equipment may be utilized. Examples include double pipe and shell and tube heat exchangers, and stirred vessels fitted with cooling coils or jackets. Sometimes, heat is generated in the process, as in extrusion which is extensively carried out in the polymer and food industry. It may also be necessary to reduce the rate at which heat is lost from a vessel or to ensure that heat is removed at a sufficient rate in equipment such as screw conveyors. In most applications, it is the rate of heat transfer within process equipment which is of principal interest. However, with thermally sensitive materials (such as foodstuffs, fermentation froths, pharmaceutical formulations), the temperature profiles must be known and maximum permissible temperatures must not be exceeded.

Because of their high consistencies, non-Newtonian materials are most frequently processed under conditions of laminar flow. Furthermore, shear stresses are generally so high that viscous generation of heat can rarely be neglected, and the temperature dependence of the rheological properties adds to the complexity of the mass, momentum and energy balance equations. Numerical techniques are often needed to obtain solutions, even for highly idealized conditions of flow.

Much of the research activity in this area has related to heat transfer to inelastic non-Newtonian fluids in laminar flow in circular and non-circular ducts. In recent years, some consideration has also been given to heat transfer to/from non-Newtonian fluids in vessels fitted with coils and jackets, but little information is available on the operation of heat exchange equipment with non-Newtonian fluids. Consequently, this chapter is concerned mainly with the prediction of heat transfer rates for flow in circular tubes. Heat transfer in external (boundary layer) flows is discussed in Chapter 7, whereas the cooling/heating of non-Newtonian fluids in stirred vessels is dealt with in Chapter 8. First of all, however, the thermo-physical properties of the commonly used non-Newtonian materials will be described.

6.2 Thermo-physical properties

The most important thermo-physical properties of non-Newtonian fluids are thermal conductivity, density, specific heat, surface tension and coefficient of thermal expansion. While the first three characteristics enter into virtually all heat transfer calculations, surface tension often exerts a strong influence on boiling heat transfer and bubble dynamics in non-Newtonian fluids, as seen in Chapter 5. Likewise, the coefficient of thermal expansion is important in heat transfer by free convection.

Only a very limited range of measurements of physical properties has been made, and for dilute and moderately concentrated aqueous solutions of commonly used polymers including carboxymethyl cellulose, polyethylene oxide, carbopol, polyacrylamide, density, specific heat, thermal conductivity, coefficient of thermal expansion and surface tension differ from the values for water by no more than 5–10% [Porter, 1971; Cho and Hartnett, 1982; Irvine, Jr. *et al.*, 1987]. Thermal conductivity might be expected to be shear rate dependent, because both apparent viscosity and thermal conductivity are dependent on structure. Although limited measurements [Loulou *et al.*, 1992] on carbopol solutions confirm this, the effect is small. For engineering design calculations, there will be little error in assuming that all the above physical properties of aqueous polymer solutions, except apparent viscosity, are equal to the values for water.

Some values of these properties of polymer melts are also available [Brandrup and Immergut, 1989; Domininghaus, 1993]. Unfortunately though, no simple predictive expressions are available for their estimation. Besides, values seem to be strongly dependent on the method of preparation of the polymer, the molecular weight distribution, etc., and therefore extrapolation from one system to another, even under nominally identical conditions, can lead to significant errors. For industrially important slurries and pastes exhibiting strong non-Newtonian behaviour, the thermo-physical properties (density, specific heat and thermal conductivity) can deviate significantly from those of its constituents. Early measurements [Orr and Dallavale, 1954] on aqueous suspensions of powdered copper, graphite, aluminium and glass beads suggest that both the density and the specific heat can be approximated by the weighted average of the individual constituents, i.e.

$$\rho_{sus} = \phi \rho_s + (1 - \phi)\rho_L \tag{6.1}$$

$$C_{p_{sus}} = \phi C_{p_s} + (1 - \phi)C_{p_L} \tag{6.2}$$

where ϕ is the volume fraction of the solids, and the subscripts s and L refer to the values for the solid and the liquid medium respectively.

The thermal conductivity of these systems, on the other hand, seems generally to be well correlated by the following expression [Tareef, 1940; Orr and

Dallavale, 1954; Skelland, 1967]:

$$k_{\text{sus}} = k_L \left[\frac{1 + 0.5(k_s/k_L) - \phi(1 - k_s/k_L)}{1 + 0.5(k_s/k_L) + 0.5\phi(1 - k_s/k_L)} \right] \tag{6.3}$$

Thermal conductivities of suspensions up to 60% (by weight) in water and other suspending media are well represented by equation (6.3). It can readily be seen that even for a suspension of highly conducting particles ($k_L/k_s \to 0$), the thermal conductivity of a suspension can be increased by several folds. Furthermore, the resulting increase in apparent viscosity from such addition would more than offset the effects of increase in thermal conductivity on the rate of heat transfer.

For suspensions of mixed particle size, the following expression due to Bruggemann [1935] is found to be satisfactory:

$$\frac{(k_{\text{sus}}/k_s) - 1}{(k_L/k_s) - 1} = \left(\frac{k_{\text{sus}}}{k_L} \right)^{1/3} (1 - \phi) \tag{6.4}$$

The scant experimental data [Rajaiah *et al.*, 1992] for suspensions ($\phi \le 0.3$) of alumina (0.5–0.8 μm) particles in a paraffin hydrocarbon are in line with the predictions of equation (6.4). An exhaustive review of the thermal conductivity of structured media including polymer solutions, filled and unfilled polymer melts, suspensions and foodstuffs has been published by Dutta and Mashelkar [1987]. Figure 6.1 shows the predictions of equations (6.3) and (6.4) for a range of values of (k_L/k_s); the two predictions are fairly close, except for the limiting value of $k_L/k_s = 0$.

Example 6.1

Estimate the value of thermal conductivity at 20°C of 25% (by vol) for aqueous suspensions of (a) alumina, $k_s = 30$ W/mK (b) thorium oxide, $k_s = 14.2$ W/mK (c) glass beads, $k_s = 1.20$ W/mK.

Solution

Here $\phi = 0.25$; thermal conductivity of water, $k_L = 0.60$ W/mK.

The values of the thermal conductivity of various suspensions are calculated using equations (6.3) and (6.4) for the purposes of comparison.

Suspension	Value of k_{sus}, W/mK	
	equation (6.3)	equation (6.4)
alumina	1.92	2.2
thorium oxide	1.85	2.06
glass beads	1.29	1.20

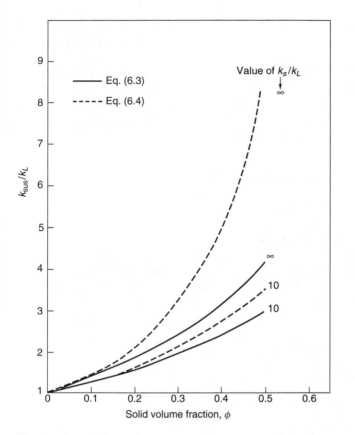

Figure 6.1   *Effect of concentration on thermal conductivity of suspensions*

The values obtained by the two methods are seen to be close and the difference, being 10%, is well within the limits of experimental error in such measurements.□

Of all the physico-chemical properties, it is the rheology which shows the strongest temperature dependence. For instance, the decrease in apparent viscosity at a fixed shear rate is well represented by the Arrhenius-type exponential expression; the pre-exponential factor and the activation energy are then both functions of shear rate. It is thus customary to denote the temperature dependence using rheological constants such as the power-law consistency coefficient and flow behaviour index. It is now reasonably well established that the flow behaviour index, n, of suspensions, polymer melts and solutions is nearly independent of temperature, at least over a range of 40–50°C, whereas the consistency coefficient exhibits an exponential dependence on temperature, i.e.

$$m = m_0 \exp(E/RT) \tag{6.5}$$

where the coefficient m_0 and E, the activation energy of viscous flow, are evaluated using experimental results for the temperature range of interest. Similarly, in the case of Bingham plastic model, both the plastic viscosity and the yield stress decrease with temperature in a similar fashion, but each with different values of the pre-exponential factors and the activation energies. Temperature dependencies of the other rheological characteristics such as the primary and the secondary normal stress differences, extensional viscosity, etc. have been discussed in detail by Ferry [1980].

6.3 Laminar flow in circular tubes

Heat transfer through a liquid in streamline flow takes place by conduction. As mentioned earlier, the consistency of most non-Newtonian materials is high so that turbulent conditions do not usually occur and free convection also is seldom significant.

When a shear-thinning power-law fluid enters a pipe heated on the outside, the fluid near the wall will be both at a higher temperature and subject to higher shear rates than that at the centre, and therefore its viscosity will be lower. It thus follows that for a given volumetric flow rate the velocity of the fluid near the wall will be greater, and that near the centre correspondingly less, as compared with an unheated fluid (Figure 6.2). Thus the velocity profile is flattened when the fluid is heated and, conversely, sharpened when it is cooled. If the fluid has a high apparent viscosity, frictional heating may be sufficient to modify the temperature and velocity profiles and the analysis of the flow problem then becomes very complex (Section 6.8).

As in the case of Newtonian flow, it is necessary to differentiate between the conditions in the entry region and in the region of fully (thermally) developed flow.

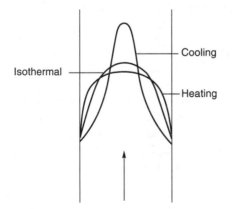

Figure 6.2   *Effect of heat transfer on velocity distribution*

Since the power-law and the Bingham plastic fluid models are usually adequate for modelling the shear dependence of viscosity in most engineering design calculations, the following discussion will therefore be restricted to cover just these two models; where appropriate, reference, however, will also be made to the applications of other rheological models. Theoretical and experimental results will be presented separately. For more detailed accounts of work on heat transfer in non-Newtonian fluids in both circular and non-circular ducts, reference should be made to one of the detailed surveys [Cho and Hartnett, 1982; Irvine, Jr. and Karni, 1987; Shah and Joshi, 1987; Hartnett and Kostic, 1989; Hartnett and Cho, 1998].

6.4 Fully-developed heat transfer to power-law fluids in laminar flow

The heating of a viscous fluid in laminar flow in a tube of radius R (diameter, D) will now be considered. Prior to the entry plane ($z < 0$), the fluid temperature is uniform at T_1; for $z > 0$, the temperature of the fluid will vary in both radial and axial directions as a result of heat transfer at the tube wall. A thermal energy balance will first be made on a differential fluid element to derive the basic governing equation for heat transfer. The solution of this equation for the power-law and the Bingham plastic models will then be presented.

Consider the differential control volume shown in Figure 6.3. The velocity profile is assumed to be fully developed in the direction of flow, i.e. $V_z(r)$. Furthermore, all physical properties including m and n for a power-law fluid and plastic viscosity and yield stress for a Bingham plastic fluid, are assumed to be independent of temperature.

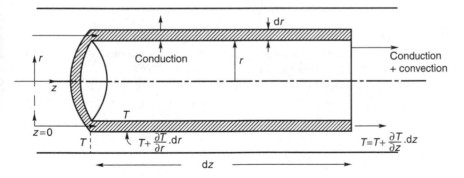

Figure 6.3   *Schematics for heat balance in a tube*

At steady state, the temperature of the fluid, T, apparent viscosity will be a function of both r and z, and the rate of transfer of heat into the control volume is:

$$(2\pi r\,dz \cdot q_r)|_r \;+\; (2\pi r\,drq_z)|_z \;+\; (2\pi rV_z\rho C_p(T - T_1)dr)|_z \qquad (6.4)$$

(conduction in   (conduction in   (convection in
r-direction)   z-direction)   z-direction)

Similarly, the rate of transfer of heat out of the control volume is given by:

$$(2\pi(r + dr)dzq_r)|_{(r+dr)} + (2\pi r\,drq_z)|_{(z+dz)}$$

$$+\, (2\pi rV_z\rho C_p(T - T_1)dr)|_{(z+dz)} \qquad (6.5)$$

where q_r and q_z respectively are the heat fluxes due to conduction in the r- and z-directions.

Under steady state conditions, expressions 6.4 and 6.5 must be equal and slight re-arrangement yields:

$$\rho C_p V_z(r)\frac{\partial T}{\partial z} + \frac{\partial q_z}{\partial z} + \frac{1}{r}\frac{\partial}{\partial r}(rq_r) = 0 \qquad (6.6)$$

Since heat transfer in r-direction is solely by conduction,

$$q_r = -k\frac{\partial T}{\partial r} \qquad (6.7)$$

Similarly, the conduction heat flux in z-direction

$$q_z = -k\frac{\partial T}{\partial z} \qquad (6.8)$$

Combining equations (6.6) to (6.8),

$$\rho C_p V_z(r)\frac{\partial T}{\partial z} = k\frac{\partial^2 T}{\partial z^2} + \frac{k}{r}\frac{\partial}{\partial r}\left(r\frac{\partial T}{\partial r}\right) \qquad (6.9)$$

It is implicit in equation (6.9) that thermal conductivities are isotropic. This is satisfactory for homogeneous systems (e.g. polymer solutions), but not always so for filled polymer melts [Dutta and Mashelkar, 1987].

Generally, conduction in the z-direction is negligible in comparison with the convection, and the term $\partial^2 T/\partial z^2$ in equation (6.9) may thus be dropped to give:

$$\rho C_p V_z(r)\frac{\partial T}{\partial z} = \frac{k}{r}\frac{\partial}{\partial r}\left(r\frac{\partial T}{\partial r}\right) \qquad (6.10)$$

Two important wall conditions will now be considered: (1) constant wall temperature (e.g. when steam is condensing on the outside of the tube), and (2) constant heat flux (e.g. when an electrical heating coil surrounds the pipe). The solutions for these two conditions are quite different, and the two cases will therefore be dealt with separately.

6.5 Isothermal tube wall

6.5.1 Theoretical analysis

Let the tube wall be at a uniform temperature $T_0(z > 0)$ and the entry temperature of the fluid be T_1 at $z = 0$. The solution of equation (6.10) is simplified by using a dimensionless temperature, θ, defined by:

$$\theta = \frac{T - T_0}{T_1 - T_0} \tag{6.11}$$

On substitution in equation (6.10):

$$V_z(r)\frac{\partial \theta}{\partial z} = \frac{\alpha}{r}\frac{\partial}{\partial r}\left(r\frac{\partial \theta}{\partial r}\right) \tag{6.12}$$

where $\alpha = k/\rho C_p$ is the thermal diffusivity of the fluid. This basic equation for temperature, $\theta(r, z)$, must be solved subject to the following boundary conditions:

At the tube wall, $r = R$, $\theta = 0$ for all values of $z > 0$ (6.13a)

At the centre of tube, $r = 0$, $\dfrac{\partial \theta}{\partial r} = 0$ for all values of z (6.13b)

and at the tube inlet, $z = 0$, $\theta = 1$,   for all values of r (6.13c)

The solution depends on the form of the velocity distribution. Unfortunately, the closed form solutions are only possible for the following three forms of $V_z(r)$:

(i) Piston or plug flow

This type of flow is characterised by the uniform velocity across the cross-section of the tube, i.e. $V_z(r) = V_0$, the constant value. This condition applies near the tube entrance, and is also the limiting condition of $n = 0$ with power-law model, i.e. infinite pseudoplasticity. In view of its limited practical utility, though this case is not discussed here, but detailed solutions are given in several books, e.g. see Skelland [1967]. However, Metzner *et al.* [1957] put forward the following expression for Nusselt number under these conditions (for Gz > 100):

$$\mathrm{Nu} = \frac{h_m D}{k} = \frac{4}{\pi}(2 + \mathrm{Gz}^{1/2}) \tag{6.14}$$

where h_m is the heat transfer coefficient averaged over the length L, of the tube, and Gz is the Graetz number defined by:

$$Gz = \frac{\dot{m}C_p}{kL} \tag{6.15}$$

where $\dot{m}$ is the mass flow rate of the fluid.

(ii) Fully developed power-law fluid flow

As seen in Chapter 3, the fully developed laminar velocity profile for power law fluids in a tube is given by

$$V_z(r) = V\left(\frac{3n+1}{n+1}\right)\left[1 - \left(\frac{r}{R}\right)^{(n+1)/n}\right] \tag{3.7}$$

which, upon substitution in equation (6.12), yields:

$$V\left(\frac{3n+1}{n+1}\right)\left[1 - \left(\frac{r}{R}\right)^{(n+1)/n}\right]\frac{\partial\theta}{\partial z} = \frac{\alpha}{r}\frac{\partial}{\partial r}\left(r\frac{\partial\theta}{\partial r}\right) \tag{6.16}$$

where V is the mean velocity of flow.

This differential equation can be solved by the *separation of variables* method by letting

$$\theta = R(r)Z(z) \tag{6.17}$$

where R is a function of r only and Z is a function of z only. Equation (6.16) then reduces to:

$$\frac{V}{\alpha}\left(\frac{3n+1}{n+1}\right)\frac{1}{Z}\frac{dZ}{dz} = \frac{1}{rR\left\{1 - (r/R)^{(n+1)/n}\right\}}\frac{d}{dr}\left(r\frac{dR}{dr}\right) \tag{6.18}$$

Both sides of equation (6.18) must be equal to a constant, say, $-\beta^2$, i.e.

$$\frac{1}{Z}\frac{dZ}{dz} = -\beta^2\left(\frac{n+1}{3n+1}\right)\cdot\frac{\alpha}{V}$$

which can be integrated to obtain the Z component of θ as

$$Z = \exp\left\{-\frac{\alpha(n+1)}{V(3n+1)}\cdot\beta^2 z\right\} \tag{6.19}$$

The variation of θ in the radial direction is given by the right-hand side of equation (6.18), with slight rearrangement as:

$$\frac{d^2R}{dr^2} + \frac{1}{r}\frac{dR}{dr} + \beta^2\left[1 - \left(\frac{r}{R}\right)^{(n+1)/n}\right]R = 0 \tag{6.20}$$

This equation is of the Sturm – Liouville type, and its complete solution for $n = 1$, 1/2 and 1/3 and for small values of the Graetz number is available [Lyche and Bird, 1956]. Detailed tabulations of the eigen values of equation (6.20) are given in the original paper, and a representative temperature profile for Gz = 5.24 is shown in Figure 6.4, for a range of values of the power-law flow behaviour index. The temperature profile is seen to become progressively flatter as the flow index n is reduced (increasing degree of shear-thinning behaviour). However, difficulties in evaluating the series expansions at high Graetz numbers limited their predictions to the range $1.57 \leq \text{Gz} \leq 31.4$.

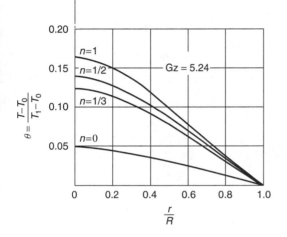

Figure 6.4 *Temperature distribution for power-law fluids in a tube [Bird et al., 1987]*

The resulting values of the mean Nusselt number, Nu_∞, under thermally fully-developed conditions are 3.66, 3.95 and 4.18 respectively for n values of 1, 1/2 and 1/3 [Bird *et al.*, 1987].

In addition, the so-called Leveque approximation [Leveque, 1928] has also been extended and applied to power-law fluids. The key assumption in this approach is that the temperature boundary layer is confined to a thin layer of the fluid adjacent to tube wall. This is a reasonable assumption for high flow rates and for short tubes, i.e. large values of Gz. A linear velocity gradient can then be assumed to exist within this thin layer:

$$V_z(r) = -\beta_v(R - r) \tag{6.21}$$

where β_v is the velocity gradient close to the tube wall. For power-law fluids, it can be evaluated using the velocity distribution given by equation (3.7) as:

$$\beta_v \simeq \left.\frac{dV_z(r)}{dr}\right|_{r=R} = -\frac{V}{R}\left(\frac{3n+1}{n}\right) \tag{6.22}$$

and the velocity profile near the wall is therefore given approximately by:

$$V_z(r) = \frac{V}{R}\left(\frac{3n+1}{n}\right)(R-r) \tag{6.23}$$

When substituted in equation (6.10), this yields:

$$\frac{V}{R}\left(\frac{3n+1}{n}\right)(R-r)\frac{\partial\theta}{\partial z} = \frac{\alpha}{r}\frac{\partial}{\partial r}\left(r\frac{\partial\theta}{\partial r}\right) \tag{6.24}$$

Furthermore, since the heat transfer is confined to a thin layer near the wall, the effects of curvature can be neglected. Denoting the distance from the wall as y, i.e. $y = R - r$, the right-hand side of equation (6.24) can be further simplified to give:

$$\left(\frac{3n+1}{n}\right)\xi\frac{\partial\theta}{\partial\zeta} = \frac{\partial^2\theta}{\partial\xi^2} \tag{6.25}$$

where $\xi = y/R$; $\zeta = \alpha z/VR^2 = \pi(z/L)(1/Gz)$. The boundary conditions (6.13a) and (6.13b) may be re-written as $\xi = 0$, $\theta = 0$ and $\xi \to \infty$, $\theta = 1$ respectively. Under these conditions, equation (6.25) can be solved by a *combination of variables* method as

$$\theta = \theta(\chi) \tag{6.26}$$

with $$\chi = \frac{\xi}{\left[9\zeta\left(\dfrac{n}{3n+1}\right)\right]^{1/3}} \tag{6.27}$$

When substituted in equation (6.25), this gives:

$$\frac{d^2\theta}{d\chi^2} + 3\chi^2\frac{d\theta}{d\chi} = 0 \tag{6.28}$$

The first integration gives:

$$\frac{d\theta}{d\chi} = A_0 e^{-\chi^3}$$

where A_0 is a constant of integration. This equation can be again integrated:

$$\theta = A_0 \int e^{-\chi^3} d\chi + B_0$$

The two constants A_0 and B_0 are evaluated by applying the boundary conditions for dimensionless temperature θ. When $\chi = 0$, $\theta = 0$ and $\chi \to \infty$, $\theta = 1$.

Thus, substituting the value of $\theta = 0$, when $\chi = 0$, it gives $B_0 = 0$. Similarly, substituting $\chi \to \infty$ and $\theta = 1$:

$$A_0 = \frac{1}{\int_0^{\infty} e^{-\chi^3} d\chi}.$$

The definite integral in the denominator is identified to be a gamma function, $\Gamma(4/3)$. Therefore, the dimensionless temperature θ is obtained as:
The dimensionless temperature θ is obtained as:

$$\theta = \frac{1}{\Gamma(4/3)} \int_0^{\chi} e^{-\chi^3} d\chi \qquad (6.29)$$

The local value of the Nusselt number, in turn, is deduced by writing the rate of transfer of heat at the tube wall as:

$$h(T_1 - T_0) = -k \frac{\partial T}{\partial r}\bigg|_{r=R} \qquad (6.30)$$

which can be re-arranged in terms of dimensionless variables:

$$
\begin{aligned}
\text{Nu}_z = \frac{h_z D}{k} &= 2 \frac{\partial \theta}{\partial \xi}\bigg|_{\xi=0} \\
&= 2 \frac{d\theta}{d\chi}\bigg|_{\chi=0} \left(\frac{\partial \chi}{\partial \xi}\right) \qquad (6.31)
\end{aligned}
$$

Substitution from equations (6.27) and (6.29) in equation (6.31) gives:

$$
\begin{aligned}
\text{Nu}_z &= \frac{2}{\Gamma(4/3)} \left[\left(\frac{3n+1}{n}\right) \frac{VR^2}{9\alpha Z}\right]^{1/3} \\
&= 1.167 \left(\frac{3n+1}{4n}\right)^{1/3} \text{Gz}_z^{1/3} \qquad (6.32)
\end{aligned}
$$

where Gz_z relates to an arbitrary axial distance z instead of the tube length L. The average value of the Nusselt number is obtained as:

$$\text{Nu} = \frac{1}{L} \int_0^L \text{Nu}_z \, dz$$

which when evaluated yields

$$\text{Nu} = 1.75 \Delta^{1/3} \text{Gz}^{1/3} \qquad (6.33)$$

$$\text{where} \quad \Delta = \left(\frac{3n+1}{4n}\right) \qquad (6.34)$$

When n is put equal to unity, equation (6.34) correctly reduces to its Newtonian form, $Nu = 1.75 \, Gz^{1/3}$.

Hirai [1959] and Schechter and Wissler [1960] have extended this approach to include Bingham plastic fluids; the factor Δ is then given by:

$$\Delta = \frac{3}{(3 - \phi - \phi^2 - \phi^3)} \tag{6.35}$$

where $\phi = \tau_0^B / \tau_w$, the ratio of the yield stress to the wall shear stress. Indeed, Pigford [1955] has asserted that equation (6.33) is applicable to any type of time-independent fluid provided that n' the apparent flow behaviour index replaces n in equation (6.33). For power-law shear-thinning fluids, equations (6.32) and (6.33) seem to be valid for $Gz > 100$.

6.5.2 *Experimental results and correlations*

The experimental studies on heat transfer to/from purely viscous fluids in laminar flow in circular tubes have been critically reviewed in many publications [Porter, 1971; Cho and Hartnett, 1982; Irvine, Jr. and Karni, 1987; Shah and Joshi, 1987; Hartnett and Kostic, 1989]. Metzner *et al.* [1957] found it necessary to modify equation (6.33) to account for the temperature dependence of the consistency index as:

$$Nu = 1.75 \Delta^{1/3} Gz^{1/3} \left(\frac{m_b'}{m_w'} \right)^{0.14} \tag{6.36}$$

where m_b' and m_w', respectively, are the apparent consistency coefficients $\{(= m((3n' + 1)/4n')^{n'}\}$ at the bulk temperature of the fluid and at the wall temperature. They asserted that this equation applies to all types of time-independent fluids provided that the local value of n' is used in evaluating Δ and in the viscosity correction term. Extensive comparisons between the predictions of equation (6.36) and experimental results suggest that it is satisfactory for $1 \geq n' > 0.1$ and $Gz > 20$. Subsequently [Metzner and Gluck, 1960], equation (6.36) has been modified to take into account free-convection effects:

$$Nu = 1.75 \Delta^{1/3} \left[Gz + 12.6 \left\{ (Pr \, Gr)_w \left(\frac{D}{L} \right) \right\}^{0.4} \right]^{1/3} \left(\frac{m_b'}{m_w'} \right)^{0.14} \tag{6.37}$$

where $Gr = \dfrac{\beta \Delta T D^3 \rho^2 g}{\mu_{eff}^2}$ \qquad\qquad (6.37a)

$$Pr = \frac{C_p \mu_{eff}}{k} \tag{6.37b}$$

The thermo-physical properties including the effective viscosity are evaluated at the wall conditions of shear rate and temperature. For a power-law fluid therefore the effective viscosity is evaluated at the shear rate of $\{(3n + 1)/4n\}(8V/D)$. However, Oliver and Jenson [1964] found that equation (6.37) underpredicted their results on heat transfer to carbopol solutions in 37 mm diameter tubes and that there was no effect of the (L/D) ratio. They correlated their results as $(0.24 \leq n' \leq 0.87)$:

$$\text{Nu} = 1.75[\text{Gz} + 0.0083(\text{Pr}\,\text{Gr})_w^{0.75}]^{1/3} \left(\frac{m'_b}{m'_w}\right)^{0.14} \tag{6.38}$$

where the effective viscosity used in evaluating the Prandtl and Grashof numbers is evaluated at the wall conditions.

The limited data on heat transfer to Bingham plastic suspensions of thoria [Thomas, 1960] in laminar flow seem to be well correlated by equations (6.33) and (6.35), except that a slightly different numerical constant must be used. Skelland [1967] has put it in a more convenient form as:

$$j_H = \left(\frac{h_m}{\rho C_p V}\right)\text{Pr}^{2/3}\left(\frac{\mu_w}{\mu_b}\right)^{0.14}\left(\frac{L}{D}\right)^{1/3}\Delta^{1/3} = 1.86\text{Re}_B^{-1/3} \tag{6.39}$$

where the Reynolds number is based on the mean plastic viscosity of the suspension. The density and heat capacity were taken as weighted averages, and the thermal conductivity was estimated using equation (6.1).

Example 6.2

A 0.2% aqueous carbopol solution at 25°C is flowing at a mass flow rate of 200 kg/h through a 30 mm diameter copper tube prior to entering a 2 m long heated section. The initial unheated section is sufficiently long for the velocity profile to be fully established. The heated section is surrounded by a jacket in which steam condenses at a pressure of 70 kPa (saturation temperature 90°C). Estimate the mean heat transfer coefficient and the bulk temperature of the fluid leaving the heating section. Compare the predicted values given by the various equations/correlations presented in the preceding section.

Physical properties of carbopol solution:
Density and specific heat as for water, 1000 kg/m^3 and 4.2 kJ/kg K respectively.
Thermal conductivity = 0.56 W/mK
Power-law index, $n = 0.26$, applies in the range $15 \leq T \leq 85°C$; power-law consistency coefficient, m, (Pa·s^n) $= 26 - 0.0566\,T$ ($288 \leq T \leq 363$ K) where T is in K.

Solution

The Graetz number for the given conditions will be calculated first.

$$Gz = \frac{\dot{m}C_p}{kL} = \frac{(200/3600) \times 4.2 \times 1000}{0.56 \times 2} = 208$$

Since this value is greater than 100, and it is not necessary to take account of the temperature dependence of viscosity, equation (6.33) may be used:

$$Nu = 1.75\Delta^{1/3}Gz^{1/3}$$

$$= 1.75 \left(\frac{3 \times 0.26 + 1}{4 \times 0.26}\right)^{1/3} 208^{1/3}$$

$$= 12.40$$

and the heat transfer coefficient, $h = 12.40 \times \dfrac{k}{D}$

$$= \frac{12.40 \times 0.56}{0.03} = 232 \, W/m^2K$$

The arithmetic mean of the temperature difference between the wall and the fluid may, as a first approximation, be used to calculate the rate of heat transfer:

At inlet, $\Delta T_1 = T_s - T_1 = 90-25 = 65°C$

At exit, $\Delta T_2 = T_s - T_2 = 90 - T_2$

∴ $\Delta T_{mean} = \dfrac{\Delta T_1 + \Delta T_2}{2} = \dfrac{65 + 90 - T_2}{2} = 77.5 - \dfrac{1}{2}T_2$

Neglecting the resistance of the film on the steam side and of the copper tube wall (high thermal conductivity), the overall heat transfer coefficient, $U \approx h = 232 \, W/m^2K$
From a heat balance on the fluid:

$$\dot{m}C_p(T_2 - 25) = U \cdot A \cdot \Delta T$$

Substituting values:

$$\frac{200}{3600} \times 4200 \times (T_2 - 25) = 232 \times 3.14 \times 0.03 \times 2 \times \left(77.5 - \frac{1}{2}T_2\right).$$

Solving: $T_2 = 36.1°C$
The temperature of the fluid leaving the heating section is therefore 36.1°C.

It is important next to establish the influence of temperature dependence of the consistency index. Equation (6.36) will be used since the values of both n and the Graetz number lie within its range of validity. In this example, since the value of n is constant,

$$\frac{m_b'}{m_w'} = \left(\frac{m_b}{m_w}\right)$$

Strictly speaking, as the outlet temperature of the fluid is not known, a trial and error solution is required. Assuming the outlet fluid temperature to be 36.1°C, as calculated above, the mean fluid temperature is $(25 + 36.1)/2$, i.e. 30.55°C and the value of m is calculated as:

$$m_b = 26 - 0.0566T = 26 - 0.0566(273 + 30.55)$$

$$= 8.82 \, \text{Pa·s}^n.$$

Similarly, $m_w = 26 - 0.0566(273 + 90)$

$$= 5.45 \, \text{Pa·s}^n$$

Substituting the values in equation (6.36):

$$\text{Nu} = 1.75 \left(\frac{3 \times 0.26 + 1}{4 \times 0.26} \right)^{1/3} 208^{1/3} \left(\frac{8.82}{5.45} \right)^{0.14} = 13.26$$

and $h = 248 \, \text{W/m}^2\text{K}.$

Under these conditions, the viscosity correction is small (only about 7%). The outlet temperature of the fluid in this case is found to be 36.8°C which is sufficiently close to the assumed value of 36.1°C for a second iteration not to be needed.

Finally, the contribution of free convection may be evaluated using equations (6.37) and (6.38).

The wall shear rate for a power-law fluid is:

$$\dot{\gamma}_{\text{wall}} = \left(\frac{3n + 1}{4n} \right) \left(\frac{8V}{D} \right)$$

The mean velocity of flow, $V = \dfrac{4\dot{m}}{\rho \pi D^2} = \dfrac{4 \times 200/3600}{1000 \times 3.14 \times (30 \times 10^{-3})^2}$

$$= 0.0785 \, \text{m/s}$$

$$\therefore \; \dot{\gamma}_{\text{wall}} = \left(\frac{3 \times 0.26 + 1}{4 \times 0.26} \right) \left(\frac{8 \times 0.0785}{30 \times 10^{-3}} \right) = 35.8 \, \text{s}^{-1}$$

The effective viscosity at the wall shear rate and temperature is:

$$\mu_{\text{eff}} = m(\dot{\gamma}_{\text{wall}})^{n-1}$$

The value of m at wall temperature, (90°C) is 5.45 Pa·s^n

$$\therefore \qquad\qquad \mu_{\text{eff}} = 5.45(35.8)^{0.26-1} = 0.386 \, \text{Pa·s}$$

Grashof number, $\text{Gr} = \dfrac{g\beta\Delta T D^3 \rho^2}{\mu_{\text{eff}}^2}$

The coefficient of thermal expansion of the carbopol solution is assumed to be same as that for water, and the mean value in this temperature range is 0.000302 K^{-1}.

Substituting values:

$$\text{Gr} = \frac{0.000302 \times 9.81 \times (90 - 25) \times (30 \times 10^{-3})^3 \times 1000^2}{0.386^2}$$

$$= 31.7$$

and Prandtl number, $\ \mathrm{Pr} = \dfrac{C_p \mu_{\mathrm{eff}}}{k} = \dfrac{4200 \times 0.386}{0.56}$

$$= 2890$$

Now substituting the above values in equation (6.37):

$$\mathrm{Nu} = 1.75 \times \left(\frac{0.26 \times 3 + 1}{4 \times 0.26} \right)^{1/3}$$

$$\times \left[208 + 12.6 \left\{ 2890 \times 31.7 \times \frac{30 \times 10^{-3}}{2} \right\}^{0.4} \right]^{1/3} \left(\frac{8.82}{5.45} \right)^{0.14}$$

$$= 17$$

$$\therefore\ h = \frac{17 \times 0.56}{30 \times 10^{-3}} = 317 \,\mathrm{W/m^2 K}$$

The resulting outlet temperature of the polymer solution in this case is 39.8°C which is significantly different from the assumed value of 36.1°C used in evaluating the viscosity correction term. However, the latter has an exponent of 0.14 which gives rise to a change in the value of h of less than 0.2%.

Similarly, equation (6.38) gives:

$$\mathrm{Nu} = 13.85 \quad \text{and} \quad h = 258 \,\mathrm{W/m^2 K}$$

The outlet fluid temperature is then about 36.2°C.

Using the minimum value of the consistency coefficient, which will apply in the wall region where the temperature is 90°C:

$$\mathrm{Re}_{MR} = \frac{\rho V^{2-n} D^n}{8^{n-1} \left(\dfrac{3n + 1}{4n} \right)^n m}$$

$$= \frac{1000 \times 0.0785^{2-0.26} \times (30 \times 10^{-3})^{0.26}}{8^{0.26-1} \left(\dfrac{0.26 \times 3 + 1}{4 \times 0.26} \right)^{0.26} \times 5.45}$$

$$= 3.57 \ll 2100$$

The flow is thus laminar.

Equations (6.33), (6.36) and (6.38) all give comparable results while equation (6.37) yields a much higher value of h. From design view point, it is desirable to establish upper and lower bounds on the value of the heat transfer coefficient and hence on the required heat transfer area. There does not appear to be much information available as to when natural convection effects become significant. Using the criterion that it should be so for $\mathrm{Gr}/\mathrm{Re}_{MR}^2 > 1$, as for Newtonian fluids, natural convection effects will be important.□

6.6 Constant heat flux at tube wall

6.6.1 Theoretical treatments

In this case, the energy balance given by equation (6.10) is still applicable; however, the boundary conditions are amended as follows:

At the tube wall, $\qquad r = R, \quad -k\dfrac{\partial T}{\partial r} = q_\omega$ (constant) $\quad$ (for $z \geq 0$)

$$(6.40a)$$

At the centre of the tube, $\quad r = 0, \quad \dfrac{\partial T}{\partial r} = 0 \quad$ (for $z \geq 0$) $\qquad (6.40b)$

and the inlet condition, $\qquad z = 0, \; T = T_1, \quad$ for all values of $r \leq R$.

$$(6.40c)$$

The use of the Leveque approximation in this instance leads to the following expression for the mean Nusselt number over the entry length [Bird, 1959]:

$$\text{Nu} = 2.11 \Delta^{1/3} \text{Gz}^{1/3} \qquad\qquad (6.41)$$

Similarly, the Nusselt number under fully developed thermal conditions is given by [Bird, 1959]:

$$\text{Nu}_\infty = \frac{8(3n + 1)(5n + 1)}{(31n^2 + 12n + 1)} \qquad\qquad (6.42)$$

On the other hand, Grigull [1956] estimated the value of Nu_∞ by modifying the value for Newtonian fluids by applying a factor of $\Delta^{1/3}$:

i.e. $\quad \text{Nu}_\infty = 4.36\Delta^{1/3} \qquad\qquad (6.43)$

where $\Delta = [(3n + 1)/4n]$

For $n = 1$, both equations (6.42) and (6.43) reduce to the correct value $\text{Nu}_\infty = 48/11 = 4.36$. For $n = 0.1$, the two predictions differ by about 6.5%. Interestingly enough, the correction is smaller for shear-thickening fluids, e.g. for $n \to \infty$, $\Delta^{1/3} = 0.908$.

Similar results for square and triangular ducts are also available in the literature [Irvine, Jr. and Karni, 1987].

6.6.2 Experimental results and correlations

For the constant heat flux condition at the tube wall, most investigators have expressed the local values of the Nusselt number as a function of the corresponding Graetz number. Cho and Hartnett [1982] concluded that for small values of temperature difference between the liquid and the tube wall, the radial variation in apparent viscosity is not significant, and most available experimental results are well correlated by the approximate isothermal solution, equation (6.41), due to Bird [1959]. Indeed, even the data obtained with weakly visco-elastic solutions of polyacrylamide and polyethylene oxide are in line with equation (6.41). However, when an appreciable temperature difference exists, the temperature dependence of the viscosity and natural convection effects can no longer be neglected. For these conditions, Cho and Hartnett [1982] have recommended using the following equation due to Mahalingam *et al.* [1975]:

$$Nu_z = 1.46\Delta^{1/3}[Gz + 0.0083(Gr\,Pr)_w^{0.75}]^{1/3}\left(\frac{m_b}{m_w}\right)^{0.14} \tag{6.44}$$

where the subscripts, z, b, w refer respectively to an arbitrary value of z (length in the direction of flow), bulk and wall conditions; the Grashof and Prandtl numbers are defined using the apparent viscosity at the shear rate of $(8\Delta V/D)$ in equations (6.37a) and (6.37b). Equation (6.44) covers the following ranges of conditions: $0.24 \leq n \leq 1$; $2.5 \leq q_0 \leq 40\,kW/m^2$; $200 \leq Gz \leq 10\,000$ and $750 \leq Gr \leq 4000$. Furthermore, Mahalingam *et al.* [1975] have suggested that natural convection effects are negligible for $(Gr/Re_{MR}) < 1$.

Example 6.3

The polymer solution referred to in Example 6.2 is to be heated from 15°C to 25°C in a 25 mm diameter 1800 mm long steel tube at the rate of 100 kg/h. The steel tube is wrapped with a electrical heating wire so that a constant heat flux is maintained on the tube wall. Estimate the rate of heat transfer to the solution, the heat flux at the wall and the temperature of the tube wall at the exit end of the tube. The physical properties are given in Example 6.2.

Solution

Initially, the temperature dependence of the fluid consistency coefficient m will be neglected so that equation (6.41) can be used. For this case,

$$Gz = \frac{(100/3600) \times 4200}{0.56 \times 1.8} = 115.7$$

For $n = 0.26$ (applicable for $T \not> 85°C$),

$$\Delta = (3n + 1)/(4n) = (3 \times 0.26 + 1)/(4 \times 0.26) = 1.71$$

$T = 15°C \longrightarrow$ $\longrightarrow T = 25°C$
$m = 100$ kg/h

$\longleftarrow$ —————— $L = 1.8$ m ——————$\longrightarrow$

Using equation (6.41),

$$\text{Nu} = \frac{hD}{k} = 2.11 \Delta^{1/3} \text{Gz}^{1/3}$$

or $\quad h = 2.11 \times 1.71^{1/3} \times 115.7^{1/3} \times \dfrac{0.56}{25 \times 10^{-3}} = 271 \text{ W/m}^2\text{K}.$

From a heat balance,

$$Q = \dot{m} C_p (T_{\text{out}} - T_{\text{in}})$$

$$= \frac{100}{3600} \times 4200 \times (25 - 15) = 1167 \text{ W}$$

Also, $\quad q_w \cdot \pi DL = Q$

$$\therefore q_w = \frac{1167}{\pi \times 0.025 \times 1.8} = 8253 \text{ W/m}^2 \text{ or } 8.25 \text{ kW/m}^2$$

$$= h(T_w - T_b)$$

For $T_b = 25°\text{C}$,

$$T_w = \frac{8253}{271} + 25 = 55.5°\text{C}. \tag{eq. (i)}$$

The role of natural convection can now be ascertained by using equation (6.44). Initially, the above estimate of the wall temperature will be used to evaluate the power-law consistency coefficient. For the wall temperature of 55.5°C, $m = 26 - 0.0566(273 + 55.5) = 7.41 \text{ Pa·s}^n$ and for the bulk temperature of 25°C, $m = 9.13 \text{ Pa·s}^n$.

$$\dot{\gamma}_{\text{wall}} = \left(\frac{3n+1}{4n} \right) \left(\frac{8V}{D} \right),$$

where $\quad V = \dfrac{100}{3600 \times 1000} \times \dfrac{4}{\pi \times 0.025^2} = 0.057 \text{ m/s}$

$$\therefore \dot{\gamma}_{\text{wall}} = \left(\frac{3n+1}{4n} \right) \left(\frac{8V}{D} \right)$$

$$= \left(\frac{3 \times 0.26 + 1}{4 \times 0.26} \right) \left(\frac{8 \times 0.057}{25 \times 10^{-3}} \right) = 15.92 \text{ s}^{-1}$$

$$\therefore \mu_{\text{eff wall}} = m(\dot{\gamma}_{\text{wall}})^{n-1} = 7.41(15.92)^{0.26-1} = 0.96 \text{ Pa·s}$$

At the wall conditions,

$$\text{Pr} = \frac{C_p \mu_{\text{eff}}}{k} = \frac{4200 \times 0.96}{0.56} = 7170$$

$$\text{Gr} = \frac{\beta g \Delta T D^3 \rho^2}{\mu_{\text{eff}}^2}$$

$$= \frac{0.000302 \times 9.81 \times (55.5 - 25) \times (25 \times 10^{-3})^3 1000^2}{0.96^2}$$

$$= 1.53$$

Note that this value is outside the range of the validity of equation (6.44), so its use here is only tentative.

$$\text{Nu} = \frac{hD}{k} = 1.46(1.71)^{1/3}[115.7 + 0.0083(7170 \times 1.53)^{0.75}]^{1/3} \left(\frac{9.13}{7.41}\right)^{0.14}$$

$$= 9$$

$$\therefore h = \frac{9 \times 0.56}{25 \times 10^{-3}} = 201 \, \text{W/m}^2\text{K}$$

At the end of the tube,

$$q_w = h(T_w - T_b) = 201(T_w - 25)$$

For $q_w = 8.25 \, \text{kW/m}^2$; $T_w = 66°\text{C}$

This value is substantially different from the assumed value of 55.5°C [from equation (i)]. Another iteration is carried out by assuming $T_w = 64°\text{C}$, at which $m = 6.93 \, \text{Pa·s}^n$
Again, at $\dot{\gamma}_{\text{wall}} = 15.92 \, \text{s}^{-1}$

$$\mu_{\text{eff}} = m(\dot{\gamma}_{\text{wall}})^{n-1} = 6.93(15.92)^{0.26-1} = 0.894 \, \text{Pa·s}$$

The new values of the Prandtl and Grashof numbers are

$$\text{Pr} = 6705 \qquad \text{Gr} = 2.25$$

Using equation (6.44), Nu = 9.12

$$\therefore h = \frac{9.12 \times 0.56}{0.025} = 204 \, \text{W/m}^2\text{K}$$

At the tube exit, $q_w = h(T_w - T_b)$
Substituting values, $q_w = 8250 \, \text{W/m}^2$; $h = 204 \, \text{W/m}^2\text{K}$ and $T_b = 25°\text{C}$,

$$T_w = 65.4°\text{C}$$

Though this value is fairly close to the assumed value of 64°C, a second iteration yields $T_w = 65°\text{C}$. Thus, if the natural convection effects are ignored, the wall temperature at the end is found to be 55.5°C [see equation (i)] which is some 10°C lower than the value of 65°C. In a sense, this discrepancy also reflects the inherent inaccuracy of the predictive equations in this field. The value of Re_{MR} (using the apparent viscosity near the wall) is calculated and this confirms that the flow is streamline:

$$\text{Re}_{MR} = \frac{\rho V^{2-n} D^n}{8^{n-1} \left(\frac{3n+1}{4n}\right)^n m} = \frac{1000 \times 0.057^{2-0.26} \times 0.025^{0.26}}{8^{0.26-1} \times 1.71^{0.26} \times 6.93}$$

$$= 1.53 \ \square$$

6.7 Effect of temperature-dependent physical properties on heat transfer

The theoretical treatments considered so far have been based on the assumption that the thermo-physical properties are constant (i.e. independent of temperature and therefore the velocity profiles do not change over the heat transfer section of the tube. Christiansen and Craig [1962] investigated the effect of temperature-dependent power-law viscosity on the mean values of Nusselt number for streamline flow in tubes with constant wall temperature. They postulated that the flow behaviour index, n was constant and that the variation of the consistency coefficient, m, with temperature could be represented by equation (6.45) giving:

$$\tau_{rz} = m_0 \left[\left(-\frac{dV_z}{dr} \right) \exp \left(\frac{E}{RT} \right) \right]^n \tag{6.45}$$

where m_0 and E are the pre-exponential factor and the activation energy of viscous flow; their values can be estimated by making rheological measurements at different temperatures covering the range of application. All other physical properties were assumed to be temperature independent. Figure 6.5 shows representative results for the mean values of Nusselt number, Nu, as a function of the Graetz number for $n = 1/3$.

The sign and magnitude of the dimensionless parameter, $\psi = (E/\mathbf{R})(1/T_1 - 1/T_0)$ depends upon the direction of heat transfer and

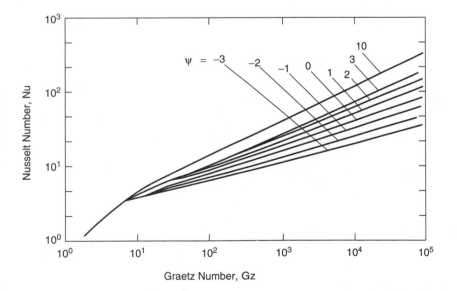

Figure 6.5   *Computed Nusselt number versus Graetz number for $n = 1/3$*

the sensitivity of m to temperature. For a cold fluid entering the tube, ψ would be positive. This analysis predicts an enhancement in the rate of heat transfer which is qualitatively consistent with the limited data presented by Christiansen and Craig [1962]. The thermal conductivity must be evaluated at the mean film temperature, i.e. $\frac{1}{2}[T_0 + \frac{1}{2}(T_1 + T_2)]$ where T_0, T_1, T_2 are the temperatures of the tube wall and of the fluid at inlet and outlet respectively. Finally, the effect of the flow behaviour index is found to be small.

On the other hand, Forrest and Wilkinson [1973, 1974] modelled the temperature-dependence of the power-law consistancy coefficient as:

$$m = m_0\{1 + \beta_w(T - T_1)\}^{-n} \tag{6.46}$$

where β_w is the temperature coefficient of viscosity. For the heating and cooling of power-law fluids in streamline flow in tubes with walls at a constant temperature, representative results from their study are shown in Figure 6.6, for a range of values β_w and $\phi = (T_0/T_1)$, which are qualitatively similar to those shown in Figure 6.5.

The enhancement of heat transfer at a isothermal tube wall due to temperature-dependent viscosity has been correlated by Kwant *et al.* [1973] as follows:

$$\frac{\mathrm{Nu}_{VP}}{\mathrm{Nu}} = 1 + 0.271 \ln\phi + 0.023(\ln\phi)^2 \tag{6.47}$$

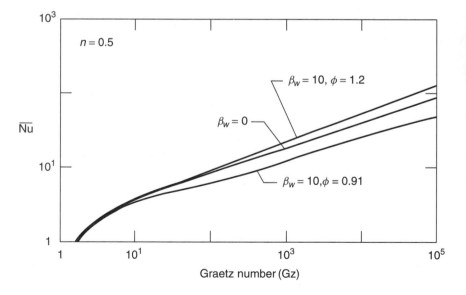

Figure 6.6 *Nusselt number versus Graetz number for heating and cooling of power-law fluids ($n = 0.5$)*

where Nu_{VP} and Nu are, respectively, the Nusselt numbers for the cases where consistencies are temperature dependent and independent respectively. Physical properties are evaluated at the mean film temperature.

To summarise, the heat transfer rate is enhanced as a result of the temperature dependence of the power-law consistency coefficient during heating ($\phi \geq 1$ or $\psi \geq 0$) while it is reduced during cooling of the fluid ($\phi < 1$, $\psi < 0$). This effect is much more pronounced for the case of constant wall temperature than that for the constant wall heat flux condition. The work of other investigators on the role of temperature-dependent viscosity has recently been reviewed by Lawal and Mujumdar [1987].

6.8 Effect of viscous energy dissipation

In the flow of all fluids, mechanical energy is degraded into heat and this process is called viscous dissipation. The effect may be incorporated into the thermal energy balance by adding a source term, S_V, (per unit volume of fluid) to the right hand side of equation (6.10). Its magnitude depends upon the local velocity gradient and the apparent viscosity of the fluid. Although, in general, the viscous dissipation includes contributions from both shearing and normal stresses, but under most conditions the contribution of shearing components outweighs that of normal stress components. Thus, it can readily be seen that the viscous dissipation term, S_V, is equal to the product of the shear stress and the shear rate. Thus, for example, in the streamline flow of a power-law fluid in a circular tube, it is of the form:

$$S_V = \tau_{rz}\left(-\frac{dV_z}{dr}\right) = m\left(-\frac{dV_z}{dr}\right)^{n+1} \tag{6.48}$$

For fully developed isothermal flow, the velocity gradient is obtained by differentiating the expression for velocity (equation 3.7) to give:

$$-\frac{dV_z}{dr} = \frac{V}{R}\left(\frac{3n+1}{n}\right)\left(\frac{r}{R}\right)^{1/n} \tag{6.49}$$

and thus the source term, S_V, becomes:

$$S_V = m\left\{\frac{V}{R}\left(\frac{3n+1}{n}\right)\right\}^{n+1}\left(\frac{r}{R}\right)^{(n+1)/n} \tag{6.50}$$

where V is the mean velocity of flow.

Evidently, for the limiting case of a shear-thinning fluid (Newtonian fluid $n = 1$), the velocity gradient is a maximum and hence, the viscous heat generation is also maximum. The effect of this process on the developing temperature profile can be illustrated by considering the situation depicted in Figure 6.7.

Here, a Newtonian fluid initially at a uniform temperature, T_0, enters a tube whose walls are also maintained at the same temperature T_0. As the fluid proceeds down the tube, heat is produced within the fluid at the rate given by equation (6.50). Near the entrance of the tube, the maximum temperature is obtained closed to the wall. Due to the conduction of heat in the radial direction (towards the centre), the temperature profile evolves progressively, eventually the maximum temperature being reached at the centre of the tube (Figure 6.7). Obviously, viscous dissipation effects will be significant for the flow of high consistency materials (such as polymer melts) and/or where velocity gradients are high (such as in lubrication and rolling processes).

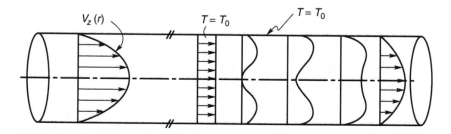

Figure 6.7   *Velocity (left) and temperature (right) profiles for the flow of a Newtonian fluid in a tube*

Viscous dissipation can be quantified in terms of the so-called Brinkman number, Br, which is defined as the ratio of the heat generated by viscous action to that dissipated by conduction. Thus for streamline flow in a circular tube (on the basis of per unit volume of fluid):

$$\text{heat generated by viscous action} \sim m(V/R)^{n+1}$$

and the heat conducted in radial direction $\sim (k\Delta T)/(R^2)$

In this case, the Brinkman number, Br is:

$$\text{Br} = \frac{m(V/R)^{n+1}}{k(\Delta T/R^2)} \tag{6.51}$$

Clearly, the viscous dissipation effects will be significant whenever the Brinkman number is much greater than unity.

While the rigorous solutions of the thermal energy equation are quite complex, some useful insights can be gained by qualitative considerations of the results of Forrest and Wilkinson [1973, 1974] and Dinh and Armstrong [1982] amongst others or from the review papers [Winter, 1977; Lawal and Mujumdar, 1987].

6.9 Heat transfer in transitional and turbulent flow in pipes

Although turbulent flow gives higher values of the heat transfer coefficient, it frequently cannot be achieved in practice with non-Newtonian polymer solutions and particulate suspensions. However, the turbulent flow of the so-called drag reducing solutions has been the subject of several studies and Cho and Hartnett [1982] have reviewed the literature on heat transfer to these dilute solutions. They have clearly shown that the results reported by different workers often do not agree and may differ by more than an order of magnitude under nominally identical conditions of flow. The rheological and heat transfer characteristics of these fluids are known to be extremely sensitive to the method of preparation, chemistry of solvent, etc. Furthermore these solutions show appreciable shear degradation and few reliable data are available on their heat transfer coefficients [Cho and Hartnett 1982]. Yoo [1974] has collated most of the available data on heat transfer to purely viscous fluids in turbulent region and has proposed the following correlation:

$$\text{St} = \frac{h}{\rho V C_p} = 0.0152(\text{Re}_{MR})^{-0.155}\text{Pr}^{-2/3} \tag{6.52}$$

where the effective viscosity used in the Reynolds and Prandtl numbers is evaluated at the wall shear rate of $[(3n + 1)/4n](8V/D)$ for a power-law fluid.

Equation (6.52) correlates the data over wide ranges ($0.24 \leq n \leq 0.9$ and $3000 \leq \text{Re}_{MR} \leq 90\,000$) with an average error of $\pm 3\%$. On the other hand, the heat transfer data for particulate suspensions ($0.42 \leq n \leq 0.89$) seems to be well represented by

$$j_H = \frac{h}{\rho V C_p}\text{Pr}^{2/3}\left(\frac{\mu_w}{\mu_b}\right)^{0.14} = 0.027\text{Re}_B^{-0.2} \tag{6.53}$$

where $\text{Re}_B = \rho V D/\mu_B$.

Other similar correlations are available in the literature [Quadeer and Wilkinson], none of which, however, has been validated using independent data.

Thus, it is concluded that although adequate information is available on laminar heat transfer to purely viscous fluids in circular tubes, further work is needed in turbulent regime, particularly with visco-elastic fluids.

6.10 Further reading

Carreau, P.J., Dekee, D. and Chhabra, R.P., *Rheology of Polymeric Systems: Principles and Applications.* Hanser, Munich (1997).

Chhabra, R.P., in *Advances in the Rheology and Flow of Non-Newtonian Fluids* (edited by Siginer, D., Dekee, D. and Chhabra, R.P. Elsevier, Amsterdam) (1999) in press.

Cho, Y.I. and Hartnett, J.P., *Adv. Heat Transf.* **15** (1982) 59.
Hartnett, J.P. and Kostic, M., *Adv. Heat Transf.* **19** (1989) 247.
Kakac, S., Shah, R.K. and Aung, W., (eds), *Handbook of Single Phase Convective Heat Transfer.* Wiley, New York (1987).
Lawal, A. and Mujumdar, A.S., *Adv. Transport Process.* **5** (1987) 352.
Rohsenow, W.M., Hartnett, J.P. and Cho, Y.I. (eds.) *Handbook of Heat Transfer*, 3rd edn., McGraw-Hill, New York (1998).

6.11 References

Bird, R.B., *Chem. Ing. Tech.* **31** (1959) 569.
Bird, R.B., Armstrong, R.C. and Hassager, O., *Dynamics of Polymeric Liquids. Vol. 1 Fluid Dynamics*, 2nd edn. Wiley, New York (1987).
Brandrup, J. and Immergut, E.H., (editors), *Polymer Handbook*, 3rd edn. Wiley, New York (1989).
Bruggemann, D.A.G., *Ann. Phys.* (Leipzig) **24** (1935) 636.
Cho, Y.I. and Hartnett, J.P., *Adv. Heat Transf.* **15** (1982) 59.
Christiansen, E.B. and Craig, S.E., *AIChEJ.* **8** (1962) 154.
Dinh, S.M. and Armstrong, R.C., *AIChEJ.* **28** (1982) 294.
Domininghans, H., *Plastics for Engineers: Materials, Properties and Applications.* Hanser, Munich (1993).
Dutta, A. and Mashelkar, R.A., *Adv. Heat Transf.* **18** (1987) 161.
Ferry, J.D., *Visco-elastic Properties of Polymers.* 3rd edn. Wiley, New York (1980).
Forrest, G. and Wilkinson, W.L., *Trans. Inst. Chem. Eng.* **51** (1973) 331.
Forrest, G. and Wilkinson, W.L., *Trans. Inst. Chem. Eng.* **52** (1974) 10.
Grigull, U., *Chem. Ing. Tech.* **28** (1956) 553 & 655.
Hartnett, J.P. and Cho, Y.I., in *Handbook of Heat Transfer* edited by Rohsenow, W.M., Hartnett, J.P. and Cho, Y.I., 3rd edn., McGraw Hill, New York (1998).
Hartnett, J.P. and Kostic, M., *Adv. Heat Transf.* **19** (1989) 247.
Hirai, E., *AIChEJ.* **5** (1959) 130.
Irvine, T.F. Jr., and Karni, J., in *Handbook of Single Phase Convective Heat Transfer* (edited by Kakac, S. Shah, R.K. and Aung, W., Wiley, New York) (1987) Chapter 5.
Irvine, T.F. Jr., Kim, I., Cho, K. and Gori, F., *Exp. Heat Transf.* **1** (1987) 155.
Kwant, P.B., Zwaneveld, A. and Dijkstra, F.C., *Chem. Eng. Sci.* **28** (1973) 1303.
Lawal, A. and Mujumdar, A.S., *Adv. Transport Process.* **5** (1987) 352.
Leveque, J., *Ann. Mines.* **13** (1928) 201, 305, 381.
Loulou, T., Peerhossaini, H. and Bardon, J.P., *Int. J. Heat Mass Transf.* **35** (1992) 2557.
Lyche, B.C. and Bird, R.B., *Chem. Eng. Sci.* **6** (1956) 34.
Mahalingam, R., Tilton, L.O. and Coulson, J.M., *Chem. Eng. Sci.* **30** (1975) 921.
Metzner, A.B. and Gluck, D.F., *Chem. Eng. Sci.* **12** (1960) 185.
Metzner, A.B., Vaughn, R.D. and Houghton, G.L., *AIChEJ.* **3** (1957) 92.
Oliver, D.R. and Jenson, V.G., *Chem. Eng. Sci.* **19** (1964) 115.
Orr, C. and Dallavalle, J.M., *Chem. Eng. Prog. Sym. Ser.* **50** #9 (1954) 29.
Pigford, R.L., *Chem. Eng. Prog. Sym. Ser.* **51** #17 (1955) 79.
Porter, J.E., *Trans. Inst. Chem. Engrs.* **49** (1971) 1.
Quader, A.K.M.A. and Wilkinson, W.L., *Int. J. Multiphase Flow.* **7** (1981) 545.
Rajaiah, J., Andrews, G., Ruckenstein, E. and Gupta, R.K., *Chem. Eng. Sci.* **47** (1992) 3863.
Schechter, R.S. and Wissler, E.H., *AIChEJ.* **6** (1960) 170.
Shah, R.K. and Joshi, S.D., in *Handbook of Single Phase Convective Heat Transfer* (edited by Kakac, S., Shah, R.K. and Aung, W. Wiley, New York) (1987) Chapter 5.
Skelland, A.H.P., *Non-Newtonian Flow and Heat Transfer.* Wiley, New York (1967).

Tareef, B.M., *Colloid, J., (U.S.S.R.)* **6** (1940) 545.
Thomas, D.G., *AIChEJ.* **6** (1960) 632.
Yoo, S.S., *PhD Thesis, University of Illinois at Chicago Circle*. IL (1974).
Winter, H.H., *Adv. Heat Transf.* **13** (1977) 205.

6.12 Nomenclature

Dimensions
in $\mathbf{M,L,T}\ \theta$

Br	Brinkman number, equation (6.51) (−)	$\mathbf{M^0L^0T^0}$
C_p	specific heat (J/kg K)	$\mathbf{L^2T^{-2}}\ \theta^{-1}$
D	pipe diameter (m)	$\mathbf{L}$
E	activation energy (J/mole)	$\mathbf{ML^2T^{-2}}$
Gr	Grashof number, equation (6.37a) (−)	$\mathbf{M^0L^0T^0}$
Gz	Graetz number, equation (6.15) (−)	$\mathbf{M^0L^0T^0}$
g	acceleration due to gravity (m/s^2)	$\mathbf{LT^{-2}}$
h	heat transfer coefficient (W/m^{2}K)	$\mathbf{MT^{-3}}\ \theta^{-1}$
j_H	heat transfer factor, equation (6.39) (−)	$\mathbf{M^0L^0T^0}$
k	thermal conductivity (W/mK)	$\mathbf{MLT^{-3}}\ \theta^{-1}$
L	length (m)	$\mathbf{L}$
m	power-law consistency coefficient (Pa·s^n)	$\mathbf{ML^{-1}T^{n-2}}$
m'	apparent power-law consistency coefficient (Pa·s$^{n'}$)	$\mathbf{ML^{-1}T^{n'-2}}$
m_0	pre-exponential factor, equation (6.45) (Pa·s^n)	$\mathbf{ML^{-1}T^{n-2}}$
$\dot{m}$	mass flow rate (kg/s)	$\mathbf{MT^{-1}}$
n	power law flow behaviour index (−)	$\mathbf{M^0L^0T^0}$
n'	apparent power law flow behaviour index (−)	$\mathbf{M^0L^0T^0}$
Nu	Nusselt number ($= hD/k$) (−)	$\mathbf{M^0L^0T^0}$
Pr	Prandtl number, equation (6.37b) (−)	$\mathbf{M^0L^0T^0}$
q	heat flux (W/m^2)	$\mathbf{MT^{-3}}$
R	pipe radius (m)	$\mathbf{L}$
$\mathbf{R}$	universal gas constant (J/kmol.K)	$\mathbf{ML^2T^{-2}}\ \theta^{-1}$
r	radial coordinate (m)	$\mathbf{L}$
Re_B	Reynolds number ($= \rho VD/\mu_B$) (−)	$\mathbf{M^0L^0T^0}$
Re_{MR}	Metzner-Reed Reynolds number ($= \rho V^{2-n}D^n/m(\Delta)^n 8^{n-1}$) (−)	$\mathbf{M^0L^0T^0}$
St	Stanton number, equation (6.52) (−)	$\mathbf{M^0L^0T^0}$
S_V	energy produced by viscous dissipation per unit volume of fluid (J/m^3)	$\mathbf{ML^{-1}T^{-2}}$
$V_z(r)$	local velocity in z-direction (m/s)	$\mathbf{LT^{-1}}$
V	average velocity of flow (m/s)	$\mathbf{LT^{-1}}$
T	temperature (K)	θ
T_1	fluid temperature at inlet (K)	θ
T_0	constant temperature of tube wall (K)	θ
z	axial coordinate (m)	$\mathbf{L}$

Greek letters

α	thermal diffusivity (m^2/s)	$\mathbf{L^2T^{-1}}$
β	coefficient of expansion (K^{-1})	θ^{-1}
Δ	correction factor ($= (3n+1)/4n$) (−)	$\mathbf{M^0L^0T^0}$

θ	dimensionless temperature ($-$)	$\mathbf{M^0 L^0 T^0}$
μ	viscosity (Pa·s)	$\mathbf{ML^{-1} T^{-1}}$
ρ	density (kg/m^3)	$\mathbf{ML^{-3}}$
ϕ	volume fraction ($-$) or $= T_1/T_0$ ($-$)	$\mathbf{M^0 L^0 T^0}$
ψ	dimensionless factor [$= (E/\mathbf{R})(1/T_1 - 1/T_0)$] ($-$)	$\mathbf{M^0 L^0 T^0}$

Subscripts

A	apparent
b	bulk condition
L	liquid
m	mean
o	outside film
s	solid
sus	suspension
vp	variable property
z	arbitrary z value
W	wall condition

Chapter 7

Momentum, heat and mass transfer in boundary layers

7.1 Introduction

When an incompressible liquid flows steadily over a solid surface, the liquid close to the surface experiences a significant retardation. The liquid velocity is zero at the surface (provided that the no-slip boundary condition holds) and gradually increases with distance from the surface, and a velocity profile is established. The velocity gradient is steepest adjacent to the surface and becomes progressively less with distance from it. Although theoretically, the velocity gradient is a continuous function becoming zero only at infinite distance from the surface, the flow may conveniently be divided into two regions.

(1) A boundary layer close to the surface where the fluid is retarded and a velocity gradient exists, and in which shear stresses are significant.
(2) The region outside the boundary layer where the liquid is all flowing at the free stream velocity.

It is thus evident that the definition of the boundary layer thickness is somewhat arbitrary. It is often defined as the distance (normal to flow) from the surface at which the fluid velocity reaches some proportion (e.g. 90%, 99% or 99.9% etc) of the free stream velocity; 99% is the most commonly used figure. As will be seen later, difficulties arise in comparing differently defined boundary layer thicknesses, because as the free stream velocity is approached, the velocity gradient becomes very low and a small difference in the velocity criterion will correspond to a very large difference in the resulting value of the boundary layer thickness. A thorough understanding of the flow in the boundary layer is of considerable importance in a range of chemical and processing applications because the nature of flow influences, not only the drag at a surface or on an immersed object, but also the rates of heat and mass transfer when temperature or concentration gradients exist.

At the outset, it is convenient to consider an incompressible fluid flowing at a constant free stream velocity V_0 over a thin plate oriented parallel to the flow, Figure 7.1.

Although this is a simplified case, it facilitates the discussion of more complex two and three dimensional boundary layers. The plate is assumed

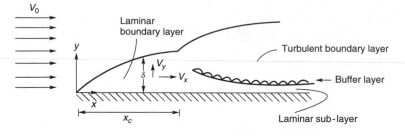

Figure 7.1   *Schematic development of boundary layer*

to be sufficiently wide in the z-direction for the flow conditions to be uniform across any width W of the plate. In addition, the extent of the fluid in y-direction is assumed to be sufficiently large for the velocity of the fluid remote from the plate to be unaffected, and to remain constant at the value of V_0. From Bernoulli's equation, the consequence of this is that the pressure gradient is zero in the direction of flow.

Although there is conflicting evidence in the literature concerning the validity of the no slip boundary condition for non-Newtonian materials, it will be assumed that it is satisfied, so that the fluid in contact with the surface ($y = 0$) can be considered to be at rest. At a distance x from the leading edge, the x-direction velocity V_x will increase from zero at the surface and approach the free stream velocity V_0 asymptotically. At the leading edge ($x = 0$, $y = 0$), the liquid will have been subjected to the retarding force exerted by the surface for only an infinitesimal time and the effective boundary layer thickness will therefore be zero. The liquid will then experience retardation for progressively longer periods of time as it flows over the surface. The retarding effects will affect fluid at greater depths, and the boundary layer thickness, δ, will therefore increase. The velocity gradient (dV_x/dy) at the surface ($y = 0$) decreases as the boundary layer thickens. Near the leading edge, the boundary layer thickness is small, the flow is laminar and the shear stresses arise solely from viscous shearing effects. However, when the boundary layer thickness exceeds a critical value, the flow becomes turbulent. However, the transition from laminar to turbulent flow is not as sharply defined as in a conduit and is strongly influenced by small protuberances at the leading edge and by surface roughness and irregularities which may possibly give rise to an early transition. The flow parameter describing the nature of the flow is a Reynolds number, defined for power-law fluids as $\rho \delta^n V_0^{2-n}/m (= \mathrm{Re}_\delta)$. Since the boundary layer thickness (δ) is a function of x, it is more convenient to define the local Reynolds number, Re_x, as:

$$\mathrm{Re}_x = \frac{\rho x^n V_0^{2-n}}{m} \tag{7.1}$$

The relation between Re_x and Re_δ will be developed later.

The transition for Newtonian fluids occurs at $Re_x \simeq 10^5$ [Schlichting, 1968].

Even when the flow in the body of the boundary layer is turbulent, flow remains laminar in the thin layer close to the solid surface, the so-called laminar sub-layer. Indeed, the bulk of the resistance to momentum, heat and mass transfer lies in this thin film and therefore interphase heat and mass transfer rates may be increased by decreasing its thickness. As in pipe flow, the laminar sub-layer and the turbulent region are separated by a buffer layer in which viscous and inertial effects are of comparable magnitudes, as shown schematically in Figure 7.1.

The frictional drag force on a submerged object will depend on the flow conditions in the boundary layer. The analysis of boundary layer flow of non-Newtonian materials will now be based on a direct extension of that applicable to Newtonian fluids. Detailed accounts are given in the literature [Schlichting, 1968], but the salient features of this approach, which is based on Prandtl's analysis, will be re-capitulated. If the flow can be regarded as uni-directional (x-direction), it implies that the effects of velocity components normal to the surface may be neglected, except at very low Reynolds number where the boundary layer thickens rapidly. A further simplifying assumption is to neglect the existence of the buffer layer and to assume that there is a sharp interface between the laminar sub-layer and the turbulent region. If there is a negligible pressure gradient in the direction of flow, i.e. $(\partial p / \partial x) \to 0$, it follows from the application of Bernoulli equation to the free stream outside the boundary layer that its velocity must be constant. The pressure gradient may in practice be either positive or negative. For $(\partial p / \partial x) > 0$, the so-called adverse pressure gradient will tend to retard the flow and cause the boundary layer to thicken rapidly with the result that *separation* may occur. Conversely, the effect of a negative pressure gradient is to reduce the boundary layer thickness.

An integral form of the general momentum balance will be obtained by applying Newton's second law of motion to a control volume of fluid. The evaluation of the resulting integral necessitates a knowledge of the velocity profile and appropriate assumptions of its form must be made for both laminar and turbulent flow conditions.

7.2 Integral momentum equation

Schilichting [1968] points out that the differential equations for flow in boundary layers require numerical solutions even when the flow is laminar and fluid behaviour Newtonian. However, reasonably good estimates of drag on a plane surface can be obtained by using the integral momentum balance approach due to von Karman, as illustrated in this section.

Consider the steady flow of an incompressible liquid of density ρ over an immersed plane surface. Remote from the surface, the free stream velocity

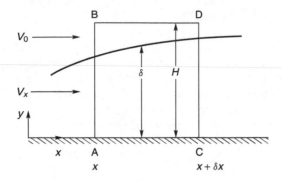

Figure 7.2   *Control volume for momentum balance on the fluid in boundary layer flow*

of liquid is V_0. A boundary layer of thickness δ develops near the surface, as shown schematically in Figure 7.1. Now consider the equilibrium of the control volume bounded by the planes AB and CD, as shown in Figure 7.2. The velocity component normal to the surface is assumed to be negligible ($V_y \ll V_x$). The rate at which the fluid is entering the control volume through the boundary AB is given by

$$\dot{m}_{AB} = \int_0^H \rho V_x W \, dy \tag{7.2}$$

Similarly, the total rate of momentum transfer through plane AB

$$= W \int_0^H \rho V_x^2 \, dy \tag{7.3}$$

In passing from plane AB to CD, the mass flow changes by:

$$W \frac{d}{dx} \left[\int_0^H \rho V_x \, dy \right] dx \tag{7.4}$$

and the momentum flux changes by:

$$W \frac{d}{dx} \left[\int_0^H \rho V_x^2 \, dy \right] dx \tag{7.5}$$

A mass flow of fluid equal to the difference between the flows at planes C–D and A–B must therefore occur through plane B–D. Since plane B–D lies outside the boundary layer, this entering fluid must have a velocity V_0 in the x-direction. Because the fluid in the boundary layer is being retarded, there will be a smaller flow at plane C–D than at A–B, and hence the amount of fluid entering through plane B–D is negative; that is to say fluid leaves the control volume through the plane B–D.

Thus, the rate of transfer of momentum through plane BD into the control volume

$$= WV_0 \frac{d}{dx} \left[\int_0^H \rho V_x \, dy \right] dx \tag{7.6}$$

The net rate of change of momentum in the x-direction associated with the fluid in the control volume must be equal to the rate of addition of momentum from outside (through plane B–D) together with the net force acting on the control volume. With zero pressure gradient, there are no pressure forces and the only external force is that due to the shear stress, τ_{wx}, acting on the surface ($y = 0$). As this is a retarding force, τ_{wx} is negative.

Thus, the net force acting on the control volume $= \tau_{wx} W \, dx$. Hence, the momentum balance on the control volume:

$$W \frac{d}{dx} \left[\int_0^H \rho V_x^2 \, dy \right] dx = WV_0 \frac{d}{dx} \left[\int_0^H \rho V_x \, dy \right] dx + \tau_{wx} W \, dx$$

Since V_0 does not vary with x;

$$\frac{d}{dx} \int_0^\delta \rho (V_0 - V_x) V_x \, dy = -\tau_{wx} dx \tag{7.7}$$

Note that the upper limit of integration has been changed to δ because the integrand is zero in the range $\delta \le y \le H$.

This expression, (equation (7.7)), known as the integral momentum equation, is valid for both laminar and turbulent flow, and no assumption has also been made about the nature of the fluid or its behaviour. In order to integrate equation (7.7), the relation between V_x and y must be known.

Laminar flow will now be considered for both power-law and Bingham plastic model fluids, followed by a short discussion for turbulent flow.

7.3 Laminar boundary layer flow of power-law liquids over a plate

For the laminar flow of a power-law fluid, the only forces acting within the fluid are pure shearing forces, and no momentum transfer occurs by eddy motion. A third degree polynomial approximation may be used for the velocity distribution:

$$V_x = a + by + cy^2 + dy^3 \tag{7.8}$$

The constants a to d are evaluated by applying the following boundary conditions:

At $y = 0$, $V_x = 0$ (no slip)   (7.9a)

shear stress $\rightarrow$ constant,

i.e. $\left(\dfrac{\partial V_x}{\partial y}\right)^n \rightarrow$ constant

$$\therefore \frac{\partial^2 V_x}{\partial y^2} = 0 \qquad (7.9b)$$

At $y = \delta$, $V_x = V_0$ (7.9c)

$$\frac{\partial V_x}{\partial y} = 0 \qquad (7.9d)$$

Condition (7.9d) is necessary to ensure the continuity of the velocity at $y = \delta$. Condition (7.9b) can be explained by the following physical reasoning. The shearing stress in the fluid increases towards the plate, where the velocity is zero. Thus, the change in the momentum of the fluid in the x-direction must be very small near the plate and the shear stress in the fluid therefore must approach a constant value so that equation (7.9b) is applicable.

Evaluating the constants and incorporating them into equation (7.8):

$$\frac{V_x}{V_0} = \frac{3}{2}\left(\frac{y}{\delta}\right) - \frac{1}{2}\left(\frac{y}{\delta}\right)^3 \qquad (7.10)$$

as for Newtonian fluids.

Thus, the shearing stress acting at any position on the plate can now be evaluated by substituting from equation (7.10) into equation (7.7):

$$-\tau_{wx} = \rho V_0^2 \frac{d}{dx} \int_0^\delta \left\{ 1 - \frac{3}{2}\left(\frac{y}{\delta}\right) - \frac{1}{2}\left(\frac{y}{\delta}\right)^3 \right\} \left\{ \frac{3}{2}\left(\frac{y}{\delta}\right) - \frac{1}{2}\left(\frac{y}{\delta}\right)^3 \right\} dy$$

(7.11)

After integration and simplification, equation (7.11) gives:

$$-\tau_{wx} = \frac{39}{280} \rho V^2 \frac{d\delta}{dx} \qquad (7.12)$$

For a power-law liquid,

$$\tau_{yx} = -m \left(\frac{dV_x}{dy}\right)^n$$

and $\tau_{wx} = \tau_{yx}|_{y=0} = -m \left(\frac{dV_x}{dy}\bigg|_{y=0}\right)^n \qquad (7.13)$

From the velocity distribution given by equation (7.10),

$$\frac{dV_x}{dy}\bigg|_{y=0} = \frac{3V_0}{2\delta} \tag{7.14}$$

Now combining equations (7.13) to (7.14) and eliminating τ_{wx}:

$$\frac{39}{280}\rho V_0^2 \frac{d\delta}{dx} = m\left(\frac{3V_0}{2\delta}\right)^n \tag{7.15}$$

which upon further integration with respect to x, and noting that $\delta = 0$ at $x = 0$, gives:

$$\frac{\delta}{x} = F(n)\,Re_x^{-1/(n+1)} \tag{7.16}$$

$$\text{where} \quad F(n) = \left[\frac{280}{39}(n+1)\left(\frac{3}{2}\right)^n\right]^{1/(n+1)} \tag{7.17}$$

$$\text{and} \quad Re_x = \frac{\rho V_0^{2-n} x^n}{m}$$

Putting $n = 1$ and $m = \mu$ for a Newtonian fluid, equation (7.16) reduces to the well known result:

$$\frac{\delta}{x} = 4.64\,Re_x^{-1/2} \tag{7.18}$$

The rate of thickening of the boundary layer for power-law fluids is given by:

$$\frac{d\delta}{dx} = \left(\frac{1}{n+1}\right)F(n)\,Re_x^{-1/(n+1)}$$

$$\text{or} \quad \frac{d\delta}{dx} \propto x^{-(n/(n+1))} \tag{7.19}$$

which suggests that, for a pseudoplastic fluid ($n < 1$), the boundary layer thickens more rapidly with distance along the surface as compared with a Newtonian liquid, and conversely for a shear-thickening fluid ($n > 1$).

7.3.1 Shear stress and frictional drag on the plane immersed surface

As seen above, the shear stress in the fluid at the surface ($y = 0$) is given by:

$$\tau_{yx}|_{y=0} = -m\left(\frac{dV_x}{dy}\bigg|_{y=0}\right)^n$$

$$= -m \left(\frac{3V_0}{2\delta} \right)^n \tag{7.20}$$

Substitution for δ from equation (7.17) and with slight re-arrangement:

$$\tau_{yx}|_{y=0} = - \left(\frac{3}{2F(n)} \right)^n \rho V_0^2 \, Re_x^{-1/(n+1)} \tag{7.21a}$$

The shear stress acting on the plate will be equal and opposite to the shear stress in the fluid at the surface, i.e. $\tau_{wx} = -\tau_{yx}|_{y=0}$.

Thus, $$\tau_{wx} = \left(\frac{3}{2F(n)} \right)^n \rho V_0^2 \, Re_x^{-1/(n+1)} \tag{7.21b}$$

Equation (7.21b) gives the local value of the shearing stress on the plate which can be averaged over the length of the plate to obtain a mean value, τ_w:

$$\tau_w = \frac{1}{L} \int_0^L \tau_{wx} \, dx$$

$$= \frac{\rho V_0^2}{L} \left\{ \frac{3}{2F(n)} \right\}^n \int_0^L Re_x^{-1/(n+1)} \, dx$$

$$= \rho V_0^2 (n+1) \left(\frac{3}{2F(n)} \right)^n Re_L^{-1/(n+1)} \tag{7.22}$$

where $Re_L = \rho V_0^{2-n} L^n / m$.

The total frictional drag force F_d on one side of the plate of length L and width W is then obtained as:

$$F_d = \tau_w \cdot (LW) = \left(\frac{3}{2F(n)} \right)^n (n+1) \rho V_0^2 \, Re_L^{-1/(n+1)} WL \tag{7.23}$$

Introducing a dimensionless drag coefficient C_D defined as:

$$C_D = \frac{F_d}{\left(\frac{1}{2} \rho V_0^2 \right) (WL)}$$

Equation (7.23) becomes:

$$C_D = 2(n+1) \left(\frac{3}{2F(n)} \right)^n Re_L^{-1/(n+1)}$$

$$= C(n) \, Re_L^{-1/(n+1)} \tag{7.24}$$

where $C(n) = 2(n+1)(3/2F(n))^n$ $\tag{7.24a}$

For the special case of Newtonian fluids ($n = 1$), equation (7.24) reduces to the well known result of $C_D = 1.292\,\mathrm{Re}_L^{-1/2}$.

The values of the laminar boundary layer thickness and of the frictional drag are not very sensitive to the form of approximations used for the velocity distribution, as illustrated by Skelland [1967] for various choices of velocity profiles. The resulting values of C(n) are compared in Table 7.1 with the more refined values obtained by Acrivos *et al.* [1960] who solved the differential momentum and mass balance equations numerically; the two values agree within 10% of each other. Schowalter [1978] has discussed the extension of the laminar boundary layer analysis for power-law fluids to the more complex geometries of two- and three-dimensional flows.

Table 7.1   *Values of* C(n) *in equation (7.24a)*

n	Value of C(n)	
	Numerical	Equation (7.24a)
0.1	2.132	1.892
0.2	2.094	1.794
0.3	1.905	1.703
0.5	1.727	1.554
1.0	1.328	1.292
1.5	1.095	1.128
2	0.967	1.014
3	0.776	0.872
4	0.678	0.79
5	0.613	0.732

7.4 Laminar boundary layer flow of Bingham plastic fluids over a plate

Outside the boundary layer region, the velocity gradient is zero, and thus the shearing forces must also be zero in the case of Newtonian and power-law fluids, as seen in the preceding section. In contrast, in the case of Bingham plastic fluids, the shear stress approaches the yield stress of the fluid at the outer edge of the boundary layer which must eventually decay to zero over a relatively short distance and thus, once again, there is no shearing force present in the bulk of the fluid outside the boundary layer. It should be noted that in this case the position of the boundary layer is defined precisely, as a point where the shear stress in the fluid equals the yield stress of the fluid.

However, to take account of the yield stress acting at the edge of the boundary layer, the integral momentum balance equation (7.7) must be modified as:

$$\rho \frac{d}{dx} \left[\int_0^\delta (V_0 - V_x) V_x \, dy \right] = -(\tau_{wx} - \tau_0^B) \tag{7.25}$$

Assuming that equation (7.11) still provides a satisfactory representation of the laminar boundary layer flow in the region $0 \le y \le \delta$, substituting equation (7.11) into (7.25) yields:

$$\frac{39}{280} \rho V_0^2 \frac{d\delta}{dx} = -(\tau_{wx} - \tau_0^B) \qquad (\tau_{wx} > \tau_0^B) \tag{7.26}$$

For a Bingham plastic, the fluid the shear stress at the surface, that is, at $y = 0$ is given by:

$$\tau_{yx}|_{y=0} = -\tau_0^B + \mu_B \frac{dV_x}{dy}\bigg|_{y=0} \tag{7.27}$$

The velocity gradient at the surface,

$$\frac{dV_x}{dy}\bigg|_{y=0} = \frac{3V_0}{2\delta} \qquad \text{(equation 7.15)}$$

Substituting in equation (7.27):

$$\tau_{yx}|_{y=0} = -\tau_{wx} = -\left(\tau_0 + \frac{3}{2} \frac{\mu_B V_0}{\delta} \right) \tag{7.28a}$$

$$\text{or} \qquad -(\tau_{wx} - \tau_0^B) = \frac{3}{2} \frac{\mu V_0}{\delta} \tag{7.28b}$$

Combining equations (7.26) and (7.28):

$$\delta \, d\delta = \left(\frac{140}{13} \right) \frac{\mu_B}{\rho V_0} \cdot dx$$

Integration with the initial condition that at $x = 0$, $\delta = 0$, yields:

$$\frac{\delta}{x} = \frac{4.64}{\sqrt{Re_x}} \tag{7.29}$$

where Re_x is now defined as $\rho V_0 x / \mu_B$.

It is interesting to note that equation (7.29) is identical to that for a Newtonian fluid (see equation 7.17 with $n = 1$). In spite of this similarity with Newtonian fluids, the frictional drag on a plate submerged in a Bingham plastic fluid deviates from that in a Newtonian fluid owing to the existence of the fluid yield stress in equation (7.28), as will be seen in the next section.

7.4.1 Shear stress and drag force on an immersed plate

The shear stress in the fluid adjacent to the plate $(y = 0)$ is given by equation (7.28a), i.e.

$$\tau_{yx}|_{y=0} = -\left(\tau_0^B + \frac{3}{2}\frac{\mu_B V_0}{\delta}\right) \tag{7.28a}$$

Substitution for δ from equation (7.29):

$$\tau_{yx}|_{y=0} = -\left(\tau_0^B + \frac{0.323}{x}V_0\mu_B\sqrt{Re_x}\right) \tag{7.30}$$

The shear stress acting on the plate will be equal and opposite of this value, i.e. $\tau_{wx} = -\tau_{yx}|_{y=0}$, thus

$$\tau_{wx} = \left(\tau_0^B + \frac{0.323}{x}V_0\mu_B\sqrt{Re_x}\right) \tag{7.31}$$

The average shear stress acting over the entire plate may be calculated as:

$$\tau_w = \frac{1}{L}\int_0^L \tau_{wx}\,dx = \frac{1}{L}\int_0^L (\tau_0^B + 0.323V_0^{3/2}\rho^{1/2}\mu_B^{1/2}x^{-1/2})\,dx$$

or

$$\tau_w = \tau_0^B + 0.646\left(\frac{\rho V_0^2}{\sqrt{Re_L}}\right) \tag{7.32}$$

Introducing a drag coefficient C_D, as defined earlier, equation (7.32) can be re-written as:

$$C_D = Bi + \frac{1.292}{\sqrt{Re_L}} \tag{7.33}$$

where Bi, the Bingham number is defined as $2\tau_0^B/\rho V_0^2$.

Again, for Newtonian fluids, $Bi = 0$ and equation (7.33) reduces to the expected form.

Example 7.1

A polymer solution (of density $1000\,\text{kg/m}^3$) is flowing parallel to a plate ($300\,\text{mm} \times 300\,\text{mm}$); the free stream velocity is $2\,\text{m/s}$. In the narrow shear rate range, the rheology of the polymer solution can be adequately approximated by both the power-law ($m = 0.3\,\text{Pa·s}^n$ and $n = 0.5$) and the Bingham plastic model ($\tau_0 = 2.28\,\text{Pa·s}$ and $\mu_B = 7.22\,\text{mPa·s}$). Using each of these models, estimate and compare the values of the shear stress and the boundary layer thickness $150\,\text{mm}$ away from the leading edge, and the total frictional force on each side of the plate.

Solution

(a) Power-law model calculations

First, check the value of the Reynolds number at $x = L = 0.3\,\mathrm{m}$,

$$\mathrm{Re}_L = \frac{\rho V_0^{2-n} L^n}{m}$$

$$= \frac{(1000)(2)^{2-0.5}(0.3)^{0.5}}{0.3} = 5164$$

The flow is likely to be streamline over the whole plate length (see Section 7.5). The value of the local shear stress at $x = 0.15\,\mathrm{m}$ can be calculated using equation (7.21):

$$\tau_{wx}|_{x=0.15\,\mathrm{m}} = -\tau_{yx}|_{y=0,x=0.15\,\mathrm{m}} = \left(\frac{3}{2F(n)}\right)^n \rho V_0^2 \mathrm{Re}_x^{-1/(n+1)}$$

where $F(n) = \left[\frac{280}{39}(n+1)\left(\frac{3}{2}\right)^n\right]^{1/(n+1)}$

For $n = 0.5$, $F(n) = \left[\frac{280}{39}(1.5)(3/2)^{0.5}\right]^{1/1.5} = 5.58$

Substitution in $\tau_{wx}|_{x=0.15\,\mathrm{m}}$ yields:

$$\tau_{wx}|_{x=0.15\,\mathrm{m}} = \left(\frac{3}{2 \times 5.58}\right)^{0.5} \times 1000 \times 2^2 \left\{\frac{(1000)(2)^{2-0.5}(0.15)^{0.5}}{0.3}\right\}^{-1/1.5}$$

$$= 8.75\,\mathrm{Pa}$$

The boundary layer thickness δ at $x = 0.15\,\mathrm{m}$ is calculated using equation (7.17):

$$\frac{\delta}{x} = F(n)\,\mathrm{Re}_x^{-1/(n+1)}$$

or $\delta|_{x=0.15\,\mathrm{m}} = (0.15)(5.58)\left\{\frac{(1000)(2)^{2-0.5}(0.15)^{0.5}}{0.3}\right\}^{-1/1.5}$

$$= 3.53 \times 10^{-3}\,\mathrm{m} \quad \text{or} \quad 3.53\,\mathrm{mm}$$

Finally, the frictional drag force on the plate is calculated using equation (7.24):

$$C_D = C(n)\,\mathrm{Re}_L^{-1/(n+1)}$$

where $C(n) = 2(n+1)\left(\frac{3}{2F(n)}\right)^n$

For $n = 0.5$, $C(n) = 2 \times 1.5 \times \left(\frac{3}{2 \times 5.58}\right)^{0.5} = 1.554$

$$\therefore C_D = 1.554(5164)^{-1/1.5} = 0.0052$$

Drag force on one side of plate $F_D = C_D \frac{1}{2} \rho V_0^2 \cdot WL$

$$F_D = 0.0052 \times \tfrac{1}{2} \times 1000 \times 2^2 \times 0.3 \times 0.3 = 0.94\,\text{N}$$

(b) Bingham plastic model calculations

The maximum value of the Reynolds number occurs at the trailing edge of the plate, i.e.

$$\text{Re}_L = \frac{\rho V_0 L}{\mu_B} = \frac{1000 \times 2 \times 0.3}{7.22 \times 10^{-3}} = 83\,100$$

Using the same transition criterion as for Newtonian fluids, namely, $\text{Re}_L < {\sim}10^5$, the flow in the whole of the boundary layer is likely to be laminar.

The value of the local shear stress at $x = 0.15\,\text{m}$ is calculated using equation (7.31):

$$\tau_{wx}|_{x=0.15\,\text{m}} = \tau_0^B + \frac{0.323}{x} V_0 \mu_B \sqrt{\text{Re}_x}$$

$$= 2.28 + \frac{0.323}{0.15}(2)(7.22 \times 10^{-3}) \left(\frac{1000 \times 2 \times 0.15}{7.22 \times 10^{-3}} \right)^{1/2}$$

$$= 8.62\,\text{Pa}$$

Note that, although about a quarter of the shear stress stems from the yield stress, the total value is comparable with that based on the use of the power-law model, as seen in Part (a) above.

The boundary layer thickness δ is calculated using equation (7.29):

$$\delta = 4.64x\,\text{Re}_x^{-1/2} = 4.64 \times 0.15 \left(\frac{1000 \times 2 \times 0.15}{7.22 \times 10^{-3}} \right)^{-1/2}$$

$$= 3.41 \times 10^{-3}\,\text{m} \quad \text{or} \quad 3.41\,\text{mm}$$

This value also closely corresponds with that calculated in part (a).

Finally, the drag coefficient is evaluated using equation (7.33):

$$C_D = \text{Bi} + \frac{1.292}{\sqrt{\text{Re}_L}} = \frac{2\tau_0^B}{\rho V_0^2} + \frac{1.292}{\sqrt{\text{Re}_L}}$$

$$= \frac{2 \times 2.28}{(1000)(2)^2} + \frac{1.292}{(83\,100)^{1/2}} = 0.00562$$

and drag force on one side of the plate,

$$F_D = C_D \cdot \tfrac{1}{2}\rho V_0^2 WL = 0.00562 \times \tfrac{1}{2} \times 1000 \times 2^2 \times 0.3 \times 0.3$$

$$= 1.015\,\text{N}$$

This value is slightly higher than that obtained in part (a) above.$\square$

7.5 Transition criterion and turbulent boundary layer flow

7.5.1 Transition criterion

As mentioned in Section 7.1, the nature of the flow in the boundary layer is influenced by several variables including the degree of turbulence in the free stream, the roughness of the immersed surface and the value of the Reynolds number, defined by equation (7.1). While no experimental data are available on the limiting values of the Reynolds number for the flow of non-Newtonian materials in boundary layers, Skelland [1967] has suggested that the limiting value of the Reynolds number decreases with increasing shear-thinning behaviour. Based on heuristic considerations, he defined an effective viscosity for a power-law fluid in such a manner that the shear stress at the surface equals that in Newtonian fluids at the transition point, i.e. at $\mathrm{Re}_x \sim 10^5$. Thus, equating the values of the shear stress at the surface for a Newtonian fluid of hypothetical viscosity to that for the power-law fluid (equation 7.21), Skelland [1967] obtained the following transition criterion for power law fluids:

$$\left(\frac{0.323}{\mathrm{B}(n)}\right)^2 \mathrm{Re}_c^{2/(n+1)} < 10^5 \tag{7.34}$$

where $\mathrm{B}(n) = (1.5/F(n))^n$, $F(n)$ is given by equation (7.17a) and $\mathrm{Re}_c = \rho V_0^{2-n} x_c^n / m$ where x_c is the distance (from the leading edge) at which the flow ceases to be streamline. For Newtonian fluids $(n = 1)$, $\mathrm{B}(n) = 0.323$ and $\mathrm{Re}_c = 10^5$ which is the value at the transition point. For a liquid of flow behaviour index, $n = 0.5$, the limiting value of the Reynolds number is 1.14×10^4 which is an order of magnitude smaller than the value for a Newtonian fluid.

7.5.2 Turbulent boundary layer flow

As mentioned previously, even when the flow becomes turbulent in the boundary layer, there exists a thin sub-layer close to the surface in which the flow is laminar. This layer and the fully turbulent regions are separated by a buffer layer, as shown schematically in Figure 7.1. In the simplified treatments of flow within the turbulent boundary layer, however, the existence of the buffer layer is neglected. In the laminar sub-layer, momentum transfer occurs by molecular means, whereas in the turbulent region eddy transport dominates. In principle, the methods of calculating the local values of the boundary layer thickness and shear stress acting on an immersed surface are similar to those used above for laminar flow. However, the main difficulty stems from the fact that the viscosity models, such as equations (7.13) or (7.27), are applicable only for laminar flow and thus appropriate expressions for the shear stress at the surface must be known if they are to be inserted in equation (7.12). For Newtonian fluids, Blasius [1913] circumvented this difficulty by inferring the values of the shear stress at the wall from the turbulent friction factor–Reynolds number relationship for flow in circular tubes. Curvature effects are likely to be negligible because the laminar sub-layer thickness is generally much smaller than the diameter of the pipe. Schlichting [1968] and Coulson and Richardson [1999] have discussed turbulent boundary layer flow in depth. Skelland [1967] has extended this approach to power-law liquids by replacing the Blasius equation by a modified form, equation (3.38). Furthermore, because non-Newtonian materials generally have high apparent viscosities, turbulent conditions are not often encountered in flow over immersed surfaces. Therefore, this topic will not be pursued further here because the turbulent flow of non-Newtonian fluids is not at all well understood and there are no well-established and validated equations for shear stress, even for flow in pipes.

7.6 Heat transfer in boundary layers

When the fluid and the immersed surface are at different temperatures, heat transfer will take place. If the heat transfer rate is small in relation to the thermal capacity of the flowing stream, its temperature will remain substantially constant. The surface may be maintained at a constant temperature, or the heat flux at the surface may be maintained constant; or surface conditions may be intermediate between these two limits. Because the temperature gradient will be highest in the vicinity of the surface and the temperature of the fluid stream will be approached asymptotically, a thermal boundary layer may therefore be postulated which covers the region close to the surface and in which the whole of the temperature gradient is assumed to lie.

Thus a momentum and a thermal boundary layer will develop simultaneously whenever the fluid stream and the immersed surface are at different temperatures (Figure 7.3). The momentum and energy equations are coupled, because the physical properties of non-Newtonian fluids are normally temperature-dependent. The resulting governing equations for momentum and heat transfer require numerical solutions. However, if the physical properties of the fluid do not vary significantly over the relevant temperature interval, there is little interaction between the two boundary layers and they may both be assumed to develop independently of one another. As seen in Chapter 6, the physical properties other than apparent viscosity may be taken as constant for commonly encountered non-Newtonian fluids.

In general, the thermal and momentum boundary layers will not correspond. In the ensuing treatment, the simplest non-interacting case will be considered

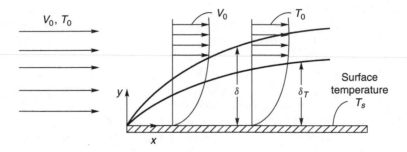

Figure 7.3 *Schematic representation of momentum and thermal boundary layers*

with physical properties taken as being temperature-independent. The temperature of the bulk fluid will be taken to be T_0 (constant) and that of the immersed plate to be T_s (also constant). For convenience, the temperature scale will be chosen so that the surface temperature is zero, giving the boundary condition $T_s = 0$, corresponding to zero velocity in the momentum balance equation.

An integral procedure similar to that adopted previously for momentum boundary layers will be used here to obtain the expression for the rate of heat exchange between the fluid and the plane surface. A heat balance will be made over a control volume (Figure 7.4) which extends beyond the limits of both the momentum and thermal boundary layers.

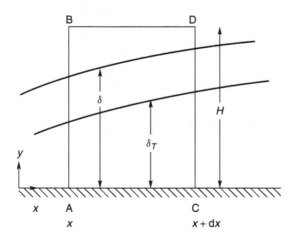

Figure 7.4 *Control volume for heat balance*

At steady state and with no source or sink present in the control volume, the heat balance for the control volume ABCD can be stated as follows:

The sum of heat convected in through planes AB and BD and that added by conduction at wall AC must be equal to the heat convected out at plane CD (7.35)

The rate at which heat enters the control volume at plane AB is:

$$W \int_0^H \rho V_x C_p T \, dy \qquad (7.36a)$$

Similarly, the rate of transfer of heat at the plane CD:

$$W \int_0^H \rho V_x C_p T \, dy + W \frac{d}{dx} \left[\int_0^H \rho V_x C_p T \, dy \right] dx \qquad (7.36b)$$

It has already been seen that the mass rate of transfer of fluid across the plane BD is given by equation (7.4):

$$W \frac{d}{dx} \left(\int_0^H \rho V_x dy \right) dx \qquad (7.4)$$

The enthalpy of this stream is:

$$C_p T_0 W \frac{d}{dx} \left(\int_0^H \rho V_x \, dy \right) dx \qquad (7.37)$$

Substituting for the terms in equation (7.35):

$$\rho W C_p \int_0^H V_x T \, dy + \rho C_p W T_0 \frac{d}{dx} \left(\int_0^H V_x \, dy \right) dx + q_w W \, dx$$

$$= \rho W C_p \int_0^H V_x T \, dy + \rho W C_p \frac{d}{dx} \left(\int_0^H V_x T \, dy \right) dx \qquad (7.38)$$

where q_w is the heat flux at the wall and is equal to $-k \, (dT/dy)|_{y=0}$ which upon substitution in equation (7.38) gives:

$$\frac{d}{dx} \int_0^{\delta_T} V_x (T_0 - T) \, dy = \alpha \frac{dT}{dy} \bigg|_{y=0} \qquad (7.39)$$

where α is the thermal diffusivity ($= k/\rho C_p$) of the fluid. Since both V_x and V_y are zero at the surface, the only mechanism of heat transfer which is relevant is conduction. Also, the upper limit of the integration in equation (7.39) has been changed to δ_T, the thermal boundary layer thickness, because for $y \geq \delta_T$, $T = T_0$ and the integrand is identically zero.

The relation between V_x and y has already been obtained for laminar flow in the boundary layer. A similar relation between the temperature of the fluid,

T, and y for laminar flow of power-law fluids in the boundary layer will now be derived.

7.6.1 Heat transfer in laminar flow of a power-law fluid over an isothermal plane surface

Consideration is now given to the flow of an incompressible power-law liquid of temperature T_0 over a plane surface maintained at a higher temperature T_s. At any given distance from the leading edge, the temperature of the fluid progressively decreases with distance y from the surface reaching T_0 at the extremity of the thermal boundary layer, $y = \delta_T$. The temperature at a distance y from the surface can be approximated by a polynomial of the form:

$$T = a + by + cy^2 + dy^3 \tag{7.40}$$

The boundary conditions for the temperature field are:

$$\text{At} \quad y = 0, \ T = T_s \tag{7.41a}$$

$$y = \delta_T, T = T_0 \quad \text{and} \quad \frac{\mathrm{d}T}{\mathrm{d}y} = 0 \tag{7.41b}$$

The fourth boundary condition for temperature can be deduced as follows: the rate of heat transfer per unit area of the surface is given by:

$$q|_{y=0} = -k\frac{\mathrm{d}T}{\mathrm{d}y}\bigg|_{y=0} \tag{7.41c}$$

If the temperature of the fluid element adhering to the wall is to remain constant at $T = T_s$, the rate of heat transfer into and out of the element must be the same, i.e. the temperature gradient must be constant and therefore:

$$\frac{\mathrm{d}}{\mathrm{d}y}\left(\frac{\mathrm{d}T}{\mathrm{d}y}\bigg|_{y=0}\right) \quad \text{or} \quad \frac{\mathrm{d}^2T}{\mathrm{d}y^2}\bigg|_{y=0} = 0 \tag{7.41d}$$

The four unknown constants in the assumed temperature profile, equation (7.40), can now be evaluated by using equation (7.41a) to (7.41d) to give:

$$a = T_s; \quad b = \frac{3}{2}\left(\frac{T_0 - T_s}{\delta_T}\right)$$

$$c = 0; \quad d = -\frac{(T_0 - T_s)}{2\delta_T^3} \tag{7.42}$$

Substitution of these values into equation (7.40) yields

$$\frac{\theta}{\theta_0} = \frac{T - T_s}{T_0 - T_s} = \frac{3}{2}\left(\frac{y}{\delta_T}\right) - \frac{1}{2}\left(\frac{y}{\delta_T}\right)^3 \tag{7.43}$$

It should be noted that equation (7.10) for velocity distribution and equation (7.43) for temperature distribution in the laminar boundary layer are identical in form, because basically the same forms of equation and boundary conditions have been used for the velocity and temperature profiles.

It will now be assumed that the thermal boundary layer is everywhere the thinner, i.e. $\delta_T < \delta$. The conditions under which this approximation is justified are examined later. Now substituting in equation (7.39) for V_x from equation (7.11), T and $(dT/dy)|_{y=0}$ from equation (7.43):

$$V_0 \theta_0 \frac{d}{dx} \left[\int_0^{\delta_T} \left\{ \frac{3}{2} \left(\frac{y}{\delta} \right) - \frac{1}{2} \left(\frac{y}{\delta} \right)^3 \right\} \left\{ 1 - \frac{3}{2} \left(\frac{y}{\delta_T} \right) + \frac{1}{2} \left(\frac{y}{\delta_T} \right)^3 \right\} dy \right]$$

$$= \frac{3}{2} \alpha \frac{\theta_0}{\delta_T} \tag{7.44}$$

Integration with respect to y between the limits $y = 0$ to $y = \delta_T$ yields:

$$\theta_0 V_0 \frac{d}{dx} \left\{ \delta \left(\frac{3}{20} \varepsilon^2 - \frac{3}{280} \varepsilon^4 \right) \right\} = \frac{3}{2} \frac{\alpha \theta_0}{\delta \varepsilon} \tag{7.45}$$

where $\varepsilon = (\delta_T/\delta) < 1$. Neglecting the ε^4 term in equation (7.45) leads to

$$\frac{V_0}{10} \frac{d}{dx} (\delta \varepsilon^2) = \frac{\alpha}{\delta \varepsilon} \tag{7.46}$$

which may be expressed as:

$$\frac{V_0}{10} \left(2\varepsilon^2 \delta^2 \frac{d\varepsilon}{\delta x} + \delta \varepsilon^3 \frac{d\delta}{dx} \right) = \alpha \tag{7.47}$$

Now introducing a new variable $z = \varepsilon^3$ and substituting for δ and $d\delta/dx$ from equations (7.17) and (7.19) respectively:

$$\frac{dz}{dx} + \frac{3}{2x(n+1)} z = \frac{15\alpha \operatorname{Re}_x^{2/(n+1)}}{V_0 \, x^2 (F(n))^2} \tag{7.48}$$

where $\operatorname{Re}_x = (\rho V_0^{2-n} x^n)/m$.

The solution of the differential equation (7.48) is given by:

$$z = \varepsilon^3 = \left(\frac{\delta_T}{\delta} \right)^3 = \frac{15\alpha}{(F(n))^2 V_0} \cdot \frac{\operatorname{Re}_x^{2/(n+1)}}{x} \cdot \frac{2(n+1)}{(2n+1)} + C_1 x^{-3/(2(n+1))}$$

$$\tag{7.49}$$

Substituting for δ from equation (7.16):

$$\delta_T^3 = \frac{15\alpha F(n)}{V_0(2n+1)} 2(n+1)x^2 \operatorname{Re}_x^{-1/(n+1)}$$

$$+ C_1(F(n))^3 x^{3(2n+1)/2(n+1)} \operatorname{Re}_x^{-3/(n+1)} \tag{7.50}$$

For the general case where the heating of the plate begins at $x = x_0$, the constant C_1 is evaluated using the condition $\delta_T = 0$ at $x = x_0$ yielding the expression for δ_T as:

$$\frac{\delta_T}{x} = \left[\frac{30(n+1)}{(2n+1)} F(n)\right]^{1/3} \left(\frac{\alpha}{xV_0}\right)^{1/3} \operatorname{Re}_x^{\frac{1}{-3(n+1)}} \left[1 - \left(\frac{x_0}{x}\right)^{(2n+1)/2(n+1)}\right]^{1/3} \tag{7.51a}$$

For the special case where the whole plate being heated, i.e. $x_0 = 0$, the thermal boundary layer thickness δ_T is given by:

$$\frac{\delta_T}{x} = \left[\frac{30(n+1)}{(2n+1)} F(n)\right]^{1/3} \left(\frac{\alpha}{xV_0}\right)^{1/3} \operatorname{Re}_x^{\frac{-1}{3(n+1)}} \tag{7.51b}$$

A heat transfer coefficient, h, may be defined such that:

$$h(T_s - T_0) = -k\frac{dT}{dy}\bigg|_{y=0} \tag{7.52}$$

The temperature gradient is found by differentiating the temperature profile, equation (7.43):

$$\frac{dT}{dy}\bigg|_{y=0} = \frac{3(T_0 - T_s)}{2\delta_T} \tag{7.53}$$

and the heat transfer coefficient is obtained by combining equations (7.52) and (7.53):

$$h = \left(\frac{3}{2}\right)\left(\frac{k}{\delta_T}\right) \tag{7.54}$$

Substitution for δ_T from equation (7.51) into (7.54) leads to

$$\operatorname{Nu}_x = \frac{hx}{k} = \frac{3}{2}\left[\frac{30F(n)(n+1)}{(2n+1)}\right]^{-(1/3)} \operatorname{Pr}_x^{1/3} \operatorname{Re}_x^{(n+2)/(3(n+1))} \tag{7.55}$$

where the Prandtl number, Pr_x, is defined as:

$$\operatorname{Pr}_x = \frac{C_p}{k} m \left(\frac{V_0}{x}\right)^{n-1} \tag{7.56}$$

The Prandtl number for a power-law fluid is not simply a combination of physical properties of the fluid but also depends on the point value of an apparent viscosity $m(V_0/x)^{n-1}$ which is a function of both x and y. The choice of linear dimension in this definition is somewhat arbitrary and no single definition may necessarily be appropriate for all applications. However, for $n = 1$, equation (7.55) reduces to the corresponding Newtonian expression, i.e.

$$\mathrm{Nu}_x = 0.328\,\mathrm{Pr}_x^{1/3}\,\mathrm{Re}_x^{1/2} \tag{7.57}$$

It is seen that the local value of the heat transfer coefficient varies with distance from the leading edge according to the relation:

$$h \propto x^{-(n+2)/(3(n+1))} \tag{7.58}$$

The heat transfer coefficient h has a (theoretical) infinite value at the leading edge where the thickness of the thermal boundary layer is zero and decreases progressively as the boundary layer thickens. The mean value of the heat transfer coefficient over the plate length can be obtained by integrating it from $x = 0$ to $x = L$ as:

$$
\begin{aligned}
h_m &= \frac{1}{L}\int_0^L h\,\mathrm{d}x \\
&= \frac{\alpha(n)}{L}k\int_0^L \frac{\mathrm{Pr}_x^{1/3}\,\mathrm{Re}_x^{(n+2)/(3(n+1))}}{x}\,\mathrm{d}x
\end{aligned}
\tag{7.59}
$$

where $\quad \alpha(n) = \dfrac{3}{2}\left[\dfrac{30F(n)(n+1)}{(2n+1)}\right]^{-(1/3)}$

The evaluation of the integral in equation (7.59) followed by some simplification and re-arrangement finally yields:

$$\mathrm{Nu}_m = \frac{hL}{k} = \frac{9(n+1)}{2(2n+1)}\left[\frac{30F(n)(n+1)}{(2n+1)}\right]^{-1/3}\mathrm{Pr}_L^{1/3}\,\mathrm{Re}_L^{(n+2)/(3(n+1))} \tag{7.60}$$

where $F(n)$, a function of n, is given by equation (7.17) and Pr_L and Re_L respectively are defined as:

$$\mathrm{Pr}_L = \frac{C_p}{k}m\left(\frac{V_0}{L}\right)^{n-1} \tag{7.61}$$

and

$$\mathrm{Re}_L = \frac{\rho V_0^{2-n}L^n}{m} \tag{7.62}$$

More rigorous numerical solutions for the laminar boundary layers of power-law and other time-independent fluids flowing over plane surfaces and objects of two dimensional axisymmetric shapes are given in the literature [Acrivos et al., 1960] and these predictions are in good agreement with the approximate result given by equation (7.60). Most of the theoretical developments in the field of thermal boundary layers of time-independent fluids have been critically reviewed in references [Shenoy and Mashelkar, 1982; Nakayama, 1988; Chhabra, 1993a,b, 1999]. The limited experimental results and empirical correlations are presented in a later section in this chapter.

Example 7.2

A dilute polymer solution at 25°C flows at 2 m/s over a 300 mm × 300 mm square plate which is maintained at a uniform temperature of 35°C. The average values of the power-law constants (over this temperature interval) may be taken as: $m = 0.3\,\text{Pa·s}^n$ and $n = 0.5$. Estimate the thickness of the boundary layer 150 mm from the leading edge and the rate of heat transfer from one side of the plate only. The density, thermal conductivity and heat capacity of the polymer solution may be approximated as those of water at the same temperature.

Solution

As seen in Example 7.1, the flow conditions in the boundary layer appear to be laminar over the entire length of the plate under the given operating conditions. Therefore, the thickness of the thermal boundary layer can be estimated using equation (7.51). The value of Re_x is:

$$\text{Re}_x = \frac{\rho V_0^{2-n} x^n}{m} = \frac{1000 \times 2^{2-0.5} \times 0.15^{0.5}}{0.3} = 3650.$$

For water at 25°C, the values of the thermo-physical properties are: $\rho = 1000\,\text{kg/m}^3$; $k = 0.615\,\text{W/mK}$; $C_p = 3800\,\text{J/kgK}$

$$\therefore \text{ thermal diffusivity } \alpha = \frac{k}{\rho C_p} = \frac{0.615}{1000 \times 3800}$$

$$= 1.62 \times 10^{-7}\,\text{m}^2/\text{s}$$

We have already calculated in example 7.1, $F(n) = 5.58$ when $n = 0.5$.
Substituting these values in equation (7.51):

$$\delta_T = \left[\frac{30 \times 1.62 \times 10^{-7} \times 5.58 \times 0.15^2}{2} \times \left(\frac{1.5}{2} \right) \right]^{1/3} 3650^{-1/4.5}$$

$$= 9.88 \times 10^{-4}\,\text{m}$$

or about 1 mm which is less than a third of the corresponding momentum boundary layer thickness.

The Prandtl number with L as characteristic linear dimension is evaluated:

$$\mathrm{Pr}_L = \frac{C_p m}{k} \left(\frac{V_0}{L} \right)^{n-1} = \frac{3800 \times 0.3}{0.615} \left(\frac{2}{0.3} \right)^{0.5-1} = 718$$

and $\quad \mathrm{Re}_L = \dfrac{\rho V_0^{2-n} L^n}{m} = \dfrac{1000 \times 2^{2-0.5} \times 0.3^{0.5}}{0.3} = 5164$

Substituting these values in equation (7.60):

$$\mathrm{Nu}_m = \frac{h_m \cdot L}{k} = \frac{9 \times 1.5}{2 \times 2} \left[\frac{30 \times 5.58 \times 1.5}{2} \right]^{-1/3} (718)^{1/3} (5164)^{2.5/4.5}$$

$$= 697.5$$

$$\therefore h_m = \frac{697.5 \times 0.615}{0.3} = 1430 \,\mathrm{W/m^2K}$$

The rate of heat loss from the plate is

$$Q = h_m \cdot A \cdot \Delta T = 1430 \times (0.3 \times 0.3) \times (35 - 25) = 1287 \,\mathrm{W.} \square$$

7.7 Mass transfer in laminar boundary layer flow of power-law fluids

The basic equation for mass transfer by molecular diffusion is Fick's law which may be expressed as:

$$N_A = -D_{AB} \frac{dC_A}{dy}$$

where N_A is the mass transfer rate per unit area $(\mathrm{kmol/m^2 s})$

C_A is the molar concentration of the diffusing component, and

D_{AB} is the molecular diffusity.

This equation is analogous to Fourier's law for heat transfer by conduction:

$$q = -k \frac{dT}{dy} = -\frac{k}{\rho C_p} \frac{d(\rho C_p T)}{dy} = -\alpha \frac{d(\rho C_p T)}{dy}$$

(assuming ρ and C_p to be constant).

It will be seen that there is a synergy between the two equations, $\rho C_p T$ the heat content per unit volume being equivalent to C_A, the molar concentration of the diffusing component.

The boundary layer equation for mass transfer may therefore be written by analogy with equaiton (7.39) as:

$$\frac{d}{dx} \int_0^H (C_{A_s} - C_A) V_x \, dy = D_{AB} \left. \frac{dC_A}{dy} \right|_{y=0} \tag{7.63}$$

where C_{A_s} and C_A are respectively the concentration of the solute A in the free stream (outside the boundary layer) at a distance x from the leading edge and at a distance y from the surface. H is the thickness of the control volume normal to the surface as shown in Figure 7.4; it is chosen to be greater than the thickness of both the momentum and concentration boundary layers.

Again the form of concentration profile in the diffusion boundary layer depends on the boundary conditions at the surface and in the fluid stream. For the conditions analogous to those used in consideration of thermal boundary layer, (constant concentrations both in the free stream outside the boundary layer and at the submerged surface) the concentration profile will be of similar form to that given by equation (7.43);

$$\frac{C_A - C_{A_o}}{C_{A_s} - C_{A_o}} = \frac{3}{2}\left(\frac{y}{\delta_m}\right) - \frac{1}{2}\left(\frac{y}{\delta_m}\right)^3 \tag{7.64}$$

and $$\frac{1}{C_{A_s} - C_{A_o}}\frac{dC_A}{dy} = \frac{3}{2}\left\{\frac{1}{\delta_m} - \frac{y^2}{\delta_m^3}\right\} \tag{7.64a}$$

where C_{A_o} is the solute concentration at the surface ($y = 0$) and δ_m is the thickness of the concentration boundary layer.

Substituting in equation (7.63) for the velocity V_x from equation (7.10), and for the concentration C_A and concentration gradient (d C_A/dy) from equations (7.64) and (7.64a):

$$V_0\frac{d}{dx}\left\{\delta\left(\frac{\varepsilon_m^2}{10} - \frac{1}{140}\varepsilon_m^4\right)\right\} = \frac{D_{AB}}{\delta\varepsilon_m} \tag{7.65}$$

where $\varepsilon_m = (\delta_m/\delta) < 1$. It should be noted that it has been implicitly assumed that the diffusion boundary layer is everywhere thinner than the momentum boundary layer. With this assumption, the fourth order term in ε_m in equation (7.65) can be neglected without incurring significant error. The resulting approximate solution is identical in form to equation (7.47) for heat transfer, except that the diffusivity D_{AB} replaces the thermal diffusivity α, and thus its solution for a power-law liquid, with the condition $\delta_m = 0$ at $x = 0$, is by analogy with equation (7.51b):

$$\frac{\delta_m}{x} = \left[\frac{30(n + 1)}{(2n + 1)}F(n)\right]^{1/3}\left[\frac{D_{AB}}{xV_o}\right]^{1/3}\mathrm{Re}_x^{-1/3(n+1)} \tag{7.66}$$

One can now define a mass transfer coefficient, h_D by the relation:

$$h_D(C_{A_o} - C_{A_s}) = -D_{AB}\frac{\partial C_A}{\partial y}\Big|_{y=0} = -\frac{3}{2}(C_{A_s} - C_{A_o})\frac{D_{AB}}{\delta_m} \tag{7.67}$$

Substituting for δ_m from equation (7.66) and re-arranging:

$$\text{Sh}_x = \frac{h_D x}{D_{AB}} = \frac{3}{2} \left[\frac{30(n+1)\text{F}(n)}{(2n+1)} \right]^{-1/3} \text{Sc}_x^{1/3} \, \text{Re}_x^{(n+2)/(3(n+1))} \tag{7.68}$$

where the Schmidt number, $\text{Sc}_x = \dfrac{m}{\rho D_{AB}} \left(\dfrac{V_0}{x} \right)^{n-1}$ \hfill (7.69)

Here, Sc_x is analogous to Pr_x for heat transfer (equation 7.56). Finally, the average value of the transfer coefficient over the plate length L can be obtained in the same way as for heat transfer (equation 7.59) by integration:

$$\text{Sh}_m = \frac{h_D L}{D_{AB}} = \frac{9(n+1)}{2(2n+1)} \left\{ \frac{30(n+1)\text{F}(n)}{(2n+1)} \right\}^{-1/3} \text{Sc}_L^{1/3} \, \text{Re}_L^{(n+2)/(3(n+1))}$$
$$\tag{7.70}$$

Both Mishra *et al.* [1976] and Ghosh *et al.* [1986] measured the rates of mass transfer from glass plates coated with benzoic acid to non-Newtonian solutions of carboxymethyl cellulose ($0.88 \le n \le 1$; $5 \le \text{Re}_L \le \sim 200$), and they found a satisfactory ($\pm 25\%$) agreement between their data and the predictions of equation (7.70). In this narrow range of n values, there is a little difference between the predictions for $n = 1$ and $n = 0.88$, however. Thus, the predictive equations developed for Newtonian fluids can be applied without incurring appreciable errors.

For both heat and mass transfer in laminar boundary layers, it has been assumed that the momentum boundary layer is everywhere thicker than the thermal and diffusion boundary layers. For Newtonian fluids ($n = 1$), it can readily be seen that ε varies as $\text{Pr}^{-1/3}$ and $\varepsilon_m \propto \text{Sc}^{-1/3}$. Most Newtonian liquids (other than molten metals) have the values of Prandtl number >1 and therefore the assumption of $\varepsilon < 1$ is justified. Likewise, one can justify this assumption for mass transfer provided $\text{Sc} > 0.6$. Most non-Newtonian polymer solutions used in heat and mass transfer studies to date seem to have large values of Prandtl and Schmidt numbers [Ghosh *et al.*, 1994], and therefore the assumptions of $\varepsilon \ll 1$ and $\varepsilon_m \ll 1$ are valid.

7.8 Boundary layers for visco-elastic fluids

For flow over a bluff body, the fluid elements are subjected to a rapid change in deformations near the frontal face; hence elastic effects are likely to be important in this region and the simple boundary layer approximations should not be applied to visco-elastic materials in this region. However, if elastic effects are negligibly small, the previous approach is reasonably satisfactory for visco-elastic fluids. For instance, the normal stresses developed in visco-elastic fluids will give rise to additional terms in the x-component of the momentum balance, equation (7.7) [Harris, 1977; Schowalter, 1978]. This leads to the boundary layer having a finite thickness at the leading edge ($x = 0$) which is a function of the Deborah number. The effect of fluid visco-elasticity on the boundary layer appears to persist at considerable distances from the leading edge, resulting in thickening at all points [Harris, 1977; Ruckenstein, 1994]. Although various treatments of visco-elastic boundary layers differ in detail, in all cases it is assumed that the fluid elasticity is small and the Deborah number is low in the flow regions of interest [Beard and Walters, 1964; Denn, 1967; Serth, 1973]. Unfortunately, there is a lack of experimental results for boundary layer flows of visco-elastic fluids. However, Hermes and Fredrickson [1967], in a study of the flow of a series of carboxymethyl cellulose solutions over a flat plate, do show that the visco-elasticity has a significant effect on the velocity profile in the boundary layer. More quantitative comparisons between experiments and predictions are not possible, as some of the rheological parameters inherent in the theoretical treatments are not capable of evaluation from simple rheological measurements. More recently, Ruckenstein [1994] has used dimensional considerations to infer that the effect of fluid elasticity is to reduce the dependence of heat and mass transfer coefficients on the free stream velocity. This conclusion is qualitatively consistent with the only limited heat transfer experimental results which are available [James and Acosta 1970; James and Gupta, 1971]. It is thus not yet possible at present to suggest simple expressions for the thickness of visco-elastic boundary layers analogous to those developed for inelastic power-law fluids. Good accounts of development in the field have been presented by Schowalter [1978] and others [Shenoy and Mashelkar, 1982].

7.9 Practical correlations for heat and mass transfer

In addition to the theoretical treatments just described, many workers have measured heat and mass transfer by forced convection from bodies such as plates, spheres and cylinders. The bulk of the literature in the field has been reviewed recently [Shenoy and Mashelkar, 1982; Irvine, Jr. and Karni, 1987; Nakayama, 1988; Chhabra, 1993a, b, 1999; Ghosh *et al.*, 1994], but only a selection of reliable correlations for spheres and cylinders is presented in this section which is based primarily on the reviews of Irvine and Karni [1987] and of Ghosh *et al.* [1994].

7.9.1 Spheres

For particle–liquid heat and mass transfer in non-Newtonian polymer solutions flowing over spheres fixed in tubes ($0.25 \leq d/D_t \leq 0.5$), Ghosh *et al.* [1992, 1994] invoked the usual heat and mass transfer analogy, that is, Sh $\equiv$ Nu and

$Sc \equiv Pr$ to correlate most of the literature data as follows:

$$Y = 1.428 Re_p^{1/3} \quad Re_p < 4 \tag{7.71a}$$

$$Y = Re_p^{1/2} \quad Re_p > 4 \tag{7.71b}$$

where $Y = (Nu - 2)(m_s/m_b)^{1/(3n+1)} Pr_p^{-1/3}$ for heat transfer

and $Y = (Sh - 2)Sc_p^{-1/3}$ for mass transfer.

The predicted results are shown in Figure 7.5 for both heat and mass transfer over the range of conditions : $0.32 \le n \le 0.93$; $Re_p < 200$; $29\,000 \le Sc_p \le 4.9 \times 10^5$ and $9500 \le Pr_p \le 1.9 \times 10^6$. The effective viscosity used in the Reynolds, Prandtl and Schmidt numbers (denoted by suffix to) is $m(V/d)^{n-1}$ where V is the mean velocity of flow in the tube. The average deviation between predictions and data is 17%.

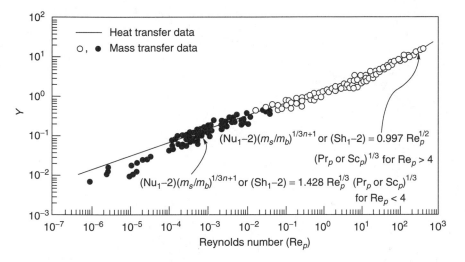

Figure 7.5 *Overall correlation for heat and mass transfer from a single sphere immersed in power law fluids*

7.9.2 Cylinders in cross-flow

The limited work on heat and mass transfer between power-law fluids and cylinders with their axis normal to the flow has been summarised recently by Ghosh *et al.* [1994] who proposed the following correlation for heat and mass transfer:

For heat transfer:

$$Nu = \frac{hd}{k} = 1.18\, Re_p^{1/3} Pr_p^{1/3} \qquad \text{for } Re_p < 10 \tag{7.72a}$$

$$= 0.76 \, \mathrm{Re}_p^{1/2} \mathrm{Pr}_p^{1/3} \qquad \text{for } \mathrm{Re}_p > 10 \tag{7.72b}$$

For mass transfer:

$$\mathrm{Sh} = \frac{h_D d}{D} = 1.18 \, \mathrm{Re}_p^{1/3} \mathrm{Sc}_p^{1/3} \qquad \text{for } \mathrm{Re}_p < 10 \tag{7.73a}$$

$$= 0.76 \, \mathrm{Re}_p^{1/2} \mathrm{Sc}_p^{1/3} \qquad \text{for } \mathrm{Re}_p > 10 \tag{7.73b}$$

where the effective viscosity is evaluated, as for spheres, with the diameter of the cylinder as the characteristic length. The overall correlations for heat and mass transfer are given in Figure 7.6.

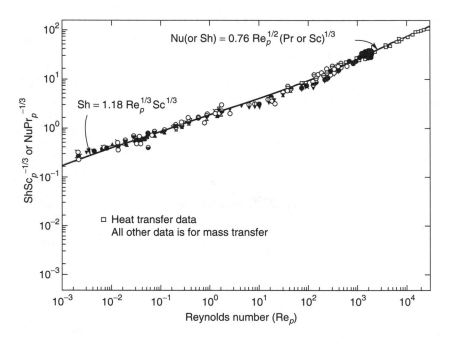

Figure 7.6   *Overall correlation for heat and mass transfer from cylinders in cross flow*

Example 7.3

A polymer solution at 25°C flows at 1.8 m/s over a heated hollow copper sphere of diameter of 30 mm, maintained at a constant temperature of 55°C (by steam condensing inside the sphere). Estimate the rate of heat loss from the sphere. The thermophysical properties of the polymer solution may be approximated by those of water, the power-law constants in the temperature interval $25 \le T \le 55°C$ are given below: $n = 0.26$ and $m = 26 - 0.0566 T$ where T is in K. What will be the rate of heat loss from a cylinder 30 mm in diameter and 60 mm long, oriented normal to flow?

Solution

At the mean film temperature of $(25 + 55)/2 = 40°C$, the values of ρ, C_p and k for water:

$$\rho = 991 \text{ kg/m}^3; \quad C_p = 4180 \text{ J/kg °C}; \quad k = 0.634 \text{ W/mK}$$

The consistency index, m:

$$m = 26 - 0.0566(273 + 40) = 8.28 \text{ Pa·s}^n$$

The effective viscosity,

$$\mu_{\text{eff}} = m \left(\frac{V}{d} \right)^{n-1}$$

$$= 8.28 \left(\frac{1.8}{30 \times 10^{-3}} \right)^{0.26-1} = 0.4 \text{ Pa·s}$$

$$\therefore \text{Re}_p = \frac{\rho V d}{\mu_{\text{eff}}} = \frac{991 \times 1.8 \times 30 \times 10^{-3}}{0.4} = 134 > 4$$

Therefore, equation (7.71b) applies in this case.

$$(\text{Nu} - 2) \left(\frac{m_s}{m_b} \right)^{1/(3n+1)} = \text{Re}_p^{1/2} \text{Pr}_p^{1/3}$$

At $T_s = 328 \text{ K} : m_s = 7.44 \text{ Pa·s}^n$

$$T_b = 313 \text{ K} : m_b = 8.28 \text{ Pa·s}^n$$

$$\text{Pr}_p = \frac{C_p \mu}{k} = \frac{4180 \times 0.4}{0.634} = 2637$$

$$\therefore \text{Nu} = \frac{hd}{k} = 2 + \text{Re}_p^{1/2} \text{Pr}_p^{1/3} \left(\frac{m_b}{m_s} \right)^{1/(3n+1)}$$

Substituting values,

$$\frac{hd}{k} = 2 + 134^{1/2} 2637^{1/3} \left(\frac{8.28}{7.44} \right)^{1/1.78}$$

$$= 172$$

$$\therefore h = \frac{172 \times 0.634}{30 \times 10^{-3}} = 3631 \text{ W/m}^2 \text{K}$$

$\therefore$ Rate of heat loss, $q = hA(T_s - T_f)$

$$= 3631 \times \pi \times (30 \times 10^{-3})^2 \times (55 - 25)$$

$$= 308 \text{ W}$$

For a cylinder, the Reynolds number of the flow is still the same and therefore equation (7.72b) applies.

$$\therefore \text{Nu} = \frac{hd}{k} = 0.76 \, \text{Re}_p^{1/2} \, \text{Pr}_p^{1/3}$$

$$= 0.76 \times (134)^{1/2} (2637)^{1/3} = 121.5$$

$$\therefore h = \frac{121.5 \times 0.634}{30 \times 10^{-3}} = 2570 \, \text{W/m}^2\text{K}$$

$$\therefore \text{Rate of heat loss, } q = hA(T_s - T_f)$$

$$= 2570 \times \pi \times 30 \times 10^{-3} \times 60 \times 10^{-3} \times (55 - 25)$$

$$= 436 \, W$$

Note that the drop in the value of the heat transfer coefficient in this case has been compensated by the increase in surface area resulting in higher rate of heat loss. Also, the heat loss from the flat ends has been neglected.□

7.10 Heat and mass transfer by free convection

In free convection, there is a new complexity in that fluid motion arises from buoyancy forces due to the density differences, and the momentum, heat and mass balance equations are therefore coupled. The published analytical results for heat transfer from plates, cylinders and spheres involve significant approximations. This work has been reviewed by Shenoy and Mashelkar [1982] and Irvine and Karni [1987]; the simple expressions (which are also considered to be reliable) for heat transfer coefficients are given in the following sections.

7.10.1 Vertical plates

For a plate maintained at a constant temperature in a power-law fluid, the mean value of the Nusselt number based on the height of the plate, L, is given as:

$$\text{Nu} = \frac{hL}{k} = C_0(n) \, \text{Gr}^{1/(2(n+1))} \text{Pr}^{n/(3n+1)} \tag{7.74a}$$

where the Grashof number, $\text{Gr} = \dfrac{\rho^2 L^{n+2}(g\beta\Delta T)^{2-n}}{m^2}$

and the Prandtl number,

$$\text{Pr} = \frac{\rho C_p}{k}\left(\frac{m}{\rho}\right)^{2/(n+1)} L^{(n-1)/(2(n+1))}(g\beta\Delta T)^{(3(n-1))/(2(n+1))} \tag{7.74b}$$

All physical properties are evaluated at the mean film temperature, except the coefficient of expansion, β, which is evaluated at the bulk fluid temperature. The constant $C_0(n)$ has values of 0.60, 0.68 and 0.72 for n values of 0.5,

1 and 1.5 respectively; a mean value of 0.66 may be used for most design calculations. The available scant data on heat transfer are in line with the predictions of equation (7.74).

For constant heat flux at the plate surface, the point values of the Nusselt number at a distance, x, from the base of the plate has been correlated empirically by:

$$\mathrm{Nu}_x = \frac{hx}{k} = C \left[\mathrm{Gr}^{(3n+2)/(n+4)} \mathrm{Pr}^n \right]^B \tag{7.75a}$$

$$\text{where } \mathrm{Gr} = \frac{\rho^2 x^4}{m^2} \left(\frac{g\beta q_w}{k} \right)^{2-n}$$

$$\mathrm{Pr} = \frac{\rho C_p}{k} \left(\frac{m}{\rho} \right)^{5/(n+4)} x^{(2(n-1))/(n+4)} \left(\frac{g\beta q_w}{k} \right)^{(3(n-1))/(n+4)} \tag{7.75b}$$

The recommended values [Irvine and Karni, 1987] of C and B are 0.6 and 0.21 respectively.

7.10.2 Isothermal spheres

Experimental results for heat transfer [Liew and Adelman, 1975; Amato and Tien, 1976], mass transfer [Lee and Donatelli, 1989] and the approximate analytical results [Acrivos, 1960; Stewart, 1971] are well represented by the following simple relationship:

$$\mathrm{Nu} = \frac{hd}{k} = 2 \left[\mathrm{Gr}^{1/(2(n+1))} \mathrm{Pr}^{n/(3n+1)} \right]^{0.682} \tag{7.76a}$$

$$\text{for } \mathrm{Gr}^{1/(2(n+1))} \cdot \mathrm{Pr}^{n/(3n+1)} < 10$$

$$\text{and } \quad \mathrm{Nu} = \frac{hd}{k} = \mathrm{Gr}^{1/(2(n+1))} \mathrm{Pr}^{n/(3n+1)} \tag{7.76b}$$

$$\text{for } 10 \le \mathrm{Gr}^{1/(2(n+1))} \mathrm{Pr}^{n/(3n+1)} \le 40$$

where the Grashof and Prandtl numbers are given by equation (7.74b) with the height of the plate, L, replaced by the sphere radius, R. For mass transfer, the Nusselt and Prandtl numbers are replaced by the corresponding Sherwood and Schmidt numbers respectively.

7.10.3 Horizontal cylinders

Gentry and Wollersheim [1974] and Kim and Wollersheim [1976] have measured the rates of heat transfer between isothermal horizontal cylinders and power-law polymer solutions. For pseudoplastic fluids, they correlated their results by the following expression:

$$\text{Nu} = \frac{hd}{k} = 1.19(\text{Gr Pr})^{0.20} \tag{7.77}$$

where the Grashof and Prandtl numbers are given by equation (7.75b) with L replaced by the diameter of the cylinder.

Example 7.4

The heat flux at the surface of an electrically heated vertical plate (250 mm × 250 mm) immersed in a polymer solution at 20°C is constant at 250 W/m². Estimate the heat transfer coefficient and the average temperature of the plate. The physical properties of the polymer solution may be approximated by those of water, and the power-law constants may be taken as: $n = 0.5$ and $m = 20 - 0.05\,T$ (Pa.s^n) in the range 288–360 K.

Solution

Since the temperature of the plate is not known, the physical properties of the polymer solution cannot immediately be evaluated. Assuming the average temperature of the plate to be 45°C, the mean film temperature, $T_f = (45 + 20)/2 = 32.5°C$ and the physical properties (of water) are:

$\rho = 993\,\text{kg/m}^3$ $\qquad\qquad$ $k = 0.625\,\text{W/mK}$

$C_p = 4180\,\text{J/kg K}$ $\qquad\quad$ β (at 20°C) $= 320.6 \times 10^{-6}(1/K)$

$m = 4.73\,\text{Pa·s}^n$ $\qquad\qquad$ $n = 0.5$

$$\therefore\ \text{Gr} = \frac{\rho^2 x^4}{m^2}\left(\frac{g\beta q_w}{k}\right)^{2-n}$$

$$= \frac{993^2 \times x^4}{(4.73)^2} \times \left(\frac{9.81 \times 320.6 \times 10^{-6} \times 250}{0.625}\right)^{2-0.5}$$

$$= 62\,189 x^4$$

where x is in metres.

$$\text{Pr} = \frac{\rho C_p}{k}\left(\frac{m}{\rho}\right)^{5/(n+4)} x^{(2(n-1))/(n+4)}\left(\frac{g\beta q_w}{k}\right)^{(3(n-1))/(n+4)}$$

$$= \frac{993 \times 4180}{0.625}\left(\frac{4.73}{993}\right)^{5/(4.5)}$$

$$\times x^{-(1/4.5)}\left(\frac{9.81 \times 320.6 \times 10^{-6} \times 250}{0.625}\right)^{-1/3}$$

$$= 16\,178 x^{-1/4.5}$$

$$\mathrm{Nu}_x = \frac{h_x x}{k} = 0.6 \left[\mathrm{Gr}^{(3n+2)/(n+4)} \mathrm{Pr}^n \right]^{0.21}$$

$$= 0.6 \left[(62\,189x^4)^{3.5/4.5} (16\,178x^{-1/4.5})^{0.5} \right]^{0.21}$$

or $h_x = 6.29x^{-0.37}$

The average value of heat transfer coefficient, $\bar{h}$:

$$\bar{h} = \frac{1}{L} \int_0^L h_x \, \mathrm{d}x = \frac{1}{L} \int_0^L 6.29x^{-0.37} \, \mathrm{d}x$$

or $\bar{h} = \dfrac{6.29}{(250 \times 10^{-3})} \times \dfrac{(250 \times 10^{-3})^{0.63}}{0.63} = 16.7 \, \mathrm{W/m^2 K}$

From heat balance,

$$h_m (T_s - T_f) = 250$$

$$\therefore \; T_s = \frac{250}{16.7} + 20 = 35°C$$

Therefore, the assumed value of 45°C is too high and another iteration must be carried out. With the surface temperature of 35.5°C, the mean film temperature is $(35.5 + 20)/2 = 27.75°C$, at which $m = 4.96 \, \mathrm{Pa \cdot s}^n$ and $\beta = 300 \times 10^{-6} \, \mathrm{K}^{-1}$ and the other properties are substantially unaltered. The new values yield $\bar{h} = 16.4 \, \mathrm{W/m^2 K}$. The plate temperature is then 35.3°C which is sufficiently close to the assumed value of 35.5°C.□

7.11 Further reading

Schlichting, H., *Boundary layer Theory*. McGraw Hill, 6th edn. New York (1968).
Schowalter, W.R., *Mechanics of non-Newtonian Fluids*. Pergamon, Oxford (1978).

7.12 References

Acrivos, A., *AIChEJ.* **6** (1960) 584.
Acrivos, A., Shah M.J. and Petersen E.E., *AIChEJ.* **6** (1960) 312.
Amato, W.S. and Tien C., *Int. J. Heat Mass Transf.* **19** (1976) 1257.
Beard, D.W. and Walters K., *Proc. Camb. Phil. Soc.* **60** (1964) 667.
Blasius, H., *Forsch. Ver. Deut. Ing.* **131** (1913).
Chhabra, R.P., *Bubbles, Drops and Particles in non-Newtonian Fluids*. CRC Press, Boca Raton, FL (1993a).
Chhabra, R.P., *Adv. Heat Transf.* **23** (1993b) 187.
Coulson, J.M. and Richardson J.F., *Chemical Engineering*, Vol. 2, 4th edn. Butterworth-Heinemann, Oxford (1991).
Denn, M.M., *Chem. Eng. Sci.* **22** (1967) 395.
Gentry, C.C. and Wollersheim, D.E. *J. Heat Transf.* (ASME). **96** (1974) 3.
Ghosh, U.K., Kumar, S. and Upadhyay, S.N., *Chem. Eng. Commun.* **43** (1986) 335.

Ghosh, U.K., Kumar, S. and Upadhyay, S.N., *Polym.-Plast. Technol. Eng.*, **31** (1992) 271.
Ghosh, U.K., Upadhyay S.N. and Chhabra, R.P., *Adv. Heat Transf.* **25** (1994) 251.
Harris, J., *Rheology and Non-Newtonian Flow*. Longman, London (1977).
Hermes, R.A. and Fredrickson, A.G., *AIChEJ.* **13** (1967) 253.
Irvine, Jr. T.F. and Karni J., in *Handbook of Single Phase Convective Heat Transfer.* (edited by S. Kakac, R.K. Shah and W. Aung, Wiley, New York) (1987) Chapter 5.
James, D.F. and Acosta, A.J., *J. Fluid Mech.* **42** (1970) 269.
James, D.F. and Gupta, O.P., *Chem. Eng. Prog. Sym. Ser.*, **67** No. 111 (1971) 62.
Kim, C.B. and Wollersheim, D.E., *J. Heat Transf. (ASME).* **98** (1976) 144.
Lee, T.-L. and Donatelli, A.A., *Ind. Eng. Chem. Res.* **28** (1989) 105.
Liew, K.S. and Adelman, M., *Can. J. Chem. Eng.* **53** (1975) 494.
Mishra, I.M., Singh B. and Mishra, P., *Indian J. Tech.* **14** (1976) 322.
Nakayama, A., in *Encyclopedia of Fluid Mechanics.* (edited by N.P. Cheremisinoff, Gulf, Houston) **7** (1988) 305.
Ruckenstein, E., *Ind. Eng. Chem. Res.* **33** (1994) 2331.
Schlichting, H., *Boundary Layer Theory*, 6th edn. McGraw Hill, New York (1968).
Schowalter, W.R., *Mechanics of non-Newtonian Fluids*, Pergamon, Oxford (1978).
Serth, R.W., *AIChEJ.* **19** (1973) 1275.
Shenoy, A.V. and R.A. Mashelkar, *Adv. Heat Transf.* **15** (1982) 143.
Skelland, A.H.P., *Non-Newtonian Flow and Heat Transfer*. Wiley, New York (1967).
Stewart, W.E., *Int. J. Heat Mass Transf.* **14** (1971) 1013.

7.13 Nomenclature

Dimensions
in $\mathbf{M, N, L, T}, \theta$

a, b, c, d	unknown constants in equation (7.8)	$\mathbf{M^0 L^0 T^0}$
Bi	Bingham number ($-$)	$\mathbf{M^0 L^0 T^0}$
C_A	solute concentration (kmol/m^3)	$\mathbf{N^{-3}}$
C_{A_o}	solute concentration at the solid surface (kmol/m^3)	$\mathbf{N^{-3}}$
C_{A_s}	solute concentration in the free stream (kmol/m^3)	$\mathbf{N^{-3}}$
C_D	drag coefficient ($-$)	$\mathbf{M^0 L^0 T^0}$
C_p	heat capacity (J/kg K)	$\mathbf{L^2 T^{-2} \theta^{-1}}$
d	sphere or cylinder diameter (m)	$\mathbf{L}$
D	tube diameter (m)	$\mathbf{L}$
D_{AB}	molecular diffusion coefficient (m^2/s)	$\mathbf{L^2 T^{-1}}$
F_d	drag force (N)	$\mathbf{MLT^{-2}}$
$F(n)$	function of n, equation (7.17a) ($-$)	$\mathbf{M^0 L^0 T^0}$
g	acceleration due to gravity (m/s^2)	$\mathbf{LT^{-2}}$
Gr	Grashof number, equation (7.74b) ($-$)	$\mathbf{M^0 L^0 T^0}$
h	heat transfer coefficient (W/m^{2}K)	$\mathbf{MT^{-3} \theta^{-1}}$
h_m	average value of heat transfer coefficient (W/m^{2}K)	$\mathbf{MT^{-3} \theta^{-1}}$
h_D	mass transfer coefficient (m/s)	$\mathbf{LT^{-1}}$
H	height of the control volume in y-direction (m)	$\mathbf{L}$
k	thermal conductivity (W/mK)	$\mathbf{MLT^{-3} \theta^{-1}}$
L	length or height of plate (m)	$\mathbf{L}$
m	power-law consistency coefficient (Pa·s^n)	$\mathbf{ML^{-1} T^{n-2}}$
n	power-law flow behaviour index ($-$)	$\mathbf{M^0 L^0 T^0}$
p	pressure (Pa)	$\mathbf{ML^{-1} T^{-2}}$

		Dimensions in **M, N, L, T,** θ
Pr	Prandtl number (–)	$\mathbf{M^0 L^0 T^0}$
q	heat flux (W/m^2)	$\mathbf{M T^{-3}}$
q_w	heat flux at wall or surface (W/m^2)	$\mathbf{M T^{-3}}$
R	sphere radius (m)	$\mathbf{L}$
Re	Reynolds number (–)	$\mathbf{M^0 L^0 T^0}$
Sc	Schmidt number (–)	$\mathbf{M^0 L^0 T^0}$
Sh	Sherwood number (–)	$\mathbf{M^0 L^0 T^0}$
T	temperature (K)	θ
T_0	free stream fluid temperature (K)	θ
T_s	temperature of surface (K)	θ
V_0	free stream fluid velocity (m/s)	$\mathbf{L T^{-1}}$
V_x	velocity in x-direction (m/s)	$\mathbf{L T^{-1}}$
W	width of plate (m)	$\mathbf{L}$
x	coordinate in the direction of flow (m)	$\mathbf{L}$
y	coordinate in the direction normal to flow (m)	$\mathbf{L}$

Greek Letters

α	thermal diffusivity (m^2/s)	$\mathbf{L^2 T^{-1}}$
β	coefficient of expansion (1/K)	θ^{-1}
δ	momentum boundary layer thickness (m)	$\mathbf{L}$
δ_m	diffusion boundary layer thickness (m)	$\mathbf{L}$
δ_T	thermal boundary layer thickness (m)	$\mathbf{L}$
ε	ratio of δ_T to δ (–)	$\mathbf{M^0 L^0 T^0}$
ε_m	ratio of δ_m to δ (–)	$\mathbf{M^0 L^0 T^0}$
θ	temperature difference $= T - T_s$ (K)	θ
θ_0	temperature difference $= T_0 - T_s$ (K)	θ
μ_B	Bingham plastic viscosity (Pa·s)	$\mathbf{M L^{-1} T^{-1}}$
ρ	fluid density (kg/m^3)	$\mathbf{M L^{-3}}$
τ_0^B	Bingham yield stress (Pa)	$\mathbf{M L^{-1} T^{-2}}$
τ_w	wall shear stress (Pa)	$\mathbf{M L^{-1} T^{-2}}$

Subscripts

k	value using effective viscosity $m(V/d)^{n-1}$
m	mean value
x	value at an arbitrary point

Chapter 8
Liquid mixing

8.1 Introduction

Mixing is one of the most common operations in the chemical, biochemical, polymer processing, and allied industries. Almost all manufacturing processes entail some sort of mixing, and the operation may constitute a considerable proportion of the total processing time. The term 'mixing' is applied to the processes used to reduce the degree of non-uniformity or gradient of a property such as concentration, viscosity, temperature, colour and so on. Mixing can be achieved by moving material from one region to another. It may be of interest simply as means of reaching a desired degree of homogeneity but it may also be used to promote heat and mass transfer, often where a system is undergoing a chemical reaction.

At the outset, it is useful to consider some common examples of problems encountered in industrial mixing operations, since this will not only reveal the ubiquitous nature of the process, but will also provide an appreciation of some of the associated difficulties. One can classify mixing problems in many ways, such as the flowability of the final product in the mixing of powders, but it is probably most satisfactory to base this classification on the phases present; liquid–liquid, liquid–solid, gas–liquid, etc. This permits a unified approach to the mixing problems in a range of industries.

8.1.1 Single-phase liquid mixing

In many instances, two or more miscible liquids must be mixed to give a product of a desired specification, as for example, in the blending of petroleum fractions of different viscosities. This is the simplest type of mixing as it does not involve either heat or mass transfer, or indeed a chemical reaction. Even such simple operations can, however, pose problems when the two liquids have vastly different viscosities, or if density differences are sufficient to lead to stratification. Another example is the use of mechanical agitation to enhance the rates of heat and mass transfer between a liquid and the wall of a vessel, or a coil. Additional complications arise in the case of highly viscous Newtonian and non-Newtonian materials.

8.1.2 Mixing of immiscible liquids

When two immiscible liquids are stirred together, one liquid becomes dispersed as droplets in the second liquid which forms a continuous phase. Liquid–liquid extraction, a process using successive mixing and settling stages, is one important example of this type of mixing. The two liquids are brought into contact with a solvent that selectively dissolves one of the components present in the mixture. Agitation causes one phase to disperse in the other and, if the droplet size is small, a high interfacial area is created for interphase mass transfer. When the agitation is stopped, phase separation may occur, but care must be taken to ensure that the droplets are not so small that a diffuse layer is formed instead of a well-defined interface; this can remain in a semi-stable state over a long period of time and prevent the completion of effective separation. The production of stable emulsions such as those encountered in food, brewing, and pharmaceutical applications provides another important example of dispersion of two immiscible liquids. In these systems, the droplets are very small and are often stabilised by surface active agents, so that the resulting emulsion is usually stable for considerable lengths of time.

8.1.3 Gas–liquid dispersion and mixing

Numerous processing operations involving chemical reactions, such as aerobic fermentation, wastewater treatment, oxidation or chlorination of hydrocarbons, and so on, require good contact between a gas and a liquid. The purpose of mixing here is to produce a large interfacial area by dispersing bubbles of the gas into the liquid. Generally, gas–liquid mixtures or dispersions are unstable and separate rapidly if agitation is stopped, provided that a foam is not formed. In some cases, a stable foam is needed; this can be formed by injecting gas into a liquid using intense agitation, and stability can be increased by the addition of a surface-active agent.

8.1.4 Liquid–solid mixing

Mechanical agitation may be used to suspend particles in a liquid in order to promote mass transfer or a chemical reaction. The liquids involved in such applications are usually of low viscosity, and the particles will settle out when agitation ceases. There is also an occasional requirement to achieve a relatively homogeneous suspension in a mixing vessel, particularly when this is being used to prepare materials for subsequent processes.

At the other extreme, in the formation of composite materials, especially filled polymers, fine particles must be dispersed into a highly viscous Newtonian or non-Newtonian liquid. The incorporation of carbon black powder into rubber is one such operation. Because of the large surface areas involved, surface phenomena play an important role in these applications.

8.1.5 Gas–liquid–solid mixing

In some applications such as catalytic hydrogenation of vegetable oils, slurry reactors, three-phase fluidised beds, froth flotation, fermentation, and so on, the success and efficiency of the process is directly influenced by the extent of mixing between the three phases. Despite its great industrial importance, this topic has received only scant attention and the mechanisms and consequences of interactions between the phases are almost unexplored.

8.1.6 Solid–solid mixing

Mixing together of particulate solids, sometimes referred to as blending, is a very complex process in that it is very dependent, not only on the character of the particles – density, size, size distribution, shape and surface properties – but also on the differences between these properties of the components. Mixing of sand, cement and aggregate to form concrete, and of the ingredients in gun powder preparation, are longstanding examples of the mixing of solids.

Other industrial sectors employing solids mixing include food, drugs and the glass industries, for example. All these applications involve only physical contacting, although in recent years, there has been a recognition of the industrial importance of solid–solid reactions, and solid–solid heat exchangers. Unlike liquid mixing, fundamental research on solids mixing has been limited until recently. The phenomena involved are very different from those when a liquid phase is present, so solid–solid mixing will not be discussed further here. However, most of the literature on solid–solid mixing has recently been reviewed [Lindlay, 1991; Harnby *et al.*, 1992; van den Bergh, 1994].

8.1.7 Miscellaneous mixing applications

Mixing equipment may be designed not only to achieve a predetermined level of homogenity but also to improve heat transfer. For example, if the rotational speed of an impeller in a mixing vessel is selected so as to achieve a required rate of heat transfer, the agitation may then be more than sufficient for the mixing duty. Excessive or overmixing should be avoided. For example, in biological applications, excessively high impeller speeds or power input are believed by many to give rise to shear rates which may damage micro-organisms. In a similar way, where the desirable rheological properties of some polymer solutions may be attributable to structured long-chain molecules, excessive impeller speeds or agitation over prolonged periods, may damage the structure particularly of molecular aggregates, thereby altering their properties. It is therefore important to appreciate that 'over-mixing' may often be undesirable because it may result in both excessive energy consumption and impaired product quality. Equally, under-mixing also is obviously undesireable.

From the examples given here, it is abundantly clear that mixing cuts across the boundaries between industries, and there may be a need to mix virtually anything with anything – be it a gas or a solid or a liquid; it is clearly not possible to consider the whole range of mixing problems here. Instead attention will be given primarily to batch liquid mixing of viscous Newtonian and non-Newtonian materials, followed by short discussions of gas–liquid systems, and heat transfer in mechanically agitated systems. To a large extent, the rheology of the liquids concerned determines the equipment and procedures to be used in a mixing operation. However, when non-Newtonian fluids are involved, the process itself can have profound effects on the rheological properties of the product.

8.2 Liquid mixing

A considerable body of information is now available on batch liquid mixing and this forms the basis for the design and selection of mixing equipment. It also affords some physical insight into the nature of the mixing process itself. In mixing, there are two types of problems to be considered – how to design and select mixing equipment for a given duty, and how to assess whether an available mixer is suitable for a particular application. In both cases, the following aspects of the mixing process must be understood:

  (i) Mechanisms of mixing
 (ii) Scale-up or similarity criteria
(iii) Power consumption
(iv) Rate of mixing and mixing time

Each of these factors is now considered in detail.

8.2.1 Mixing mechanisms

If mixing is to be carried out in order to produce a uniform product, it is necessary to understand how mixtures of liquids move and approach uniformity of composition. For liquid mixing devices, it is necessary that two requirements are fulfilled: Firstly, there must be bulk or convective flow so that there are no dead or stagnant zones. Secondly, there must be a zone of intensive or high-shear mixing in which the inhomogeneities are broken down. Both these processes are energy-consuming and ultimately the mechanical energy is dissipated as heat; the proportion of energy attributable to each varies from one application to another. Depending upon the fluid properties, primarily viscosity, the flow in mixing vessels may be laminar or turbulent, with a substantial transition zone in between and frequently both types of flow occur simultaneously in different parts of the vessel. Laminar and turbulent flows arise from different mechanisms, and it is convenient to consider them separately.

(i) Laminar mixing

Large-scale laminar flow is usually associated with high viscosity liquids ($>\sim 10\,Pa\cdot s$) which may exhibit either Newtonian or non-Newtonian characteristics. Inertial forces therefore tend to die out quickly, and the impeller must sweep through a significant proportion of the cross-section of the vessel to impart sufficient bulk motion. Because the velocity gradients close to a moving impeller are high, the fluid elements in that region deform and stretch. They repeatedly elongate and become thinner each time they pass through the high shear rate zone. Figure 8.1 shows such a shearing sequence.

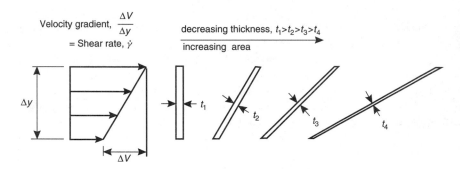

Figure 8.1 *Schematic representation of the thinning of fluid elements due to laminar shear flow*

   In addition, extensional or elongational flow usually occurs simultaneously. As shown in Figure 8.2, this can be result of convergence of the streamlines and consequential increase of velocity in the direction of flow. Since for incompressible fluids the volume remains constant, there must be a thinning or flattening of the fluid elements, as shown in Figure 8.2. Both of these mechanisms (shear and elongation) give rise to stresses in the liquid which then effect a reduction in droplet size and an increase in interfacial area, by which means the desired degree of homogeneity is obtained.

   In addition, molecular diffusion always acts in such a way as to reduce inhomogeneities, but its effect is not significant until the fluid elements have been sufficiently reduced in size for their specific areas to become large. It must be recognised, however, that the ultimate homogenisation of miscible liquids

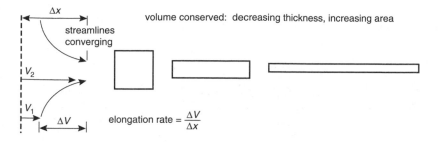

Figure 8.2 *Schematic representation of the thinning of fluid elements due to extensional flow*

can be brought about only by molecular diffusion. In the case of liquids of high viscosity, this is a slow process.

In laminar flow, a similar mixing process occurs when a liquid is sheared between two rotating cylinders. During each revolution, the thickness of an initially radial fluid element is reduced, and molecular diffusion takes over when the fluid elements are sufficiently thin. This type of mixing is shown schematically in Figure 8.3 in which the tracer is pictured as being introduced perpendicular to the direction of motion. It will be realized that, if an annular fluid element had been chosen to begin with, then no obvious mixing would have occurred. This emphasises the importance of the orientation of the fluid elements relative to the direction of shear produced by the mixer.

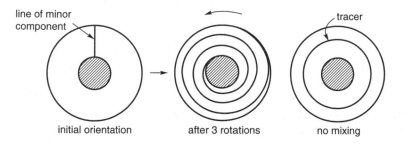

Figure 8.3 *Laminar shear mixing in a coaxial cylinder arrangement*

Finally, mixing can be induced by physically dividing the fluid into successively smaller units and then re-distributing them. In-line mixers for laminar flows rely primarily on this mechanism, as shown schematically in Figure 8.4.

Thus, mixing in liquids is achieved by several mechanisms which gradually reduce the size or scale of the fluid elements and then re-distribute them. If, for example, there are initial differences in concentration of a soluble material, uniformity is gradually achieved, and molecular diffusion becomes

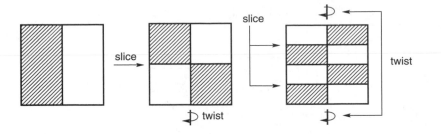

Figure 8.4   *Schematic representation of mixing by cutting and folding of fluid elements*

progressively more important as the element size is reduced. Ottino [1989] has illustrated the various stages in mixing by means of colour photographs.

(ii) Turbulent mixing

In low viscosity liquids ($<\sim$10 m Pa·s), the flows generated in mixing vessels with rotating impellers are usually turbulent. The inertia imparted to the liquid by the impeller is sufficient to cause the liquid to circulate throughout the vessel and return to the impeller. Turbulence may occur throughout the vessel but will be greatest near the impeller. Mixing by eddy diffusion is much faster than mixing by molecular diffusion and, consequently, turbulent mixing occurs much more rapidly than laminar mixing. Ultimately, homogenisation at the molecular level depends on molecular diffusion, which in general takes place more rapidly in low viscosity liquids. Mixing is fastest near the impeller because of the high shear rates and associated Reynolds stresses in vortices formed at the tips of the impeller blades; furthermore, a high proportion of the energy is dissipated here.

Turbulent flow is inherently complex, and calculation of the flow fields prevailing in a mixing vessel is not amenable to rigorous theoretical treatment. If the Reynolds number of the main flow is sufficiently high, some insight into the mixing process can be gained by using the theory of local isotropic turbulence. Turbulent flow may be considered to contain a spectrum of velocity fluctuations in which eddies of different sizes are superimposed on an overall time-averaged mean flow. In a mixing vessel, it is reasonable to suppose that the large primary eddies, of a size corresponding approximately to the impeller diameter, would give rise to large velocity fluctuations but would have a low frequency. Such eddies are anisotropic and account for much of the kinetic energy present in the system. Interaction between these primary eddies and slow moving streams produces smaller eddies of higher frequency which undergo further disintegration until, finally, viscous forces cause dissipation of their energy as heat.

The description given here is a gross over-simplification, but it does give a qualitative representation of the salient features of turbulent mixing. This whole process is similar to that of the turbulent flow of a fluid close to a boundary surface. Although some quantitative results for the scale size of eddies have been obtained and some workers [van der Molen and van Maanen, 1978; Tatterson, 1991] have reported experimental measurements on the structure of turbulence in mixing vessels, these studies have little relevance to the mixing of non-Newtonian substances which are usually processed under laminar conditions.

8.2.2 Scale-up of stirred vessels

One of the problems confronting the designers of mixing equipment is that of deducing the most satisfactory arrangement for a large unit from experiments carried out with small units. An even more common problem is the specification of small scale experiments to optimise the use of existing large scale equipment for new applications. In order to achieve similar flow patterns in two units, geometrical, kinematic and dynamic similarity and identical boundary conditions must be maintained. This problem of scale-up has been discussed in a number of books [Oldshue, 1983; Tatterson, 1991, 1994; Harnby *et al.*, 1992; Coulson and Richardson, 1999] and therefore only the salient features are re-capitulated here. It has been found convenient to relate the power used by the agitator to the geometrical and mechanical arrangement of the mixer, and thus to obtain a direct indication of the change in power arising from alteration of any of the factors relating to the mixer. A typical mixer arrangement is shown in Figure 8.5. The criteria for similarity between two systems is expressed in terms of dimensionless ratios of geometric dimensions and of forces occurring in the fluid in the mixing vessel. In the absence of heat and mass transfer and for geometrically similar systems, the resulting dimensionless groups are the Reynolds, Froude and Weber numbers for Newtonian fluids, defined respectively as:

$$\text{Reynolds number: Re} = \frac{\rho D^2 N}{\mu} \tag{8.1}$$

$$\text{Froude number: Fr} = \frac{D N^2}{g} \tag{8.2}$$

$$\text{Weber number: We} = \frac{D^3 N^2 \rho}{\sigma} \tag{8.3}$$

where D is the impeller diameter.

For heat and mass transfer in stirred vessels, additional dimensionless groups which are important include the Nusselt, Sherwood, Prandtl, Schmidt and Grashof numbers. Likewise, in the case of non-Newtonian fluids, an

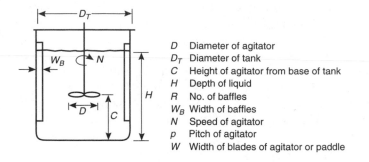

D   Diameter of agitator
D_T Diameter of tank
C   Height of agitator from base of tank
H   Depth of liquid
R   No. of baffles
W_B Width of baffles
N   Speed of agitator
p   Pitch of agitator
W   Width of blades of agitator or paddle

Figure 8.5   *Typical configuration and dimensions of an agitated vessel*

appropriate value of the apparent viscosity must be identified for use in equation (8.1). Furthermore, it may also be necessary to introduce dimensionless parameters indicative of non-Newtonian effects including a Bingham number (Bi) for viscoplastic materials and a Weissenberg number (We) for visco-elastic fluids. It is thus imperative that in order to ensure complete similarity between two systems, all the pertinent dimensionless numbers must be equal. In practice, however, this is not usually possible, owing to conflicting requirements. In general, it is necessary to specify the one or two key features that must be matched. Obviously, in the case of substances not exhibiting a yield stress, the Bingham number is redundant, as is the Weissenberg number for inelastic fluids. Similarly, the Froude number is usually important only when significant vortex formation occurs; this effect is almost non-existent for viscous materials.

   Aside from these theoretical considerations, further difficulties can arise depending upon the choice of scale-up criteria, and these are strongly dependent on the type of mixing (e.g. gas–liquid or liquid–liquid) and on the ultimate goal of the mixing process. Thus, for geometrically similar systems, the size of the equipment is determined by the scale-up factor. For power consumption, one commonly used criterion is that the power input per unit volume of liquid should be the same in two systems. In most practical situations, the equality of the Reynolds numbers ensures complete similarity of flow between two geometrically identical systems. Similarly, processes involving heat transfer in stirred vessels are often scaled up, either on the basis of equal heat transfer per unit volume of liquid, or by maintaining the same value of the heat transfer coefficient. Other commonly used criteria are equal mixing times, the same specific interfacial area in gas–liquid systems or the same value of mass transfer coefficient.

8.2.3 Power consumption in stirred vessels

From a practical point of view, power consumption is perhaps the most important parameter in the design of stirred vessels. Because of the very different flow patterns and mixing mechanisms involved, it is convenient to consider power consumption in low and high viscosity systems separately.

(i) Low Viscosity Systems

Typical equipment for low viscosity liquids consists of a baffled vertical cylindrical tank, with a height to diameter ratio of 1.5 to 2, fitted with an agitator. For low viscosity liquids, high speed impellers of diameter between one third and one half that of the vessel are suitable, running with tip speeds of around 1–3 m/s. Although work on single phase mixing of low viscosity liquids is of limited relevance to industrial applications involving non-Newtonian materials, it does, however, serve as a useful starting point for the subsequent treatment of high viscosity liquids.

For a stirred vessel of diameter D_T in which a Newtonian liquid of viscosity μ, and density ρ is agitated by an impeller of diameter D rotating at a speed of N revolutions per unit time; (the other dimensions are as shown in Figure 8.5) the power input P to the liquid depends on the independent variables (μ, ρ, D, D_T, N, g, other geometric dimensions) and may be expressed as:

$$P = \mathrm{f}(\mu, \rho, D, D_T, N, g, \text{geometric dimensions}) \tag{8.4}$$

In equation (8.4), P is the impeller power, that is, the energy per unit time dissipated within the liquid. Clearly, the electrical power required to drive the motor will be greater than P on account of transmission losses in the gear box, motor, bearings and so on.

It is readily acknowledged that the functional relationship in equation (8.4) cannot be established from first principles. However, by using dimensional analysis, the number of variables can be reduced to give:

$$\frac{P}{\rho N^3 D^5} = \mathrm{f}\left(\frac{\rho N D^2}{\mu}, \frac{N^2 D}{g}, \text{geometric ratios}\right) \tag{8.5}$$

where the dimensionless group on the left hand side is called the Power number, Po; ($\rho N D^2/\mu$) is the Reynolds number, Re and ($N^2 D/g$) is the Froude number, Fr. The geometric ratios relate to the specific impeller/vessel configuration. For geometrically similar systems, these ratios must be constant and the functional relationship between the Power number and the other groups reduces to:

$$\mathrm{Po} = \mathrm{f}(\mathrm{Re}, \mathrm{Fr}) \tag{8.6}$$

In equation (8.6), the Froude number is generally important only when vortex formation occurs, and in single phase mixing can be neglected if the value of the Reynolds number is less than about 300. In view of the detrimental effect of vortex formation on the quality of mixing, tanks are usually fitted

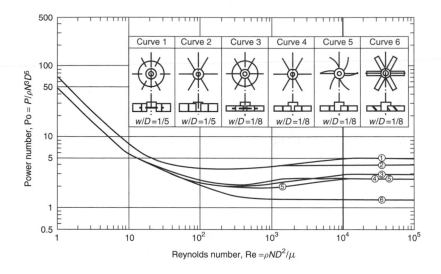

Figure 8.6 *Power number–Reynolds number correlation in Newtonian fluids for various turbine impeller designs. (Redrawn from Bates et al., 1963)*

with baffles and hence in most situations involving low viscosity Newtonian fluids, the Power number is a function of the Reynolds number and geometry only. Likewise, as the liquid viscosity increases, the tendency for vortex formation decreases and thus so does the need of installing baffles in mixing tanks ($\mu > \sim 5 \, \text{Pa·s}$). Figure 8.6 shows the functional relationship given by equation (8.6) for a range of impellers used to mix/agitate Newtonian liquids of relatively low viscosity [Bates *et al.*, 1963]. For a fixed geometrical arrangement and a single phase liquid, the data can be represented by a unique power curve. Three distinct zones can be discerned in the power curve: at small values of the Reynolds number ($<\sim 10$), laminar flow occurs and the slope of the power curve on log–log coordinates is -1. This region, which is characterised by slow mixing, is where the majority of highly viscous (Newtonian and non-Newtonian) liquids are processed. The limited experimental data suggest that the smaller the value of power-law index (<1), the larger is the value of the Reynolds number up to which the laminar flow conditions persist.

At very high values of the Reynolds number ($>\sim 10^4$), the flow is fully turbulent and inertia dominated, resulting in rapid mixing. In this region, the Power number is virtually constant and independent of Reynolds number, as shown in Figure 8.6 for Newtonian fluids and demonstrated recently for shear-thinning polymer solutions [Nouri and Hockey, 1998]; however, it depends upon the impeller – vessel configuration. Often gas–liquid, liquid–solid and liquid–liquid contacting operations are carried out in this region. Though the mixing itself is quite rapid, the overall process may be mass transfer controlled.

In between the laminar and turbulent zones, there exists a transition zone in which the viscous and inertial forces are of comparable magnitudes. No simple mathematical relationship exists between Po and Re in this flow region and, at a given value of Re, the value of Po must be read off the appropriate power curve. Though it is generally accepted that laminar flow occurs for Re < ~10, the transition from transitional to the fully turbulent flow is strongly dependent on the impeller – vessel geometry. Grenville *et al.* [1995] correlated the critical Reynolds number for Newtonian fluids at the boundary between the transitional and turbulent regime as follows:

$$\text{Re}_{cr} = 6370 \, \text{Po}_t^{-1/3} \qquad (8.7)$$

where Po_t is the value of the Power number in fully turbulent conditions.

Power curves for many different impeller geometries, baffle arrangements, and so on are available in the literature [Skelland, 1983; Harnby *et al.*, 1992; Tatterson, 1992, 1994; Ibrahim and Nienow, 1995], but it must always be remembered that though the power curve approach is applicable to any single phase Newtonian liquid, at any impeller speed, the curve will be valid for only one system geometry. Adequate information on low viscosity systems is now available for the estimation of power requirements for a given duty under most conditions of practical interest.

Example 8.1

On the assumption that the power required for mixing in a stirred tank is a function of the variables given in equation (8.4), obtain the dimensionless groups which are important in calculating power requirements for geometrically similar arrangements.

Solution

The variables in this problem, together with their dimensions, are as follows:

P	: $\mathbf{ML^2T^{-3}}$	μ : $\mathbf{ML^{-1}T^{-1}}$
ρ	: $\mathbf{ML^{-3}}$	N : $\mathbf{T^{-1}}$
g	: $\mathbf{LT^{-2}}$	D : $\mathbf{L}$
D_T	: $\mathbf{L}$	

There are seven variables and with three fundamental units ($\mathbf{M}$, $\mathbf{L}$, $\mathbf{T}$), there will be $7 - 3 = 4$ dimensionless groups.

Choosing as the recurring set ρ, N and D which themselves cannot be grouped together to form a dimensionless group, $\mathbf{M}$, $\mathbf{L}$ and $\mathbf{T}$ can now be expressed in terms of combinations of ρ, N and D. Thus:

$$\mathbf{L} \equiv D \quad \mathbf{T} \equiv N^{-1} \quad \mathbf{M} \equiv \rho D^3$$

Dimensionless group $1 = P\mathbf{M}^{-1}\mathbf{L}^{-2}\mathbf{T}^3 = P(\rho D^3)^{-1}(D)^{-2}(N^{-1})^3$

$$= P/\rho D^5 N^3$$

Dimensionless group $2 = \mu M^{-1}LT = \mu(\rho D^3)^{-1}(D)(N^{-1})$

$$= \mu/\rho D^2 N$$

Dimensionless group $3 = g L^{-1}T^2 = g(D)^{-1}(N^{-1})^2$

$$= g/N^2 D$$

Dimensionless group $4 = D_T L^{-1} = D_T/D$

Hence,

$$\frac{P}{\rho N^3 D^5} = f\left(\frac{\rho D^2 N}{\mu}, \frac{N^2 D}{g}, \frac{D_T}{D}, \ldots\right)$$

which corresponds with equation (8.5).□

As discussed above, the Froude number exerts little influence on power requirement under most conditions of practical interest, and hence for geometrically similar systems:

$$Po = f(Re)$$

At low Reynolds numbers, $Po \propto (1/Re)$ and therefore power varies as $D^3 N^2/\mu$ and it is independent of the fluid density. On the other hand, power varies as $\rho D^5 N^3$ under fully turbulent conditions when the fluid viscosity is of little importance.

(ii) High viscosity Newtonian and inelastic non-Newtonian systems

As noted previously, mixing in highly viscous liquids is slow both at the molecular scale, on account of the low values of molecular diffusivity, as well as at the macroscopic scale, due to low levels of bulk flow. Whereas in low viscosity liquids momentum can be transferred from a rotating impeller through a relatively large body of fluid, in highly viscous liquids only the fluid in the immediate vicinity of the impeller is influenced by the agitator and the flow is normally laminar.

For the mixing of highly viscous and non-Newtonian fluids, it is usually necessary to use specially designed impellers with close clearances at the vessel walls (as discussed in a later section). High-speed stirring with small impellers merely wastefully dissipates energy at the central portion of the vessel, particularly when the liquid is highly shear-thinning. The power-curve approach is usually applicable. Although highly viscous Newtonian fluids include sugar syrups, glycerol and many lubricating oils, most of the highly viscous fluids of interest in the chemical and processing industries exhibit non-Newtonian flow characteristics.

A simple relationship has been shown to exist, however, between the power consumption for time-independent non-Newtonian liquids and for Newtonian liquids in the laminar region. This link, which was first established by Metzner and Otto [1957] for pseudoplastic fluids, depends on the fact that there appears to be a characteristic average shear rate $\dot{\gamma}_{avg}$ for a mixer which characterises power consumption, and which is directly proportional to the rotational speed of impeller:

$$\dot{\gamma}_{avg} = k_s N \tag{8.8}$$

where k_s is a function of the type of impeller and the vessel configuration. If the apparent viscosity corresponding to the average shear rate defined above is used in the equation for a Newtonian liquid, the power consumption for laminar conditions is satisfactorily predicted for most inelastic non-Newtonian fluids.

The validity of the linear relationship given in equation (8.8) was subsequently confirmed by Metzner and Taylor [1960]. The experimental evaluation of k_s for a given geometry proceeds as follows:

  (i) The Power number (Po) is determined for a particular value of N.
 (ii) The corresponding value of Re is obtained from the appropriate power curve for a Newtonian liquid.
(iii) The 'equivalent' apparent viscosity is computed from the value of Re.
 (iv) The value of the corresponding shear rate is obtained, either directly from a flow curve obtained by independent viscometric experiment, or by use of an appropriate fluid model such as the power-law model.
  (v) The value of k_s is calculated for a particular impeller configuration using equation (8.8).

This procedure can be repeated for different values of N and an average value of k_s estimated. A compilation of the experimental values of k_s for a variety of impellers, turbines, propeller, paddle, anchor, and so on, has been given by Skelland [1983], and an examination of Table 8.1 suggests that for pseudoplastic fluids, k_s lies approximately in the range 10–13 for most configurations of practical interest, while slightiy larger values 25–30 have been reported for anchors and helical ribbons [Bakker and Gates, 1995]. Skelland [1983] has also given a correlation for most of the data on the agitation of inelastic non-Newtonian fluids and this is shown in Figure 8.7.

The prediction of power consumption for agitation of a given time-independent non-Newtonian fluid in a particular mixer, at a desired impeller speed, may be evaluated by the following procedure:

  (i) The average shear rate is estimated from equation (8.8).
 (ii) The corresponding apparent viscosity is evaluated, either from a flow curve, or by means of the appropriate viscosity model.
(iii) The value of the Reynolds number is calculated as $(\rho ND^2/\mu)$ and then the value of the Power number, and hence of P, is obtained from the appropriate curve in Figure 8.7.

Table 8.1 Values of k_s for Various Types of Impellers and key to Figure 8.7

Curve	Impeller	Baffles	D (m)	D_T/D	N (Hz)	k_s ($n < 1$)
A-A	Single turbine with 6 flat blades	4, $W_B/D_T = 0.1$	0.051–0.20	1.3–5.5	0.05–1.5	11.5 ± 1.5
A-A$_1$	Single turbine with 6 flat blades	None.	0.051–0.20	1.3–5.5	0.18–0.54	11.5 ± 1.4
B-B	Two turbines, each with 6 flat blades and $D_T/2$ apart	4, $W_B/D_T = 0.1$	–	3.5	0.14–0.72	11.5 ± 1.4
B-B$_1$	Two turbines, each with 6 flat blades and $D_T/2$ apart	4, $W_B/D_T = 0.1$ or none	–	1.02–1.18	0.14–0.72	11.5 ± 1.4
C-C	Fan turbine with 6 blades at 45°	4, $W_B/D_T = 0.1$ or none	0.10–0.20	1.33–3.0	0.21–0.26	13 ± 2
C-C$_1$	Fan turbine with 6 blades at 45°	4, $W_B/D_T = 0.1$ or none	0.10–0.30	1.33–3.0	1.0–1.42	13 ± 2
D-D	Square-pitch marine propellers with 3 blades (downthrusting)	None, (i) shaft vertical at vessel axis, (ii) shaft 10° from vertical, displaced $r/3$ from centre	0.13	2.2–4.8	0.16–0.40	10 ± 0.9
D-D$_1$	Same as for D-D but upthrusting	None, (i) shaft vertical at vessel axis, (ii) shaft 10° from vertical, displaced $r/3$ from centre	0.13	2.2–4.8	0.16–0.40	10 ± 0.9
D-D$_2$	Same as for D-D	None, position (ii)	0.30	1.9–2.0	0.16–0.40	10 ± 0.9
D-D$_3$	Same as for D-D	None, position (i)	0.30	1.9–2.0	0.16–0.40	10 ± 0.9
E-E	Square-pitch marine propeller with 3 blades	4, $W_B/D_T = 0.1$	0.15	1.67	0.16–0.60	10

F-F	Double-pitch marine propeller with 3 blades (downthrusting)	None, position (ii)	–	1.4–3.0	0.16–0.40	10 ± 0.9
F-F$_1$	Double-pitch marine propeller with 3 blades (downthrusting)	None, position (i)	–	1.4–3.0	0.16–0.40	10 ± 0.9
G-G	Square-pitch marine propeller with 4 blades	4, $W_B/D_T = 0.1$	0.12	2.13	0.05–0.61	10
G-G$_1$	Square-pitch marine propeller with 4 blades	4, $W_B/D_T = 0.1$	0.12	2.13	1.28–1.68	–
H-H	2-bladed paddle	4, $W_B/D_T = 0.1$	0.09–0.13	2–3	0.16–1.68	10
–	Anchor	None	0.28	1.02	0.34–1.0	11 ± 5
–	Cone impellers	0 or 4, $W_B/D_T = 0.08$	0.10–0.15	1.92–2.88	0.34–1.0	11 ± 5

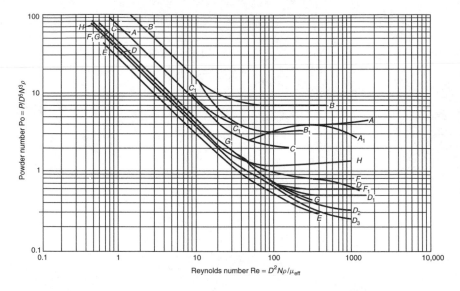

Figure 8.7   *Power curve for pseudoplastic liquids agitated by different types of impeller [from Skelland, 1983]*

Although this approach of Metzner and Otto [1957] has gained wide acceptance [Doraiswamy *et al.*, 1994], it has come under some criticism. For instance, Skelland [1967] and Mitsuishi and Hirai [1969] argued that this approach does not always yield a unique power curve for a wide range of the flow behaviour index, n. Despite this deficiency, it is safe to conclude that this method predicts power consumption with an accuracy of 25–30%. Furthermore, Godfrey [1992] has asserted that the constant k_s is independent of equipment size, and thus there are no scale-up problems.

It is not yet established, however, how strongly the value of k_s depends on the rheology, and on the geometrical arrangement of the system. For example, both Calderbank and Moo–Young [1959] and Beckner and Smith [1966] have related k_s to the impeller/vessel configuration and the power-law index, n; the dependence on, n, however is quite weak.

Data for power consumption in Bingham plastic and dilatant fluids have been reported and correlated in this manner [Metzner *et al.*, 1961; Nagata, 1975; Johma and Edwards, 1990], whereas Edwards *et al.* [1976] have studied the mixing of time-dependent thixotropic materials.

This approach has also been used in the reverse sense, using impeller power data as a means of characterising fluid rheology. Because of the indeterminate nature of the flow field produced by an impeller, the method is in principle suspect, though it can provide useful guidance in some circumstances, e.g. when working with suspensions. Even in these cases, it is imperative to

limit agitation conditions to ensure laminar flow in the vessel, say, keeping
Re ≤ 10.

As the viscosity of the liquid increases, the performance of both high speed
agitators and the close-clearance gate or anchor impellers deteriorates rapidly
and therefore impellers with high pumping capacity are then preferred. Two
such devices which have gained wide acceptance are the helical screw and
helical ribbon impellers (see Figures 8.24 and 8.26). The approach of Metzner
and Otto [1957] also provides a satisfactory correlation of power-consumption
data with such impellers [e.g. see Ulbrecht and Carreau, 1985; Carreau *et al.*,
1993; Bakker and Gates, 1995]; the value of k_s, however, is sensitive to the
geometry of the impeller.

In conclusion, it is possible to estimate the power requirements for the
agitation of single phase Newtonian and inelastic non-Newtonian fluids with
reasonable accuracy under most conditions of interest. Many useful review
articles on recent developments in this area are available in the literature
[Chavan and Mashelkar, 1980; Harnby *et al.*, 1992].

(iii) Effects of visco-elasticity

Little is known about the effect of fluid visco-elasticity on power consumption,
but early studies [Kelkar *et al.*, 1972; Chavan *et al.*, 1975; Yap *et al.*, 1976]
seem to suggest that it is negligible under laminar flow conditions. Others
have argued that the fluid visco-elasticity may increase and/or decrease the
power consumption for turbine impellers as compared with that in Newtonian
and inelastic non-Newtonian fluids under similar flow conditions [Nienow
et al., 1983; Ducla *et al.*, 1983]. Since, most polymer solutions often show
pseudoplastic as well as visco-elastic behaviour, it is not possible to distinguish the separate contributions of these non-Newtonian characteristics on
power consumption. However, the development of synthetic test fluids having
a constant apparent viscosity but a range of high degrees of visco-elasticity
(measured in terms of first normal stress difference) has helped resolve this
difficulty. The experimental results to date suggest that the extent of visco-
elastic effects is strongly dependent on the impeller geometry and operating
conditions (Reynolds and Deborah numbers, etc.). For instance, for Rushton-
type turbine impellers, the power consumption may be either greater or less
than that for Newtonian fluids depending upon the values of the Reynolds
and Deborah numbers [Oliver *et al.*, 1984; Prud'homme and Shaqfeh, 1984;
Collias and Prud'homme, 1985]. Recent work with helical ribbons and hydro-
foil impellers suggests that visco-elasticity increases the power requirement for
single phase agitation both in the laminar and transitional regimes [Carreau
et al., 1993; Özcan-Taskin and Nienow, 1995]. Thus caution must be used in
applying results obtained from any one geometry to another. No satisfactory
correlations are thus available enabling the estimation of power consumption
in visco-elastic fluids.

Finally, before concluding this section on power consumption, it should be noted that the calculation of the power requirement requires a knowledge of the impeller speed which is necessary to blend the contents of a tank in a given time, or of the impeller speed needed to achieve a given mass transfer rate in a gas–liquid system. Since a full understanding of the mass transfer/mixing mechanism is not yet available, the selection of the optimum operating speed therefore remains primarily a matter of experience.

Some guidelines for choosing an appropriate rotational speed for disc turbine blades are available for Newtonian fluids. Hicks *et al.* [1976] introduced a scale of agitation, S_A, which ranges from 1 to 10 with 1 being *mildly mixed* and 10 being *intensely mixed*. The scale of agitation is defined as [Fasano *et al.*, 1994; Bakker and Gates, 1995]:

$$S_A = 32.8 \frac{N_q N D^3}{(\pi/4D_{T_{\mathrm{eff}}})^2} \tag{8.9}$$

where N_q is the pumping number ($= Q/ND^3$), Q being the bulk flow induced by the impeller and $D_{T_{\mathrm{eff}}}$ is the effective tank diameter evaluated as $(4V_l/\pi)^{1/3}$; V_l is the volume of liquid batch. Most chemical and processing applications are characterised by scales of agitation in the range 3–6, while values of 7–10 are typical of applications requiring high fluid velocities such as in chemical reactors, fermentors and in mixing of highly viscous systems. Some guidelines for choosing a suitable value of S_A for specific applications employing turbine agitators are also available in the literature [Gates *et al.*, 1976]. Typical values of the pumping number, N_q, as a function of the Reynolds number and geometrical arrangement, are often provided by the manufacturers of mixing equipment.

Typical power consumptions (kW/m^3) are shown in Table 8.2.

Table 8.2   *Typical power consumptions*

Duty	Power (kW/m^3)
Low power	
Suspending light solids, blending of low viscosity liquids	0.2
Moderate power	
Gas dispersion, liquid–liquid contacting, heat transfer, etc.	0.6
High power	
Suspending heavy solids, emulsification, gas dispersion, etc.	2
Very high power	
Blending pastes, doughs	4

Example 8.2

A thickened lubricating oil exhibiting power-law behaviour is to be agitated using the configuration shown in Figure 8.5. Using dimensional analysis, obtain the relevant dimensionless parameters for calculating power consumption in geometrically similar equipment.

Solution

The variables in this problem, together with their dimensions, are as follows:

$$P \quad : \mathbf{ML^2T^{-3}} \qquad m : \mathbf{ML^{-1}T^{n-2}}$$
$$n \quad : \mathbf{M^0L^0T^0} \qquad N : \mathbf{T^{-1}}$$
$$g \quad : \mathbf{LT^{-2}} \qquad D : \mathbf{L}$$
$$D_T : \mathbf{L} \qquad \rho : \mathbf{ML^{-3}}$$

By Buckingham's π theorem, there will be $8 - 3 = 5$ dimensionless groups. Since n is already a dimensionless parameter, there will be four more π-groups.

Choosing ρ, N and D as the recurring set as in example 8.1; then expressing $\mathbf{M}$, $\mathbf{L}$ and $\mathbf{T}$ in terms of these variables

$$\mathbf{L} \equiv D; \quad \mathbf{M} \equiv \rho D^3 \quad \text{and} \quad \mathbf{T} \equiv N^{-1}$$

$$\pi_1 = P\mathbf{M}^{-1}\mathbf{L}^{-2}\mathbf{T}^3 = P(\rho D^3)^{-1}(D)^{-2}(N^{-1})^3$$

$$= P/\rho D^5 N^3$$

$$\pi_2 = m\mathbf{M}^{-1}\mathbf{L}\mathbf{T}^{2-n} = m(\rho D^3)^{-1}(D)(N^{-1})^{2-n}$$

$$= m/\rho D^2 N^{2-n}$$

$$\pi_3 = g\mathbf{L}^{-1}\mathbf{T}^2 = g(D)^{-1}(N^{-1})^2 = g/DN^2$$

$$\pi_4 = D_T\mathbf{L}^{-1} = D_T/D$$

Thus, the functional relationship is given:

$$\frac{P}{\rho D^5 N^3} = \mathrm{Po} = \mathrm{f}\left(\frac{\rho D^2 N^{2-n}}{m}, \frac{N^2 D}{g}, \frac{D_T}{D}, n\right)$$

where $\rho D^2 N^{2-n}/m$ is the Reynolds number which can be further re-arranged as:

$$\mathrm{Re} = \frac{\rho D^2 N}{m(N)^{n-1}}$$

On comparing this expression with the corresponding definition for Newtonian fluids, the effective viscosity of a power-law fluid is given by:

$$\mu_{\mathrm{eff}} = m(N)^{n-1}$$

This, in turn, suggests that the average shear rate, $\dot{\gamma}_{avg}$ for a mixing tank can be defined as being proportional to N, which is consistent with equation (8.8).□

Example 8.3

In a polymerisation reactor, a monomer/polymer solution is to be agitated in a baffled mixing vessel using a double turbine (6 flat blades) impeller, with the configuration B–B in Table 8.1, at a rotational speed of 2 Hz. The solution exhibits power-law behaviour with $n = 0.6$ and $m = 12\,\text{Pa·s}^{0.6}$. Estimate the power required for a 300 mm diameter impeller. The density of the solution is $950\,\text{kg/m}^3$.

Solution

Since the mixing tank is fitted with baffles, one can assume that no vortex formation will occur and the Power number is a function only of the Reynolds number.
From Table 8.1 for configuration A-A,

$$k_s = 11.5 \quad \text{and} \quad \dot{\gamma}_{avg} = k_s N = 23\,\text{s}^{-1}$$

∴ the corresponding effective viscosity of the solution.

$$\mu_{eff} = m(\dot{\gamma}_{avg})^{n-1} = 12 \times (23)^{0.6-1} = 3.42\,\text{Pa·s}$$

The Reynolds number, $\text{Re} = \dfrac{\rho D^2 N}{\mu_{eff}} = \dfrac{950 \times (300 \times 10^{-3})^2 \times 2}{3.42}$

$$= 50$$

From Figure 8.7 at $\text{Re} = 50$

$$\text{Po} = {\sim}7.4$$

and $P = \text{Po} \cdot \rho N^3 D^5$

$$= 7.4 \times 950 \times 2^3 \times (300 \times 10^{-3})^5$$

$$= 137\,\text{W}□$$

Example 8.4

It is desired to scale-up a mixing tank for the agitation of a power-law liquid. The same fluid is used in both model and large-scale equipment. Deduce the functional dependence of power consumption per unit volume of fluid on the size of the impeller and the speed of rotation.

Solution

For geometrically similar systems and in the absence of vortex formation,

$$Po = f(Re)$$

Regardless of the nature of the function, f, in order to ensure similarity between the two systems Re_1 and Re_2 in the model and large-scale equipment respectively must be equal, and Po_1 must equal Po_2. The equality of the two Reynolds numbers yields:

$$\frac{\rho_1 N_1 D_1^2}{\mu_{\text{eff1}}} = \frac{\rho_2 N_2 D_2^2}{\mu_{\text{eff2}}}$$

For a power-law fluid, $\mu_{\text{eff}} = m(k_s N)^{n-1}$. Thus,

$$\mu_{\text{eff1}} = m_1 (k_s N_1)^{n_1 - 1} \quad \text{and} \quad \mu_{\text{eff2}} = m_2 (k_s N_2)^{n_2 - 1}$$

Since the same fluid is to be used in the two cases (and assuming that the power-law constants are independent of the shear rate over the ranges encountered), $m_1 = m_2 = m$; $n_1 = n_2 = n$ and $\rho_1 = \rho_2 = \rho$, therefore the equality of Reynolds number gives:

$$\frac{N_1 D_1^2}{(N_1)^{n-1}} = \frac{N_2 D_2^2}{(N_2)^{n-1}}$$

which can be further simplified as:

$$\frac{N_1}{N_2} = \left(\frac{D_2}{D_1}\right)^{2/(2-n)}$$

From the equality of the Power numbers:

$$\frac{P_1}{\rho_1 N_1^3 D_1^5} = \frac{P_2}{\rho_2 N_2^3 D_2^5}$$

Nothing that $P_1 = P_2$, the ratio of the powers per unit volume is given by:

$$\frac{(P_1/D_1^3)}{(P_2/D_2^3)} = \left(\frac{N_1}{N_2}\right)^3 \left(\frac{D_1}{D_2}\right)^2$$

Substituting for $\frac{N_1}{N_2}$:

$$\frac{(P_1/D_1^3)}{(P_2/D_2^3)} = \left(\frac{D_1}{D_2}\right)^{\frac{2n+2}{n-2}}$$

Similarly, expressing the final result in terms of the rotational speed:

$$\frac{(P_1/D_1^3)}{(P_2/D_2^3)} = \left(\frac{N_1}{N_2}\right)^3 \left(\frac{N_1}{N_2}\right)^{n-2} = \left(\frac{N_1}{N_2}\right)^{n+1}$$

Thus, power per unit volume $(\alpha P/D^3)$ increases less rapidly with both stirrer speed and impeller diameter for a shear-thinning fluid than for a Newtonian fluid (as would be expected). The converse applies to a shear thickening fluid $(m > 1)$.

Other scale-up criteria such as equal mixing times, or equality of interfacial areas per unit volume, etc., would in general yield different relations between variables. Most significantly, complete similarity is not achievable as it would lead to conflicting requirements. In such cases, it is necessary to decide the most important similarity feature for the particular problem and then to equate the appropriate dimensionless groups. Furthermore, almost without exception, scale-up of small-scale data gives non-standard agitator speeds and power requirements for the large-scale equipment, and thus one must choose the unit nearest to that calculated from scale-up considerations.□

8.2.4 Flow patterns in stirred tanks

A qualitative picture of the flow field created by an impeller in a mixing vessel in a single phase liquid is useful in establishing whether there are stagnant or dead regions in the vessel, and whether or not particles are likely to be maintained in suspension. In addition, the efficiency of mixing equipment, as well as product quality, are influenced by the flow patterns in the mixing vessel.

Flow patterns produced in a mixing vessel are very much dependent upon the geometry of the impeller. It is thus convenient to classify the agitators used in non-Newtonian applications into three types:

   (i) those which operate at relatively high speeds, producing high shear rates in the vicinity of the impeller as well as giving good momentum transport rates throughout the whole of the liquid; typical examples include turbine impellers and propellers,
  (ii) The second type is characterised by close clearance impellers (such as gates and anchors) which extend over the whole diameter of the vessel and rely on shearing the fluid in the small gaps at the walls,
 (iii) Finally, there are slowly rotating impellers which do not produce high shear rates but rely on their effective pumping action to ensure that an adequate velocity is imparted to the fluid in all parts of the vessel; typical examples include helical screw and helical ribbon impellers (see Figures 8.24 and 8.26).

(i) Class I impellers

The flow patterns for single phase Newtonian and non-Newtonian fluids in tanks agitated by class 1 impellers have been reported in the literature by, amongst others, Metzner and Taylor [1960], Norwood and Metzner [1960], Godleski and Smith [1962] and Wichterle and Wein [1981]. The experimental methods used have included the introduction of tracer liquids, neutrally buoyant particles or hydrogen bubbles; and measurement of local velocities by

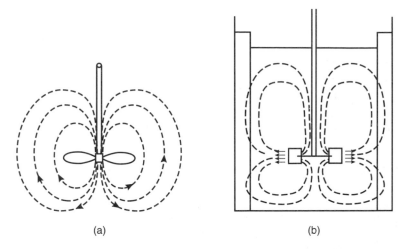

(a)   (b)

Figure 8.8   *(a) Flow pattern from propeller mixer. (b) Radial flow pattern for disc turbine*

means of pitot tubes, laser doppler velocimeters and so on. The salient features of the flow patterns produced by propellers and disc turbines are shown in Figures 8.8a and b respectively. Essentially, the propeller creates an axial flow through the impeller, which may be upwards or downwards depending upon its direction of rotation. Though the flow field is three-dimensional and unsteady, circulation patterns such as those shown in Figure 8.8a are useful in avoiding the formation of dead regions. If the propeller is mounted centrally and there are no baffles in the tank, there is a tendency for the lighter liquid to be drawn in to form a vortex, and for the degree of mixing to be reduced. These difficulties are usually circumvented by fitting baffles to the walls of the tank. The power requirement is then increased, but an improved flow pattern is obtained as seen in Figure 8.9a. Another way of minimising vortex formation is to mount the agitator off-centre to give a flow pattern similar to that depicted in Figure 8.9b.

The flat-bladed turbine impeller produces a strong radial flow outwards from the impeller (Figure 8.8b), thereby creating circulation zones in the top and bottom of the tank. The flow pattern can be altered by changing the impeller geometry and, for example, if the turbine blades are angled to the vertical, a stronger axial flow component is produced. This can be useful in applications where it is necessary to suspend solids. Furthermore, as the Reynolds number decreases (by lowering the speed of rotation and/or due to the increase in the liquid viscosity), the flow is mainly in the radial direction. The liquid velocities are very weak further away from the impeller and the quality of mixing deteriorates. A flat paddle produces a flow field with large tangential components of velocity, and this does not promote good mixing. Propellers, turbines and

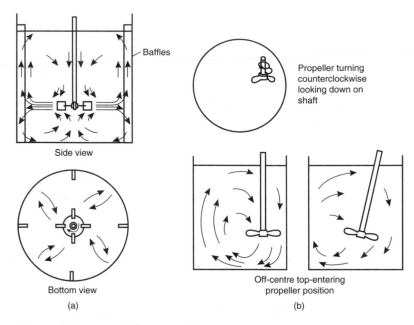

Figure 8.9   *(a) Flow pattern in vessel with vertical baffles. (b) Flow patterns with agitator offset from centre*

paddles are the principal types of impellers used for low viscosity Newtonian and pseudoplastic liquids operating in the transitional and turbulent regimes.

Generally, shear rates are highest in the region of the impeller and taper off towards the walls of the vessel. Thus, for a pseudoplastic fluid, the apparent viscosity is lowest in the impeller region and the fluid motion decreases much more rapidly for a pseudoplastic than for a Newtonian fluid as the walls are approached. Dilatant fluids display exactly the opposite behaviour. Using square cross-section vessels, Wichterle and Wein [1981] marked out zones with motion and those with no motion for pseudoplastic fluids agitated by turbine and by propeller impellers. Typical results showing stagnant zones are illustrated in Figure 8.10 in which it is clearly seen that the diameter (D_c) of the well-mixed region corresponds with the diameter of impeller D at low Reynolds numbers, but covers an increasing volume of the liquid with increasing Reynolds number or rotational speed. The authors presented the following expressions for D_c:

$$\frac{D_c}{D} = 1 \qquad \text{Re} < \frac{1}{a_0^2} \qquad\qquad (8.10a)$$

$$\frac{D_c}{D} = a_0\sqrt{\text{Re}} \qquad \text{Re} > \frac{1}{a_0^2} \qquad\qquad (8.10b)$$

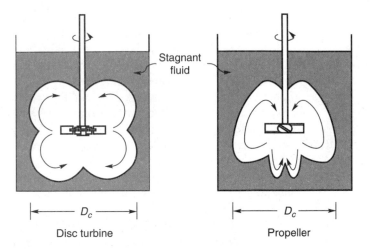

Disc turbine   Propeller

Figure 8.10 *Shape of the mixing cavity in a shear-thinning suspension. (Wichterle and Wein, 1981)*

The constant a_0 was found to be 0.3 for propellers, 0.6 for turbines and $\sim 0.375\ (\text{Po}_t)^{1/3}$ for other types where Po_t is the constant value of the Power number under fully turbulent conditions. The Reynolds number here is defined by assuming $k_s = 1$, i.e. $\rho N^{2-n} D^n / m$.

A direct link between the flow pattern and the corresponding power consumption is well illustrated by the study of Nagata *et al.* [1970] which relates to the agitation of viscoplastic media. These workers reported a cyclic increase and decrease in power consumption which can be explained as follows: Initially, the power consumption is high due to the high viscosity of the solid-like structure; however, once the yield stress is exceeded and the material begins to behave like a fluid, the power consumption decreases. The structure then becomes re-established and the solid-like zones reform leading to an increase in the power consumption; then the cycle repeats. There was a tendency for a vortex to form at the liquid surface during this cyclic behaviour. This tendency was either considerably reduced or eliminated with class II impellers. More quantitative information on flow patterns in viscoplastic materials agitated by Rushton disc turbine impellers has been obtained using X-rays and hot wire anemometry [Solomon *et al.*, 1981; Elson *et al.*, 1986]. When the stress levels induced by the impeller rotation drop below the yield stress, relative motion within the fluid ceases in viscoplastic fluids. Solomon *et al.* [1981] attempted to predict the size of the well-mixed cavern, shown in Figure 8.10. They suggested the following relation for D_c:

$$\frac{D_c}{D} = \left[\left(\frac{4\text{Po}}{\pi^3} \right) \left(\frac{\rho N^2 D^2}{\tau_0^{\ B}} \right) \right]^{1/3} \tag{8.11}$$

Equation (8.11) was stated to be applicable in the ranges of conditions as: $(\rho N^2 D^2/\tau_0{}^B) \leq 4/\pi^3 \mathrm{Po} \leq (\rho N^2 D^2/\tau_0{}^B)(D/T)^3$, i.e. when $D \leq D_c \leq D_T/D_T$.

In contrast, the influence of fluid visco-elasticity is both more striking and difficult to assess. Photographs of rotating turbine and propeller-type impellers in visco-elastic fluids suggest two distinct flow patterns [Giesekus, 1965]. In a small region near the impeller the flow is outwards, whereas elsewhere the liquid is flowing inwards towards the impeller in the equatorial plane and outwards from the rotating impeller along the axis of rotation. The two regions are separated by a streamline and thus there is no convective transport between them. A more quantitative study made by Kelkar *et al.* [1973] suggests that, irrespective of the nature of the secondary flow pattern, the primary flow pattern (i.e. tangential velocity) around a rotating body is virtually unaffected by the visco-elasticity of the fluid. Indeed, various types of flow patterns may be observed depending upon the relative magnitudes of the elastic, inertial and viscous forces, i.e. the values of the Reynolds and Deborah numbers.

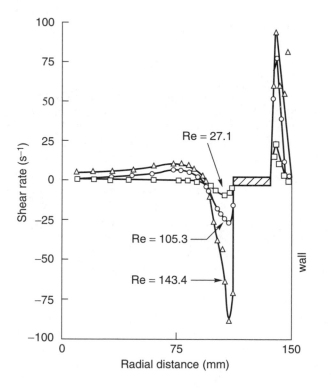

Figure 8.11   *Shear rate profiles for an anchor impeller rotating in a visco-elastic liquid. [Peters and Smith, 1967]*

(ii) Class II impellers

While anchors and gate-type impellers are known to produce poor axial circulation of the liquid in a vessel, in one study [Peters and Smith, 1967] it seems that the fluid visco-elasticity promotes axial flow. For instance, axial flow is reported to be almost 15 times greater in a visco-elastic solution as compared with a Newtonian medium. The shear rate profiles reported by these authors shown in Figure 8.11 clearly indicate that the liquid in the tank is virtually unaffected by the blade passage.

Broadly speaking, both gate and anchor agitators promote fluid motion close to the wall but leave the region near the shaft relatively stagnant, as can be seen in the typical pattern of streamlines in Figure 8.12. Furthermore, due to the only modest top to bottom turnover, vertical concentration gradients usually exist, but these may be minimised by using a helical ribbon or a screw twisted in the opposite sense, pumping the fluid downward near the shaft. Typical flow patterns for an anchor impeller are shown schematically in Figure 8.13. In such systems, the flow pattern changes with the impeller speed and the average shear rate cannot be described adequately by a linear equation such as equation (8.8). Furthermore, any rotational motion induced within the tank will produce a secondary flow in the vertical direction; the

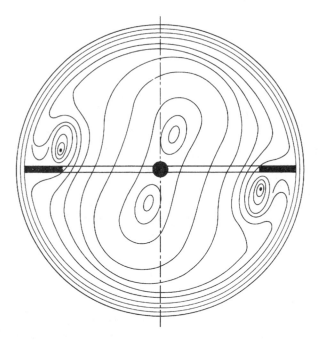

Figure 8.12   *Streamlines for a visco-elastic liquid in a tank with a gate agitator, drawn relative to the arm of the stirrer*

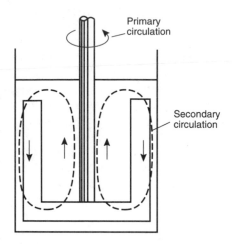

Figure 8.13   *Secondary circulation in an anchor agitated tank. [Peters and Smith, 1967]*

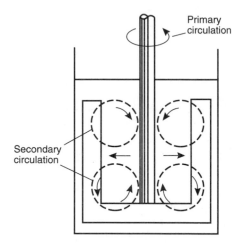

Figure 8.14   *Schematic double-celled secondary flow pattern*

liquid in contact with the tank bottom is essentially stationary while that at higher levels is rotating and will experience centrifugal forces. Consequently, the prevailing unbalanced forces within the fluid lead to the formation of a toroidal vortex. Depending upon the viscosity and type of fluid, the secondary flow pattern may be single-celled as in Figure 8.13 or double-celled, as shown schematically in Figure 8.14. Indeed, these schematic flow patterns have been substantiated by a recent experimental and numerical study for pseudoplastic fluids [Abid *et al.*, 1992].

(iii) Class III impellers

Apart from the qualitative results for a composite impeller (anchor fitted with a ribbon or screw) referred to in the preceding section, little is known about the flow patterns induced by helical ribbon and screw impellers. The salient features of the flow pattern produced by a helical ribbon impeller are shown in Figure 8.15 [Nagata *et al.*, 1956]. The primary top-to-bottom circulation, mainly responsible for mixing, is due only to the axial pumping action of the ribbons. The shear produced by the helical ribbon is localised in the regions inside and outside the blade, whereas the shear between the wall and the bulk liquid is cyclic in nature. Notwithstanding the considerable scatter of results of Bourne and Butler [1969], the velocity data shown in Figure 8.16 seem to be independent of the scale of equipment, and the nature of fluid e.g. pseudoplastic or visco-elastic. There was virtually no radial flow except in the top and bottom regions of the tank, and the vertical velocity inside the ribbon helix varied from 4 to 18% of the ribbon speed.

In addition to the primary flow pattern referred to above, secondary flow cells develop with increasing impeller speed and these are similar to those observed by Peters and Smith [1967] for class II agitators as can be seen in Figure 8.13. Carreau *et al.* [1976] have also recorded flow patterns for a helical ribbon impeller. Visco-elasticity seems to cause a considerable reduction in the

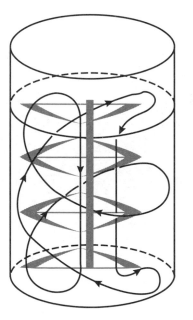

Figure 8.15 *Flow pattern produced by a helical ribbon impeller. [Nagata et al. 1956]*

axial circulation, as can be seen in Figure 8.17 where the dimensionless axial velocity is plotted for an inelastic (2% carboxymethyl cellulose and a visco-elastic (1% Separan) solutions. The axial velocities in the inelastic solution were found to be of the same order as in a Newtonian fluid.

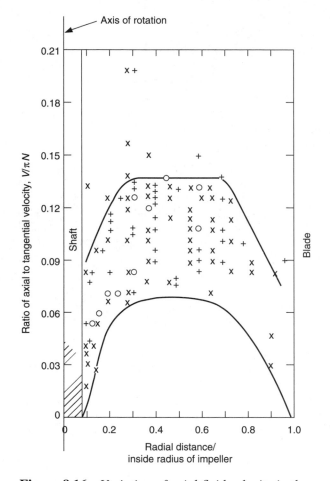

Figure 8.16 *Variation of axial fluid velocity in the core region of helical ribbon impellers pumping downwards in 27 and 730 litre tanks. The curves indicate the upper and lower bounds of data [Bourne and Butler, 1969]*

$+$ $D/D_T = 0.889$

$\times$ $D/D_T = 0.952$(small tank)

$\bigcirc$ $D/D_T = 0.954$(large tank)

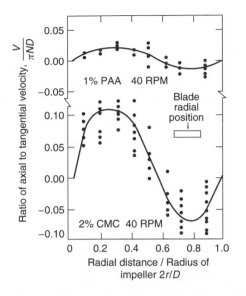

Figure 8.17 *Axial velocity distribution in a vessel agitated by a helical ribbon impeller [Carreau et al., 1976]*

On the other hand, the tangential velocities were so increased in visco-elastic fluids that the whole contents of the vessel, except for a thin layer adhering to the wall, rotated as a solid body with an angular velocity equal to that of the impeller. More definitive conclusions regarding the role of non-Newtonian rheology, especially visco-elasticity, must await additional work in this area.

Virtually nothing is known about the flow patterns produced by helical screw impellers. In a preliminary study, Chapman and Holland [1965] presented pictures of dye flow patterns for an off-centre helical screw impeller, pumping upwards with no draft tube. There seems to exist a dispersive flow between the flights of the screw, the dispersion being completed at the top of the screw. The flow into the screw impeller was from the other side of the tank, whereas the fluid in the remaining parts of the tank appeared to be virtually stagnant. Limited numerical predictions based on the assumption of three-dimensional flow induced by a helical ribbon-screw impeller show good agreement with experimental data for power consumption in Newtonian fluids [Tanguy *et al.*, 1992].

The limited results available on the flow patterns encountered in mixing devices used for thick pastes with complex rheology have been discussed by Hall and Godfrey [1968] and Kappel [1979]. One common geometry used in mixing thick pastes is that of the sigma-blade mixer (Figure 8.24). Such devices have thick S- or Z-shaped blades which look like high-pitch helical ribbon impellers. Generally, two units are placed horizontally in separate troughs inside a mixing chamber and the blades rotate in opposite directions at different speeds. Preliminary results obtained using a positive displacement mixer suggest it has advantages over helical ribbon and sigma mixers for thick pastes and extremely viscous materials [Cheng *et al.*, 1974].

Thus, the flow patterns established in a mixing tank depend critically on the vessel – impeller configuration, the rheology of the liquid batch and the operating conditions. In selecting the appropriate combination of equipment, care must be taken to ensure that the resulting flow pattern is suitable for the required application.

8.2.5 Rate and time of mixing

Before considering the question of the rate and time of mixing, it is necessary to have some means of assessing the quality of the product mixture. The wide scope and spectrum of mixing problems make it impossible to develop a single criterion for all applications. One intuitive and convenient, but perhaps unscientific, criterion is whether or not the mixture meets the required specification. Whatever the criteria used, mixing time is defined as the time needed to produce a mixture or a product of pre-determined quality, and the rate of mixing is the rate at which the mixing progresses towards the final state. When a tracer is added to a single-phase liquid in a stirred tank, the mixing time is measured as the interval between the introduction of tracer and the time when the contents of the vessel have reached the required degree of uniformity or mixedness. If the tracer is completely miscible and has the same viscosity and density as the liquid in the tank, the tracer concentration may be measured as a function of time at any point in the vessel by means of a suitable detector, such as a colour meter, or by electrical conductivity. For a given amount of tracer, the equilibrium concentration C_∞ may be calculated; this value will be approached asymptotically at any point as shown in Figure 8.18.

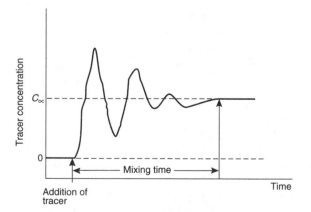

Figure 8.18 *Mixing-time measurement curve*

In practice, the mixing time will be that required for the mixture composition to come within a specified (95 or 99%) deviation from the equilibrium value C_∞, and this will be dependent upon the way in which the tracer is added and the location of the detector. It may therefore be desireable to record the tracer concentration at several locations, and to define the variance of concentration σ^2 about the equilibrium value as:

$$\sigma^2 = \frac{1}{p-1} \sum_{i=0}^{i=p} (C_i - C_\infty)^2 \tag{8.12}$$

where C_i is the tracer concentration at time t recorded by the ith detector. A typical variance curve is shown in Figure 8.19.

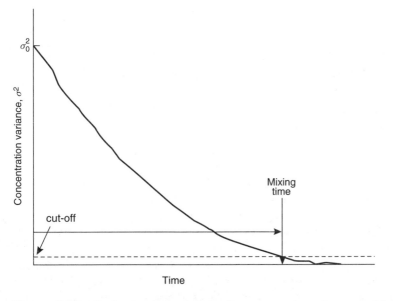

Figure 8.19   *Reduction in variance of concentration of tracer with time*

Several experimental techniques may be used, such as acid/base titration, electrical conductivity or temperature measurement, measurement of refractive index, light absorption, and so on. In each case, it is necessary to specify the manner of tracer addition, the position and number of recording points, the sample volume of the detection system, and the criterion used in locating the end point (such as a predetermined cut-off point). Each of these factors will influence the measured value of mixing time, and therefore, care must be exercised in comparing results from different investigations [Manna, 1997]. Irrespective of the technique used to measure the mixing time, the response curve may show periodic behaviour. This may most probably be due to the repeated passage of a fluid element with a locally high concentration of tracer. The time interval between any two successive peaks is known as the circulation time.

For a given experiment and configuration, the mixing time will depend upon the process and operating variables as follows:

$$t_m = f\ (\rho, \mu, N, D, g, \text{geometrical dimensions of the system}) \tag{8.13}$$

Using dimensional analysis, the functional relationship may be re-arranged as:

$$Nt_m = \theta_m = f\ (\text{Re, Fr, geometrical ratios}) \tag{8.14}$$

For geometrically similar systems and assuming that the Froude number, Fr $(= DN^2/g)$ is not important,

$$\theta_m = f(\text{Re}) \tag{8.15}$$

Broadly speaking, the dimensionless mixing times in both the laminar and fully turbulent regions are independent of the Reynolds number; a substantial transition zone exists between these two asymptotic values. Undoubtedly, the functional relationship between θ_m and Re is strongly dependent on the mixer geometry and the flow patterns produced.

For class I impellers, the limited available work [Norwood and Metzner, 1960] confirms this dependence of θ_m on Re for turbine impellers in baffled tanks, albeit these results are believed to be rather unreliable. Nonetheless, Norwood and Metzner [1960] suggested that the correlations developed for Newtonian liquids can also be used for purely inelastic shear-thinning fluids, simply by using a generalised Reynolds number based on the effective viscosity evaluated at the average shear rate given by equation (8.8). Although, it is widely reported that θ_m is quite sensitive to impeller–vessel geometry for low viscosity liquids, there have been few studies of the effect of physical properties on θ_m. Early work of Bourne and Butler [1969] suggests that the rates of mixing, and hence the mixing times, are not very sensitive to the fluid rheology for both Newtonian and inelastic non-Newtonian materials. On the other hand, Godleski and Smith [1962] reported mixing times for pseudoplastic fluids (agitated by turbines) up to 50 times greater than those expected from the corresponding results for materials exhibiting Newtonian behaviour. This emphasises the care which must be exercised in applying any generalised conclusions to a particular system. Intuitively, one might expect a similar deterioration in mixing for visco-elastic liquids, especially when phenomena such as secondary flow and flow reversal occur.

The only study relating to the use of class II impellers for non-Newtonian media is that of Peters and Smith [1967] who reported a reduction in mixing and circulation times for visco-elastic polymer solutions agitated by an anchor impeller. The decrease in mixing time can be ascribed to the enhanced axial circulation.

In contrast to this, class III impellers have generated much more interest. Circulation times with helical impellers are known not to be affected by shear-thinning behaviour [Chavan and Ulbrecht, 1972, 1973; Carreau *et al.*, 1976]. Thus, the circulation time is constant in the laminar regime (Re < ~10); it decreases in the transition zone both with increasing Reynolds number and with increasingly shear-thinning behaviour. A subsequent more detailed study [Guerin *et al.*, 1984] shows that, even though the average circulation times are not influenced significantly by shear-thinning characteristics, their distribution becomes progressively narrower.

Mixing times in inelastic systems follow a similar pattern; namely, θ_m is independent of Reynolds number in the laminar region (Re < ~10). For helical screw impellers in the intermediate zone ($10 \leq \text{Re} \leq 1000$), θ_m decreases with Reynolds number. With shear-thinning behaviour, the apparent viscosity should be evaluated at $\dot{\gamma}_{\text{avg}}$ estimated using equation (8.8). On the other hand, Carreau *et al.* [1976] have reported that non-Newtonian fluids required considerably longer mixing times as compared with Newtonian fluids at comparable circulation rates.

The results of the scant work with visco-elastic fluids are conflicting but suggest that relationship between mixing and circulation times is strongly dependent on the geometrical arrangement and that it is not yet possible to account quantitatively for the effects of visco-elasticity. Visco-elastic fluids appear to be much more difficult to mix than inelastic fluids. Reference should be made to a review paper for further details of studies in this area [Takahashi, 1988].

8.3 Gas–liquid mixing

Many gas–liquid reactions of industrial significance are carried out in agitated tank reactors, and the design requirements vary from one application to another. For instance, in effluent aeration and in some fermentation reactions, the systems are dilute and reactions are slow so that mass transfer is not likely to be a limiting factor. Energy efficiency is then the most important consideration, and large tanks giving long hold-up times are used. Chlorination and sulphonation reactions, on the other hand, are fast and the gases have high solubilities; and it is then desirable to have high rates of heat and mass transfer and short contact times. In the food industry, the non-Newtonian properties of batters and creams are of dominant importance and the flow field and temperature must be closely controlled. Irrespective of the application, a rational understanding of the fluid mechanical aspects of gas dispersion into liquids is a necessary precursor to the modelling of heat and mass transfer, and reactions in such systems.

Gas dispersion in agitated tanks may be described in terms of bubble size, gas hold-up, interfacial area and mass transfer coefficient. While gas dispersion in low viscosity systems [Smith, 1985; Tatterson, 1991; Harnby *et al.*, 1992] has been extensively studied, little is known about the analogous process in highly viscous Newtonian and non-Newtonian media, such as those encountered in polymer processing, pulp and paper manufacturing and fermentation applications.

Good dispersion of a gas into a liquid can only be achieved by using high speed (class I) agitators which unfortunately are not very effective for mixing high viscosity liquids. Hence, for gas dispersion into highly viscous media, there are two inherently conflicting requirements which may be met in practice by using a combination of two impellers mounted on a single shaft. For instance, the Rushton disc turbine with a 45° pitched blade impeller combines the advantages of the low flow and high shear of a disc turbine with the high flow and low shear produced by the second component of the assembly. Such composite agitators provide a good compromise between those agitators which cause mixing as a result of local turbulence generated by their shape, and those which give large-scale convective flows. Figure 8.20a shows a system which combines turbulence generation at the blade tips with an induced large-scale flow from the angled blade arms; two or more agitators can be mounted on the same shaft at 90° to each other. If suitably designed, such assemblies can be rotated at moderately high speeds without excessive power consumption.

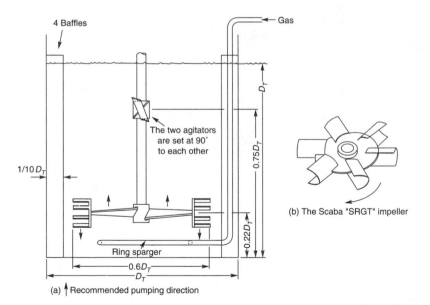

Figure 8.20   *Composite impellers used for gas–liquid dispersions: (Left) The Intermig, (right) The Scaba SRGT impeller*

In some applications, two independent agitators are employed. One is of the positive displacement type and rotates slowly to give good mixing, whereas the other operates at a high rotational speed to facilitate gas dispersion. Another more widely used impeller for dispersing gases into liquids is the so-called scaba SRGT impeller (shown in Figure 8.20b) in which half-cut pipes are used instead of plane blades in a four- or six-bladed Rushton turbine. The gas is invariably introduced through a sparger placed beneath the impeller.

Basically, the process of gas dispersion involves two competing processes: breakdown of the gas into bubbles which occurs predominantly in the high shear region near the impeller, and coalescence which takes place in quiescent regions away from the impeller. Depending upon the impeller speed and the physical properties of the liquid phase (mainly viscosity and surface tension), the gas bubbles may either re-circulate through the impeller region or escape from the system through the free surface, or they may coalesce. In addition, the dispersion of gases into highly viscous liquids differs from that in low viscosity systems because the gas bubbles have a greater tendency to follow the liquid motion when the viscosity is high. Admittedly, the exact mechanism of gas dispersion is not fully understood even for the extensively studied low viscosity systems; however, the mechanisms of dispersion in highly viscous Newtonian and non-Newtonian systems appear to be qualitatively similar. Essentially, the sparged gas gets sucked into the low pressure regions ('cavities') behind the impeller. The shape and stability of these cavities are strongly influenced by the liquid rheology, and the Froude and the Reynolds numbers for the agitator. There is, however, a minimum impeller speed required for these cavities to form. For instance, Van't Reit [1975] suggested that the Froude number should be greater than 0.1 in liquids agitated by a disc turbine. As remarked earlier, high speed agitators are, however, not effective in dispersing a gas in highly viscous liquids (Re < ~10) [Nienow *et al.*, 1983]. For Re > 1000, the minimum impeller speed required for gas dispersion rises slowly with increasing liquid viscosity, e.g. a 50% increase in impeller speed is required for a 10-fold increase in viscosity. In the intermediate zone, $10 \leq Re \leq 1000$, the formation and stability of cavities is determined by complex interactions between Reynolds and Froude numbers, and the gas flow rate. Preliminary results seem to suggest that the effects of shear-thinning viscosity are negligible. Yield stress also does not appear to play any role in this process except that some gas may be entrapped in cavities, even after the impeller has been stopped [Nienow *et al.*, 1983]. On the other hand, cavities which are much bigger and of different shapes have been observed in visco-elastic liquids [Solomon *et al.*, 1981]. Furthermore, dispersion of a gas in visco-elastic media is more difficult because the gas entrapped in a cavity does not disperse until the gas flow is stopped. The effect of visco-elasticity is even more pronounced with the composite Intermig impeller (Figure 8.20a). In this case, the gas-filled cavities may extend behind the outer twin-split blades, almost all the way back to the following blade. Qualitatively, these differences in behaviour have been attributed to large extensional viscosities of highly visco-elastic media [Nienow and Ulbrecht, 1985].

8.3.1 Power consumption

The limited work on Rushton turbines suggests that at low Reynolds numbers (<10), the Power number is almost unaltered by the introduction of gas, possibly due to the fact that no gas-filled cavities are formed at such low impeller speeds. In the intermediate region (10 < Re < 1000), gas cavities begin to form, and the Power number decreases with increasing Reynolds number, goes through a minimum value *ca*. Re ~ 300–500 and begins to increase again, as shown in Figure 8.21 for a range of polymer solutions. However, the gas flow rate does not seem to influence the value of the power number at a fixed impeller speed. The decrease in Power number arises

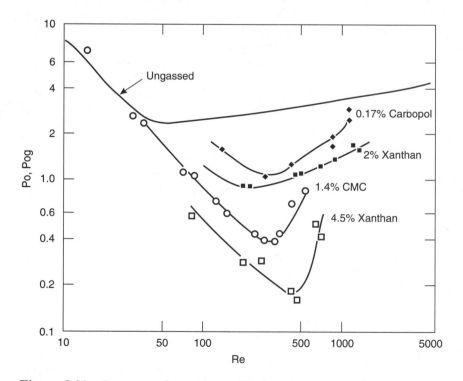

Figure 8.21   *Power number with gas-filled cavities on impeller versus Reynolds number at $Q_G = 0.5$ and 1 vvm (volumetric gas flow rate per minute per unit volume of liquid): $10 \leq$ Re ≤ 1100 (○, 1.4% CMC; □, 4.3% Xanthan gum; ■, 2% Xanthan gum; ♦, 0.17% Carbopol)*

from the combined effect of streamlining, the reduced pumping capacity of the impeller and the increased pressure in the gas-filled cavities. When the Power number is a minimum, the cavities are of maximum size and the impeller rotates in a pocket of gas without causing any dispersion. Undoubtedly, the reduction in power consumption, as well as its minimum value, are manifestations of the complex interplay between the size and structure of the cavities, the rheological characteristics of the liquid and the kinematic conditions, but the nature of these interactions is far from clear [Nienow and Ulbrecht, 1985; Tatterson, 1991; Harnby *et al.*, 1992].

Subsequent work with a helical ribbon impeller [Carreau *et al.*, 1992; Cheng and Carreau, 1994] suggests that the power consumption in the presence of a gas may either increase or decrease, depending upon the non-Newtonian flow properties of the liquid. For instance, Carreau *et al.* [1992] report reduced levels of power consumption in highly pseudoplastic liquids when gas is present (presumably due to the enhanced levels of shearing) whereas, in aerated visco-elastic liquids, the power always increases. The currently available correlations for the estimation of power under aerated conditions are too tentative to be included here [Cheng and Carreau, 1994].

Some information is also available on the performance of more widely used composite or modified impellers, for instance, Intermig and 6 bladed Scaba SRGT impellers shown in Figure 8.20 [Galindo and Nienow, 1993]. When a gas is present, the power is reduced considerably in non-Newtonian liquids agitated by a Scaba impeller. Furthermore, a point is reached where the Power number becomes independent of the gas flow rate for a fixed impeller speed. As far as scale-up is concerned, a preliminary study suggests that the power requirements for complete mixing decrease rapidly with increasing size of impeller irrespective of whether or not gas is present [Solomon *et al.*, 1981].

8.3.2 Bubble size and hold-up

Bubble size (distribution) and hold-up, together with the specific interfacial area and volumetric mass transfer coefficients, may be used to characterise the effectiveness of gas dispersion into liquids. It is important to emphasise that these parameters, and bubble coalescence, are extremely sensitive to the presence of surface-active agents. Although all these variables show spatial variation, only globally averaged values are usually reported and these are frequently found to be adequate for the engineering design calculations. For a given liquid, the mean bubble size does not show a strong dependence on the level of agitation. Ranade and Ulbrecht [1978] studied the influence of polymer addition on hold-up and gas–liquid mass transfer in agitated vessels. Even small amounts of dissolved polymer were shown to give rise to substantial reductions in both hold-up and mass transfer, albeit the degree of reduction showing strong dependence on the type and concentration of the polymer.

Qualitatively similar results have been reported for markedly shear-thinning fluids [Machon *et al.*, 1980]. This reduction has been generally ascribed to the formation of large bubbles which have a shorter residence time. The average gas hold-up was found to vary with power P_g and gas flow rate, Q_g, as:

$$\phi \propto P_g^{0.3} Q_g^{0.7} \tag{8.16}$$

or with power consumption per unit mass, ε, and the superficial gas velocity, V_g, as:

$$\phi \propto \varepsilon^{0.3} V_g^{0.7} \tag{8.17}$$

For a fixed rate of feed of gas, both hold-up and power consumption decrease with increasing polymer concentration, i.e. with increasing apparent viscosity. In these systems, bubbles have generally been found to be predominantly of two sizes, with a large population of small bubbles and a very few large bubbles. Figure 8.22 shows typical results for gas hold-up in highly viscous shear-thinning polymer solutions, obtained with a single disc turbine and a composite impeller (45° downward pumping impeller and disc turbine) for a range of (D/D_T) ratios. Finally, the point of minimum power consumption (largest amount of gas present in cavities), as expected, corresponds to the point of maximum gas holdup [Hickman, 1988].

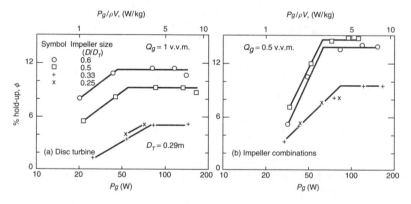

Figure 8.22   *Typical relationship between hold-up and power input when dispersing gas in a high viscosity, shear-thinning fluid. (a) A single disc turbine; (b) A combination of 45° downward pumping impeller and disc turbine*

8.3.3 Mass transfer coefficient

In view of what has been said so far regarding the mixing of highly viscous Newtonian and non-Newtonian systems, a reduction in the mass transfer coefficient, $k_L a$ is inevitable. This is indeed borne out well by the few studies available on this subject [Perez and Sandall, 1974; Yagi and Yoshida, 1975; Ranade and Ulbrecht, 1978; Nishikawa *et al.*, 1981].

Dimensionless empirical correlations relating $k_L a$ to the system geometry, and kinematic and physical variables are available in the literature [Tatterson, 1991]. For example, the following correlation due to Kawase and Moo–Young [1988] is one which embraces a very wide range of power-law parameters:

$$k_L a = \frac{0.675 \rho^{0.6} (P/V)^{(9+4n)/(10(1+n))}}{(m/\rho)^{0.5(1+n)} \sigma^{0.6}} D_L^{0.5} \left(\frac{V_g}{V_t}\right)^{0.5} \left(\frac{\mu_{\text{eff}}}{\mu_w}\right)^{-0.25} \tag{8.18}$$

Equation (8.18) is not dimensionally consistent and all quantities must be expressed in S.I. units; that is, $k_L a$, the volumetric mass transfer coefficient (s^{-1}); ρ, the density of liquid (kg/m^3); (P/V), the power input per unit volume of dispersion (W/m^3); σ, the interfacial tension (N/m); V_g, the superficial velocity of gas (m/s); V_t, the terminal velocity of a single bubble in a quiescent medium (m/s), (Kawase and Moo–Young recommended a constant value of $0.25 \, \text{m/s}$); μ_{eff}, the effective viscosity estimated using equation (8.8) (Pa·s); μ_w, the viscosity of water (Pa·s); D_L, the diffusivity of gas into liquid (m^2/s) and m is the power-law consistency coefficient (Pa·s^n). Equation (8.18) applies over the following ranges of conditions: $0.59 \le n \le 0.95$; $0.0036 \le m \le 10.8$ Pa·s^n and $0.15 \le D_T \le 0.6 \, \text{m}$.

A comprehensive discussion of other contemporary work in this field is available elsewhere [Nienow and Ulbrecht, 1985; Harnby *et al.*, 1992; Herbst *et al.*, 1992].

8.4 Heat transfer

The rate of heat transfer to process materials may be enhanced by externally applied motion both within the bulk of the material and in the proximity of heat transfer surfaces. In most applications, fluid motion is promoted either by pumping through tubes (Chapter 6) or by mechanical agitation in stirred vessels. A simple jacketed vessel is very commonly used in chemical, food, biotechnological and pharmaceutical processing applications to carry out a range of operations. In many cases, heat has to be added or removed from the contents of vessel, either to control the rate of reaction, or to bring it to completion. The removal/addition of heat is customarily accomplished by using water or steam in a jacket fitted on the outside of the vessel or in an immersed cooling coil in the tank contents. In either case, an agitator is used both to achieve uniform temperature distribution in the vessel and to improve the heat transfer rate. As is the case with power requirement and mixing time, the rate of heat transfer in stirred vessels is strongly dependent on the tank – impeller configuration, type and number of baffles, the liquid rheology, the rotational speed and the type of heat transfer surface, for example jacket or coil. In the following sections, two methods (jacket and coil) of heat removal or addition for non-Newtonian materials are discussed separately.

8.4.1 Helical cooling coils

A vessel fitted with a cooling coil and an agitator is shown schematically in Figure 8.23. In this case the thermal resistances to heat transfer arise from the fluid film on the inside of the cooling coil, the wall of the tube (usually negligible), the fluid film on the outside of the coil, and the scale that may form on either surface. The overall heat transfer coefficient, U, can thus be expressed as:

$$\frac{1}{UA} = \frac{1}{h_i A_i} + \frac{x_w}{k_w h_w} + \frac{1}{h_o A_o} + \frac{R_o}{A_o} + \frac{R_i}{A_i} \tag{8.19}$$

where the subscripts 'i', 'o' and 'w' refer to the inside, outside and the tube wall conditions respectively, and R is the scale resistance.

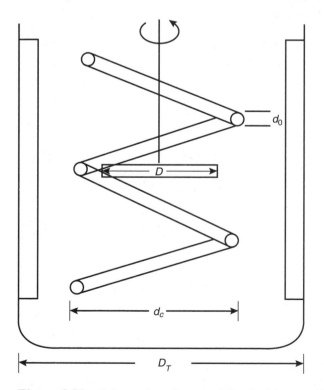

Figure 8.23   *Schematics of a vessel fitted with a cooling coil and an agitator*

(i) Inside film coefficients

The inside film coefficient, h_i, may be estimated using the correlations available in the literature such as the well known Dittus–Boelter equation for low viscosity coolants, as recommended in the literature [Coulson and Richardson, 1999]:

$$\text{Nu} = \frac{h_i d}{k} = 0.023 \left(\frac{\rho V d}{\mu} \right)^{0.8} \left(\frac{C_p \mu}{k} \right)^{0.4} \tag{8.20}$$

For more viscous liquids such as concentrated brine solutions, the Sieder–Tate equation is preferable:

$$\frac{h_i d}{k} = 0.027 \text{Re}^{0.8} \text{Pr}^{0.33} \left(\frac{\mu_b}{\mu_w} \right)^{0.14} \tag{8.21}$$

where the subscripts 'b' and 'w' relate to the bulk and wall conditions respectively. Both of these equations were based on data for straight tubes and it is necessary to apply a small correction factor for the coil configuration [Jeschke, 1925] as:

$$h_i(\text{coil}) = h_i(\text{tube}) \left[1 + 3.5 \frac{d}{d_c} \right] \tag{8.22}$$

where d is the inside diameter of the tube and d_c that of the helix, as shown in Figure 8.23.

For steam condensing inside the tube, the heat transfer coefficient is usually sufficiently high for its contribution to the overall heat transfer coefficient to be neglected.

(ii) Outside film coefficients

The value of h_o is much more difficult to determine, particularly in view of the strong interplay between the non-Newtonian rheology of the liquid and the geometry of the tank – impeller combination. Other internal fixtures such as baffles affect the flow pattern and heat transfer, as well as power consumption and mass transfer. It is thus dangerous to make detailed cross-comparisons between different studies unless the two systems are completely similar.

The results of heat transfer studies are usually expressed in terms of the relevant dimensionless groups as [Edwards and Wilkinson, 1972; Poggermann *et al.*, 1980]:

$$\text{Nu} = f \text{ (Re, Pr, Gr, geometric ratios)} \tag{8.23}$$

The effect of geometric ratios on the Nusselt number is difficult to quantify, although some investigators have incorporated the principal length parameters as ratios in their heat transfer correlations. This factor alone precludes the possibility comparisons of the results of some workers!

For geometrically similar systems, equation (8.23) simplifies to:

$$Nu = f \ (Re, Gr, Pr) \tag{8.24}$$

The Grashof number, Gr $(= g\beta\Delta T L_c^3 \rho^2 / \mu^2)$, is indicative of free convection effects which are generally significant in low viscosity liquids, but become increasingly less important with rising viscosity and increasing impeller speed [Carreau *et al.*, 1994]. For a power-law fluid, the apparent viscosity estimated using the average shear rate given by equation (8.8) is used in evaluating the Grashof number.

Owing to the inherently different flow fields produced by each class of agitators, it is convenient to consider separately the application of equation (8.24) to equipment of particular forms. Since most of the work related to heat transfer to pseudoplastic materials has been critically reviewed elsewhere [Gluz and Pavlushenko, 1966; Edwards and Wilkinson, 1972; Poggermann *et al.*, 1980; Desplanches *et al.*, 1980], only a selection of widely used correlations is given here.

(a) Class I impellers

As mentioned earlier, these impellers operate at relatively high speeds and are effective only in low to medium viscosity liquids. In most cases, the main flow in the vessel tends to be transitional and/or turbulent. For shear-thinning polymer solutions and particulate suspensions agitated by paddle, turbine and propeller type impellers, many correlations of varying complexity and form are available for the estimation of the outside film coefficient. One such correlation, based on wide ranges of conditions ($400 \leq Re \leq 10^6$; $4 \leq Pr \leq 1\,900$; $0.65 \leq \mu_{\text{eff}} \leq 283$ mPa·s), is due to Edney and Edwards [1976]:

$$Nu = \frac{h_o d_o}{k} = 0.036 \left(\frac{\rho N D^2}{\mu_{\text{eff}}}\right)^{0.64} \left(\frac{C_p \mu_{\text{eff}}}{k}\right)^{0.35} \left(\frac{\mu_{\text{eff},b}}{\mu_{\text{eff},w}}\right)^{0.2} \left(\frac{d_c}{D_T}\right)^{-0.375}$$

$$\tag{8.25}$$

where d_o is the outside (coil) tube diameter, d_c is the coil helix and D_T is the vessel diameter respectively. The effective viscosity is evaluated using equation (8.8) with $k_s = 11.6$.

(b) Class II impellers

These impellers, including gates and anchors, reach the far corners of the vessels directly instead of relying on momentum transport by bulk motion. Among others, Pollard and Kantyka [1969] carried out an extensive study on heat transfer from a coil to aqueous chalk slurries ($0.3 \leq n \leq 1$) in vessels up to 1.1 m in diameter fitted with anchor agitators; they proposed the following equation for Nusselt number:

$$\text{Nu} = \frac{hD_T}{k} = 0.077\text{Re}^{2/3}\text{Pr}^{1/3}\left(\frac{\mu_{\text{eff},b}}{\mu_{\text{eff},w}}\right)^{0.14}\left(\frac{D_T}{d_o}\right)^{0.48}\left(\frac{D_T}{d_c}\right)^{0.27} \quad (8.26)$$

Equation (8.26) applies over the range $200 \leq \text{Re} \leq 6 \times 10^5$ and the effective viscosity is estimated using equation (8.8) and the value of k_s given by:

$$k_s = \left(9.5 + \frac{9}{1 - (D/D_T)^2}\right)\left(\frac{3n+1}{4n}\right)^{n/(n-1)} \quad (8.27)$$

The indeterminate form of equation (8.27) should be noted for $n = 1$, but then k_s is redundant for Newtonian fluids.

(c) Class III impellers

This class of impellers has generated much more interest than the gates and anchors. Consequently, numerous workers [Coyle *et al.*, 1970; Nagata *et al.*, 1972; Heim, 1980; Kuriyama *et al.*, 1983; Ayazi Shamlou and Edwards, 1986; Kai and Shengyao, 1989; Carreau *et al.*, 1994] have investigated the rates of heat transfer to viscous non-Newtonian materials in vessels fitted with helical ribbons and screw-type impellers. For instance, Carreau *et al.* [1994] studied heat transfer between a coil (acting as a draft tube) and viscous Newtonian, inelastic shear-thinning and visco-elastic polymer solutions agitated by a screw impeller. The rate of heat transfer was measured for both cooling and heating of solutions. The flow rate of water inside the coil was sufficiently high for the inside film resistance to be negligible (i.e. large values of h_i). The effective viscosity of shear-thinning and visco-elastic polymer solutions was evaluated using equation (8.8) with $k_s = 16$, the value deduced from their results on power consumption. These workers proposed a single correlation for Newtonian, inelastic power-law and visco-elastic fluids as:

$$\text{Nu} = \frac{h_o d_o}{k} = 0.39\text{Re}^{0.51}\text{Pr}^{1/3}\left(\frac{d_o}{D}\right)^{0.594} \quad (8.28)$$

All physical properties are evaluated at the mean film temperature, i.e. $(T_w + T_b)/2$, and equation (8.28) encompasses the range of conditions: $3 \leq \text{Re} \leq 1300$ and $500 \leq \text{Pr} \leq 30\,000$.

8.4.2 Jacketed vessels

In many applications, it is not practicable to install cooling coils inside a tank, and heating of the contents of the vessel is achieved using condensing steam or coolant in a jacket. Indeed, this is a preferred arrangement for moderately viscous systems. The numerous studies of heat transfer to non-Newtonian fluids agitated by a range of impellers have been reviewed by Edwards and Wilkinson [1972], and more recently by Dream [1999]; most of the work to date has been carried out in relatively small vessels (<650 mm diameter). Furthermore, investigators have used diverse methods of estimating the apparent viscosity which should be used in the evaluation of the Reynolds and Prandtl numbers and it is therefore not possible to discriminate between predictive expressions. In the following sections, a selection of the more reliable correlations is presented for each class of agitators.

For a steam jacketed tank (360 mm in diameter) fitted with baffles, Hagedorn and Salamone [1967] measured the rates of heat transfer to water, glycerol and aqueous carbopol solutions over wide ranges of conditions ($0.36 \le n \le 1$; $35 \le \text{Re} \le 6.8 \times 10^5$; $\text{Pr} \le 2.4 \times 10^4$). They measured temperatures at various locations in the vessel and suggested the following general form of heat transfer correlation:

$$\text{Nu} = \frac{hD}{k} = C \, \text{Re}^{a/((n+1)+b)} \, \text{Pr}^d \left(\frac{m_b}{m_w}\right)^e \left(\frac{D_T}{D}\right)^f \left(\frac{W}{D}\right)^g n^i \quad (8.29)$$

where the apparent viscosity is evaluated at the shear rate, $\dot{\gamma}_{\text{eff}} = 11 \, N$. The values of the constants in equation (8.29) for various impellers are listed in Table 8.3.

Table 8.3   *Values of constants in equation (8.29)*

Type of Impeller	a	b	C	d	e	f	g	i	$\dfrac{D_T}{D}$
Anchor	1.43	0	0.56	0.30	0.34	–	–	0.54	1.56
Paddle	0.96	0.15	2.51	0.26	0.31	−0.46	0.46	0.56	1.75–3.5
Propeller	1.28	0	0.55	0.30	0.32	−0.40	–	1.32	2.33–3.41
Turbine	1.25	0	3.57	0.24	0.30	0	0	0.78	2–3.50

Hagedorn and Salamone [1967] reported the mean error resulting from the use of equation (8.29) to be 14% for moderately shear-thinning materials ($n > 0.69$) increasing to 20% for highly shear-thinning fluids ($n < 0.69$); both are well within the limits of experimental errors generally associated with this type of work. Similarly, Sandall and Patel [1970] and Martone and Sandall [1971] have developed correlations for the heating of pseudoplastic (carbopol solutions) and Bingham plastic slurries (chalk in water) in a steam-jacketed vessel fitted with a turbine impeller and baffles or with an anchor agitator. In their study which was limited to only one size of vessel, they were able to correlate their results in the following simple form:

$$\text{Nu} = \frac{hD_T}{k} = C \left(\frac{\rho N D^2}{\mu_{\text{eff}}} \right)^a \left(\frac{C_p \mu_{\text{eff}}}{k} \right)^b \left(\frac{\mu_{\text{eff},b}}{\mu_{\text{eff},w}} \right)^d \tag{8.30}$$

where the apparent viscosity was evaluated at $\dot{\gamma}_{\text{eff}} = k_s N$ and k_s is given by equation (8.27). Based on their data ($0.35 \leq n \leq 1$; $80 \leq \text{Re} \leq 10^5$ and $2 \leq \text{Pr} \leq 700$), they proposed $a = 2/3$ $b = 1/3$ (the values also applicable for Newtonian liquids) and $d = 0.12$. The remaining constant C showed some dependence on the type of the impeller, having values of 0.482 for turbines and 0.315 for anchors.

In conclusion, it should be emphasised that most of the currently available information on heat transfer to non-Newtonian fluids in stirred vessels relates to specific geometrical arrangements. Few experimental data are available for the independent verification of the individual correlations presented here which, therefore, must be regarded somewhat tentative. Reference should also be made to the extensive compilations [Edwards and Wilkinson, 1972; Poggermann *et al.*, 1980; Dream, 1999] of other correlations available in the literature. Although the methods used for the estimation of the apparent viscosity vary from one correlation to another, especially in terms of the value of k_s, this appears to exert only a moderate influence on the value of h, at least for shear-thinning fluids. For instance, for $n = 0.3$ (typical of suspensions and polymer solutions), a two-fold variation in the value of k_s will give rise to a 40% reduction in viscosity, and the effects on the heat transfer coefficient will be further diminished because $\text{Nu} \propto \mu_{\text{eff}}^{0.33 \text{ to } 0.7}$. Thus, an error of 100% in the estimation of μ_{eff} will result in an error of only 25–60% in the value of h which is a reasonable engineering estimate.

Example 8.5

A polymer solution is to be heated from 18°C to 27°C before use as a thinner in a wall paint. The heating is to be carried out in a stainless steel vessel (1 m diameter) fitted with an anchor agitator of diameter equal to 0.9 m which is rotated at 100 RPM. The tank which is filled up to 0.8 m depth is fitted with a helical coil (helix diameter 0.8 m) made of 25 mm od and 22 mm id copper tube (total external heat transfer area of 2 m²). Hot water at a mean temperature of 45°C (assumed to be approximately constant) is fed to the coil at a rate of 30 kg/min.

The thermal conductivity, heat capacity and density of the polymer solution can be taken as the same as for water. The values of the power-law constants are: $n = 0.36$ and $m = 26 - 0.0566\,T\,(Pa \cdot s^n)$ in the range $288 \leq T \leq 323$ K. Estimate the overall heat transfer coefficient and the time needed to heat one batch of liquid.

Solution

The inside film coefficient will be evaluated first using equations (8.20) and (8.22):
For water, $k = 0.59$ W/mK; $\mu = 0.85$ mPa·s; $C_p = 4180$ J/kgK.

$$\text{mean velocity of water, } V = \frac{30}{60 \times 1000} \times \frac{4}{\pi(0.022)^2} = 1.315 \text{ m/s.}$$

$$\therefore \text{ the Reynolds number, Re} = \frac{\rho V d}{\mu} = \frac{1000 \times 1.315 \times 0.022}{0.85 \times 10^{-3}}$$

$$= 34\,035$$

$$\text{Prandtl number, Pr} = \frac{C_p \mu}{k} = \frac{4180 \times 0.85 \times 10^{-3}}{0.59} = 6$$

From equation (8.20),

$$\text{Nu} = 0.023 \text{Re}^{0.8} \text{Pr}^{0.4} = 0.023(34\,035)^{0.8}(6)^{0.4}$$

$$\text{Nu} = 200$$

$$\text{or } h = \frac{200 \times 0.59}{0.022} = 5360 \text{ W/m}^2\text{K}$$

This value is for straight tubes and the corresponding value for the coil is given by equation (8.22):

$$h_i = 5360 \left(1 + 3.5 \times \frac{0.022}{0.80}\right) = 5876 \text{ W/m}^2\text{K. (based on the inside tube area)}$$

$$\text{or } 5170 \text{ Wm}^2\text{K (based on the outer area).}$$

Now, the outside coefficient, h_0, will be evaluated using equation (8.26).
For $D = 0.9$ m and $D_T = 1$ m, the value of k_s from equation (8.27):

$$k_s = \left[9.5 + \frac{9}{1 - \left(\frac{0.9}{1}\right)^2}\right] \left[\frac{3 \times 0.36 + 1}{4 \times 0.36}\right]^{0.36/(0.36-1)} = 46.2$$

$\therefore$ the effective shear rate, $\dot{\gamma}_{\text{eff}} = k_s N = 46.2 \times (100)/(60) = 77 \text{ s}^{-1}$
The mean temperature of the liquid is $(18 + 27)/2 = 22.5°$C.
Therefore, $m = 26 - 0.0566(273 + 22.5) = 9.27$ Pa·s^n

$$\therefore \mu_{\text{eff},b} = 9.27(77)^{0.36-1} = 0.575 \text{ Pa·s}$$

$$\text{Re} = \frac{\rho N d^2}{\mu_{\text{eff}}} = \frac{1000 \times \dfrac{100}{60} \times 0.9^2}{0.575} = 2346$$

This value is within the range of applicability of equation (8.26).

$$\text{Pr} = \frac{C_p \mu_{\text{eff}}}{k} = \frac{4180 \times 0.575}{0.59} = 4074$$

Neglecting the thermal resistance of the coil, the outside coil surface would be at a mean temperature of 45°C, at which

$$m = 26 - 0.0566(273 + 45) = 8 \, \text{Pa·s}^n$$

$$\therefore \mu_{\text{eff},w} = 8(77)^{0.36-1} = 0.5 \, \text{Pa·s}$$

Now substituting values in equation (8.26):

$$\text{Nu} = \frac{h_o \times 1}{0.59} = 0.077(2346)^{2/3}(4074)^{1/3} \left(\frac{0.575}{0.5}\right)^{0.14}$$

$$\times \left(\frac{1}{0.25}\right)^{0.48} \left(\frac{1}{0.8}\right)^{0.27}$$

or $\qquad h_o = 816 \, \text{W/m}^2\text{K}.$

Neglecting the thermal resistance of the coil wall, the overall heat transfer coefficient based on the external area of the coil, U is given by:

$$U = \frac{816 \times 5170}{816 + 5170} = 705 \, \text{W/m}^2\text{K}.$$

The mass of liquid in the tank $= \rho \dfrac{\pi D_T^2}{4} \cdot L$

$$= \frac{1000 \times \pi \times 1^2 \times 0.8}{4} = 628 \, \text{kg}.$$

Let the temperature of the solution be T_f at any time t, a heat balance on the solution gives:

$$mC_p \frac{dT_f}{dt} = UA(T_s - T_f)$$

where T_s is the mean surface temperature of the coil, assumed here to be 45°C. Substituting values and integrating for $T_f = 18°\text{C}$ to $T_f = 27°\text{C}$:

$$t = \frac{mC_p}{UA} \ln \frac{T_s - 18}{T_s - 27} = \frac{628 \times 4180}{705 \times 2} \ln \frac{45 - 18}{45 - 27}$$

$$= 755 \, \text{s} \quad \text{i.e. } 12.6 \, \text{minutes}$$

Heat losses to the surroundings have been neglected. □

8.5 Mixing equipment and its selection

The wide range of mixing equipment available commercially reflects the enormous variety of mixing duties encountered in the processing industries. It is reasonable to expect therefore that no single item of mixing equipment will be suitable for carrying out such a range of duties. Few manufacturers have taken into account non-Newtonian characteristics of the fluid, but in general those mixers which are suitable for high viscosity Newtonian fluids are also likely to be appropriate for shear-thinning fluids. This normally means the mixer should have small clearances between the moving and fixed parts, and be designed to operate at low speeds. This has led to the development of a number of distinct types of mixers over the years. Very little has been done, however, by way of standardisation of equipment, and no design codes are available. The choice of a mixer type and its design is therefore primarily governed by experience. In the following sections, the main mechanical features of commonly used types of equipment, together with their range of applications, are described briefly. Detailed descriptions of design and selection of various types of mixers have been presented by Oldshue [1983], Harnby *et al.* [1992] and Tatterson [1994]. Most equipment manufacturers also provide performance charts of their products and offer some guidelines for the selection of most appropriate configuration for a specific application.

8.5.1 Mechanical agitation

This is perhaps the most commonly used method of mixing liquids, and essentially there are three elements in such devices.

(i) Vessels

These are often vertically-mounted cylindrical tanks, up to 10 m in diameter, and height to diameter ratios of at least 1.5, and typically filled to a depth equal to about one diameter. In some gas–liquid applications, tall vessels are used and the liquid depth is then up to three tank diameters; multiple impellers fitted on a single shaft are frequently used, e.g. see Kuboi and Nienow [1982]. The base of the tanks may be flat, dished, conical, or specially contoured, depending upon factors such as ease of emptying, or the need to suspend solids, etc.

For the batch mixing of thick pastes and doughs using ribbon impellers and Z- or sigma blade mixers, the tanks may be mounted horizontally. In such units, the working volume of pastes and doughs is often relatively small, and the mixing blades are massive in construction.

(ii) Baffles

To prevent gross vortexing, which is detrimental to mixing, particularly in low viscosity systems, baffles are often fitted to the walls of the vessel. These take the form of thin strips about one-tenth of the tank diameter in width and typically four equi-spaced baffles may be used. In some cases, the baffles are mounted flush with the wall, although more usually a small clearance is left between the wall and the baffle to facilitate fluid motion in the wall region. Baffles are, however, generally not required for high viscosity fluids ($>\sim 5$ Pa·s) because the viscous stresses are sufficiently large to damp out the rotary motion. Some times, the problem of vortexing is circumvented by mounting impellers off-centre or horizontally.

(iii) Impellers

Figure 8.24 shows some of the impellers which are frequently used. Propellers, turbines, paddles, anchors, helical ribbons and screws are usually mounted on a

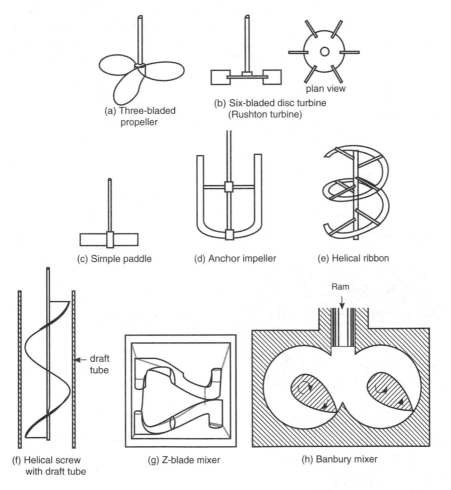

Figure 8.24 *Commonly used impellers*

central vertical shaft in a cylindrical tank, and they are selected for a particular duty, largely on the basis of liquid viscosity. It is generally necessary to move from a propeller to a turbine and then, in order, to a paddle, to an anchor or a gate, and then to a helical ribbon, and finally to a screw, as the viscosity of the liquid increases. The speed of rotation or agitation is reduced as the viscosity increases.

Propellers, turbines and paddles are generally used with relatively low viscosity liquids and operate at high rotational speeds (in the transitional–turbulent region). A typical velocity for the tip of the blades of a turbine is of the order of 3 m/s, with a propeller being a little faster and a paddle little slower. These are classed as remote-clearance impellers having diameters in the range 13 to 67% of the tank diameter. For instance, Figure 8.24b shows a standard six flat-bladed Rushton turbine, and possible variations are shown elsewhere in Figure 8.25. Thus, it is possible to have retreating blades, angled-blades, four- to twenty-bladed, hollow bladed turbines, wide blade hydro-foils

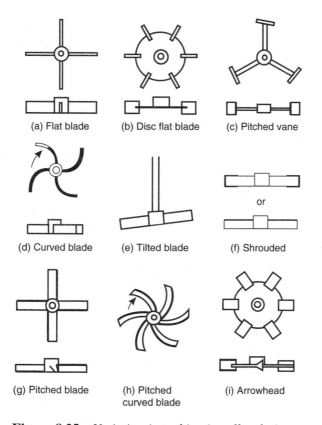

(a) Flat blade   (b) Disc flat blade   (c) Pitched vane

(d) Curved blade   (e) Tilted blade   (f) Shrouded

(g) Pitched blade   (h) Pitched curved blade   (i) Arrowhead

Figure 8.25   *Variation in turbine impeller designs*

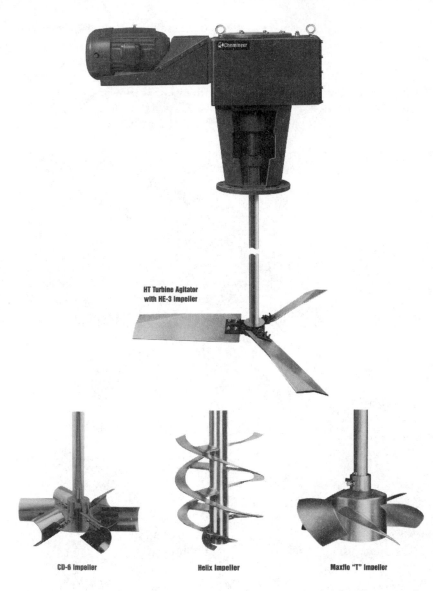

HT Turbine Agitator
with HE-3 Impeller

CD-6 Impeller   Helix Impeller   Maxflo "T" Impeller

Figure 8.26   *Specially designed impellers (a) HE-3 (b) CD-6 (c) Maxflo 'T' impeller (courtesy Chemineer, Inc, Dayton, Ohio)*

(see Figure 8.26) and so on. For dispersion of gases in liquids, turbines and modified turbines are usually employed (Figure 8.20). Commonly two or more disc turbine impellers ($D_T/2$ distance apart) are mounted on the same shaft to ensure mixing over the whole depth of the tank.

Anchors, helical ribbons and screws, are generally used for high viscosity and non-Newtonian liquids. The anchor and ribbon types are arranged with a close clearance at the vessel wall, whereas the helical screw has a smaller diameter and is often used inside a draft tube to promote liquid motion through out the vessel. Helical ribbons or interrupted ribbons are often used in horizontally-mounted cylindrical vessels. A variation of the simple helix mixer is the helicone (Figure 8.27), which has the additional advantage that

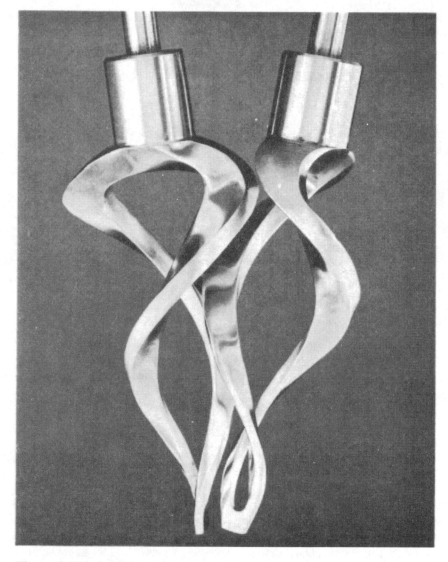

Figure 8.27   *A double helicone impeller*

the clearance between the blade and the vessel wall is easily adjusted by a small axial shift of the impeller. For some applications involving anchor-stirrers the shear stresses generated are not adequate for the breakup and dispersion of agglomerated particles. In such cases, it may be necessary to use an anchor to

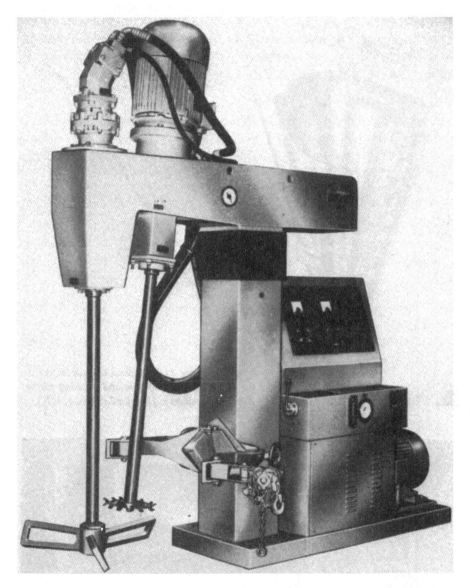

Figure 8.28   *Mastermix HVS/TS high-speed/low-speed impeller combination (Courtesy, Mastermix)*

promote general flow in the vessel in conjunction with a high shear mixing device mounted on a separate eccentric shaft and operating at high speed. A similar arrangement involving a modified paddle and a small high speed dispenser is shown in Figure 8.28.

Kneaders, Z- and sigma-blade (Figure 8.29), and Banbury mixers (sketched in Figure 8.24) are generally used for the mixing of high viscosity liquids, pastes, rubbers, doughs and so on, many of which have non-Newtonian flow characteristics. The tanks are usually mounted horizontally and two impellers are often used. The impellers are massive and the clearances between blades, as well as between the wall and the blade, are very small thereby ensuring that the entire mass of liquid is sheared. While mixing heavy pastes and doughs using a sigma blade mixer, it is usual for the two blades to rotate at different speeds–a ratio of 3 : 2 is common. Various blade profiles of different helical pitch are used. The blade design differs from that of the helical ribbons considered above in that the much higher effective viscosities (of the order of 10^4 Pa·s) require a more solid construction; the blades consequently tend to sweep a greater quantity of fluid in front of them, and the main small-scale mixing process takes place by extrusion between the blade and the wall. Partly for

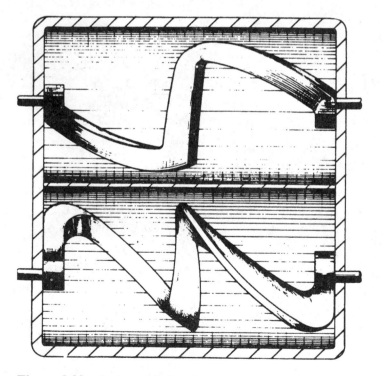

Figure 8.29   *A sigma blade mixer*

this reason, mixers of this type are usually operated only partially full, though the Banbury mixer (Fig. 8.24) used in the rubber industry is filled completely and pressurised as well. The pitch of the blades produces the necessary motion along the channel, and this gives the larger scale blending that is required in order to limit batch blending times to reasonable periods.

The motion of the material in the sigma blade mixer can be considered in three stages, as illustrated in Fig. 8.30. Material builds up in front of the blade in region A where it will undergo deformation and flow – the relative extent of these processes depending on the material properties. Materials will tend to be rolled and deformed until some is trapped in the gap B.

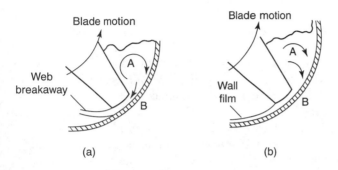

Figure 8.30   *Fluid motion in a sigma blade mixer (a) non-wetting paste (b) surface-wetting fluid*

The difference between the solid and liquid behaviour will be evident in the regions where the shear stresses are least, i.e. less than the yield stress. Whether or not there is a radially inward flow in front of the blade will depend on the magnitude of this yield stress. Unfortunately, the dynamics of this region are so complex that it is not possible to quantify this generalisation.

Once the material has entered region B it will be subject to an unsteady-state developing-flow situation; in the full-developed form the velocity gradient is linear, changing uniformly from the blade velocity at the blade surface to zero at the vessel wall.

The general situation is the same as that existing between moving planes; normally the radius of the vessel is large and the shear stress is sensibly uniform throughout the gap. Under these circumstances full developed flow will be established providing that the *channel* length (i.e. the thickness of the edge of the blade facing the wall) is of the order of two or three times the clearance between the blade and the wall. For this situation the mean velocity of the material in the gap is just half the velocity of the blade relative to the wall. A solid material may well not achieve this condition, however, since the rate at which it is drawn into the gap is largely controlled by the deformation in zone A.

On discharge from the gap the material will either break away from the wall (as in *a*) or remain adhering to it (as in *b*). In the former case, a web may be formed which will eventually break off to be incorporated with the material coming in front of the blade on its next revolution. If the extruded material remains adhering to the wall, the amount of material left on the wall is determined from mass balance considerations, at a thickness equal to about half the clearance.

8.5.2 Rolling operations

Several blending operations involving highly viscous or non-Newtonian fluids have to be carried out in a flow regime which is essentially laminar. Although mixing is less efficient in a laminar flow system than under turbulent conditions, the natural extension of elements of fluid in a shear field reduces the effective path length over which the final dispersive stage of diffusion has to take place. With materials possessing effective viscosities of the order of $100\,\mathrm{Nsm}^{-2}$, one of the more usual ways of imposing a suitable velocity gradient for a significant period is to feed the material through the gap between two counter-rotating rollers. This operation is often termed *calendering*. In principle, the rolls need not be rotating at the same speed, nor even need they be the same diameter. The simplest case to consider, however, is that where the diameters and speeds are the same. The difficulties in the analysis lie in the fact that it is not possible to solve unambiguously for the position at which the breakaway from the roll surface will occur.

Figure 8.31 shows the situation in the "nip" region between two rolls.

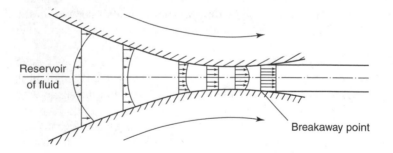

Figure 8.31   *Flow between rolls*

Rolling machines are commonly used for batch blending very viscous materials, and the method of operation is shown in Fig. 8.32. The preferential adherence of the extruded film to one of the rollers can be obtained with a small speed differential. The viscosity of the fluid is high enough to prevent centrifugal forces throwing it clear of the roller, and it returns to the feed

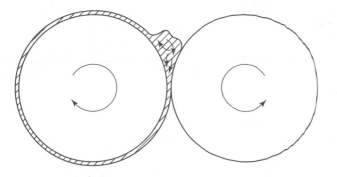

Figure 8.32   *Mixing with a rolling machine*

side of the machine where it is accumulated in the highly sheared circulating fillet of excess material. This flow is, of course, simply circumferential around the roller and through the gap, and it is necessary to ensure adequate lateral mixing along the length of the machine by separate mechanical or manual means from time to time.

In addition to the basic impeller designs shown in Figures 8.24 and 8.25, many speciality impellers developed by the manufacturers of mixing equipment are also available which supposedly give better performance than the basic configurations. For instance, the so called HE-3 high efficiency impeller shown in Figure 8.26 requires a much smaller motor than the standard turbine to achieve comparable fluid velocities in the vessel. Similarly, the concave blade disk impeller (CD-6), also shown in Figure 8.26, results in up to 100% larger values of liquid-phase mass transfer coefficients in gas–liquid systems than those obtained with the six flat bladed turbine impeller. Both these designs have been developed and/or marketed by Chemineer, Inc, Dayton, Ohio. Obviously, it is not possible to provide a complete list of designs offered by different manufacturers but it is always desirable to be aware of such developments prior to selecting equipment for a given application. Some guidelines for equipment design and selection are also available in literature [Bakker *et al.*, 1994].

8.5.2 Portable mixers

For a wide range of applications, a portable mixer which can be clamped on the top or side of the vessel is often used. This is commonly fitted with two propeller blades so that the bottom rotor forces the liquid upwards and the top rotor forces the liquid downwards. The power supplied is up to about 2 kW, though the size of the motor becomes too great at higher powers. To avoid excessive strain on the armature, some form of flexible coupling should be fitted between the motor and the main propeller shaft. Units of this kind are usually driven at a fairly high speed (15 Hz) and, for use with high viscosity and non-Newtonian materials, a reduction gear can be fairly easily fitted to the unit for low speed operation, although this increases the mass of the unit.

8.6 Mixing in continuous systems

The mixing problems considered so far have related to batch systems in which two or more materials are mixed together and uniformity is maintained by continued operation of the agitator. Consideration will now be given to some of the equipment used for continuous mixing duties.

8.6.1 Extruders

Mixing duties in the plastics industry (and to a lesser extent in food industry) are often carried out in either single or twin screw extruders. The feed to such units usually contains the base polymer in either granular or powder form, together with additives such as stabilisers, pigments, plasticisers, fire retardants, and so on. During processing in the extruder, the polymer is melted and the additives thoroughly mixed. The extrudate is delivered at high pressure and at a controlled rate from the extruder for shaping by means of either a die or a mould. Considerable progress has been made in the design of extruders in recent years, particularly by the application of finite-element methods. One of the problems is that an enormous amount of heat is generated (by viscous action), and the fluid properties may change by several orders of magnitude as a result of temperature changes. It is therefore always essential to solve the coupled equations of flow and heat transfer.

In the typical single-screw shown in Figure 8.33, the shearing which occurs in the helical channel between the barrel and the screw is not intense, and therefore this equipment does not give good mixing. Twin-screw extruders, shown in Figure 8.34, may be co- or counter-rotating and, as there are regions where the rotors are in close proximity, extremely high shear stresses may

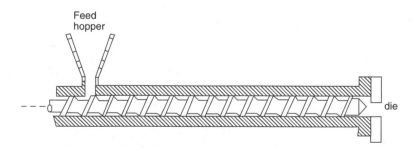

Figure 8.33   *Single-screw extruder*

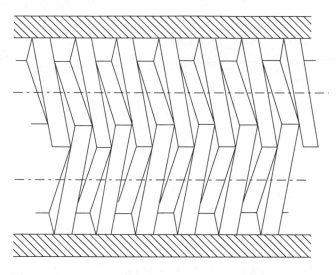

Figure 8.34 *Counter-rotating twin-screw extruder*

be generated. Clearly, twin-screw units can yield a product of better mixture quality than a single-screw unit. Detailed accounts of the design and performance of extruders are available in the literature [Schenkel, 1966; Janssen, 1978; Rauendaal, 1992].

8.6.2 Static mixers

All the mixers so far described have been of the dynamic type in the sense that moving blades are used to impart motion to the fluid and produce the mixing effect. In static mixers, sometimes called 'motionless' or in-line mixers, the materials to be mixed are pumped through a pipe containing a series of specially shaped stationary blades. Static mixers can be used with liquids of a wide range of viscosities in either the laminar or turbulent regimes, but their features are perhaps best appreciated in relation to laminar-flow mixing which is the normal condition for high-viscosity and non-Newtonian fluids.

Figure 8.35 shows a particular type of static mixer in which a series of stationary helical blades mounted in a circular pipe is used to divide, split, and twist the flowing streams (Figure 8.36). In streamline flow, the material divides at the leading edge of each of these elements and follows the channels of complex shape created by the element. At each succeeding element, the two channels are further divided, and mixing proceeds by a distributive process similar to the cutting and folding mechanism shown schematically in Figure 8.4. In principle, if each element divided the streams neatly into two, feeding two dissimilar streams to the mixer would give a striated material in which the thickness of each striation would be of the order $D_t/2^n$ where D_t

Figure 8.35   *Twisted-blade type of static mixer elements*

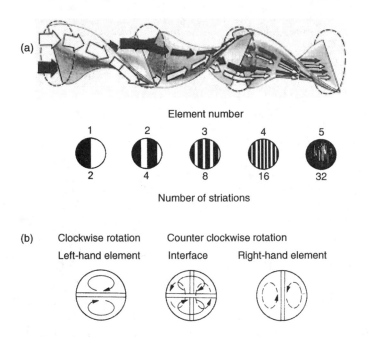

Figure 8.36   *Twisted-blade type of static mixer operating in the laminar flow regime: (a) Distributive mixing mechanism showing, in principle, the reduction in striation thickness produced. (b) Radial mixing contribution from laminar shear mechanism*

is the diameter of the tube and n is the number of elements. However, the helical elements shown in Figure 8.35 also induce further mixing by a laminar shear mechanism (illustrated in Figures 8.1 and 8.2). This, combined with the twisting action of the element, helps to promote radial mixing which is important in eliminating any radial gradients of composition, velocity and possibly temperature that might exist in the material. Figure 8.37 shows how these mixing mechanisms together produce after only 10–12 elements, a well-mixed material.

Figure 8.38 shows a Sulzer type SMX static mixer where the mixing element consists of a lattice of intermeshing and interconnecting bars contained

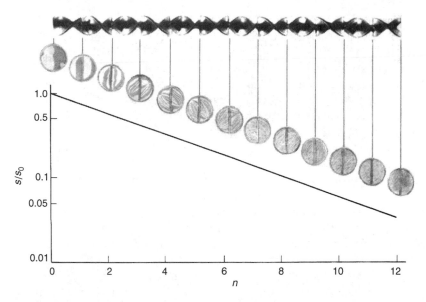

Figure 8.37 *Static mixer in laminar flow: reduction in relative standard deviation of samples indicating improvement in mixture quality with increasing number n of elements traversed*

Figure 8.38 *Static mixer for viscous Newtonian and non-Newtonian materials*

in a 80 mm diameter pipe. It is recommended for viscous Newtonian and non-Newtonian materials in laminar flow. The mixer shown is used in food processing, for example for mixing fresh cheese with whipped cream, and in polymer processing, as in dispersing TiO_2 into polymer melts for de-lustering.

Quantitatively, a variety of methods [Heywood *et al.*, 1984] has been proposed to describe the degree or quality of mixing produced in a static mixer. One such measure of mixing quality is the relative standard deviation s/s_0, where s is the standard deviation in composition of a set of samples taken at a certain stage of the mixing operation, and s_0 is the standard deviation for a similar set of samples taken at the mixer inlet. Figure 8.37 shows schematically how the relative standard deviation decreases as the number of elements, n, through which the material has passed increases, a perfectly mixed product having a zero relative standard deviation. One of the problems in using relative standard deviation or a similar index as a measure of mixing is that this depends on sample size which therefore must be taken into account in any assessment.

One of the most important considerations in comparing or selecting static mixers is the power consumed by the mixer to achieve a desired mixture quality. The pressure-drop characteristics of a mixer are most conveniently described by the ratio of mixer pressure drop to empty tube pressure drop for the same flow rate and diameter. Thus, different static mixers designs can be compared on a basis of mixing quality, pressure-drop ratio, initial cost and convenience of installation. It should be noted that pressure-drop ratio may be dependent on the rheology of the material, which may well change significantly during the course of its passage through the static mixer. Great care should therefore be taken to ensure that the material used to determine the pressure-drop criterion has a rheology which closely matches that of the material to be used in the static mixer.

Static mixers are widely used for highly viscous and non-Newtonian fluids in processes in which polymers, fibres and plastics of all kinds are manufactured, and where the material is often hot and at high pressures. However, static mixers have also achieved widespread use for mixing and blending of low viscosity liquids, liquid-liquid dispersions, and even gas–liquid dispersions. In some cases, the designs used for high viscosity liquids have also proved effective in the turbulent mixing regime for low viscosity fluids. In other cases, equipment manufacturers have developed special designs for turbulent mixing, and a wide variety of static mixer devices in now available.

8.7 Further reading

Harnby, N., Edwards, M.F. and Nienow, A.W., *Mixing in the Process Industries.* 2nd edn. Butterworth-Heinemann, Oxford (1992).

McDonough, R.J., *Mixing for the Process Industries*, Van Nostrand Reinhold, New York (1992).

Nagata, S., *Mixing: Principles and Applications*, Wiley, New York (1975).

Oldshue, J.Y., *Fluid Mixing Technology.* McGraw Hill, New York (1983).

Ottino, J.M., *The Kinematics of Mixing.* Cambridge University Press, London (1990).

Tatterson, G.B., *Fluid Mixing and Gas Dispersion in Agitated Tanks.* McGraw Hill, New York (1991).

Tatterson, G.B., *Scaleup and Design of Industrial Mixing Processes.* McGraw Hill, New York (1994).

Ulbrecht, J.J. and Patterson, G.K. (editors), *Mixing of Liquids by Mechanical Agitation.* Gordon and Breach, New York (1985).

8.8 References

Abid, M., Xuereb C. and Bertrand, J., *Chem. Eng. Res. Des.* **70** (1992) 377.

Ayazi Shamlou, P. and Edwards, M.F., *Chem. Eng. Sci.* **41** (1986) 1957.

Bakker, A. and Gates, L.E., *Chem. Eng. Prog.* **91** (Dec, 1995) 25.

Bakker, A., Morton J.R. and Berg, G.M., *Chem. Eng.* **101** (Mar, 1994) 120.

Bates, R.L., Fondy P.L. and Corpstein, R.R., *Ind. Eng. Chem. Proc. Des. Dev.* **2** (1963) 310.

Beckner, J.L. and Smith, J.M., *Trans. Inst. Chem. Eng.* **44** (1966) 224.

Bourne, J.R. and Butler, H., *Trans Inst. Chem. Eng.* **47** (1969) 11.

Calderbank, P.H. and Moo-Young, M., *Trans. Inst. Chem. Eng.* **37** (1959) 26.

Carreau, P.J., Chhabra R.P. and Cheng, J., *AIChEJ.* **39** (1993) 1421.

Carreau, P.J., Paris J. and Guerin, P., *Can. J. Chem. Eng.* **70** (1992) 1071.

Carreau, P.J., Paris J. and Guerin, P., *Can. J. Chem. Eng.* **72** (1994) 966.

Carreau, P.J., Patterson I. and Yap, C.Y., *Can. J. Chem. Eng.* **54** (1976) 135.

Chapman, F.S. and Holland, F.A., *Trans. Inst. Chem. Eng.* **43** (1965) 131.

Chavan, V.V., Arumugam M. and Ulbrecht, J., *AIChEJ.* **21** (1975) 613. Also see *Can. J. Chem. Eng.* **53** (1975) 62.

Chavan, V.V. and Mashelkar, R.A., *Adv. Transp. Process.* **1** (1980) 210.

Chavan, V.V. and Ulbrecht, J.J., *Trans. Inst. Chem. Eng.* **50** (1972) 147.

Chavan, V.V. and Ulbrecht, J.J., *Ind. Eng. Chem. Proc. Des. Dev.* **12** (1973) 472. Corrigenda ibid **13** (1974) 309.

Cheng, D.C.-H., Schofield C. and Jane, R.J., *Proc. First Eng. Conf. on Mixing & Centrifugal Sep., BHRA Fluid Eng.* Paper #C2–15 (1974).

Cheng, J. and Carreau, P.J., *Chem. Eng. Sci.* **49** (1994) 1965.

Collias, D.J. and Prud'homme, R.K., *Chem. Eng. Sci.* **40** (1985) 1495.

Coulson, J.M. and Richardson, J.F., *Chemical Engineering.* Vol 1, 6th edn. Butterworth-Heinemann, Oxford (1999).

Coyle, C.K., Hirschland, H.E., Michel B.J. and Oldshue, J.Y., *Can. J. Chem. Eng.* **48** (1970) 275.

Desplanches, H., Llinas, J.R. and Chevalier, J.L., *Can. J. Chem. Eng.* **58** (1980) 160.

Doraiswamy, D., Grenville R.K. and Etchells III, A.W., *Ind. Eng. Chem. Res.* **33** (1994) 2253.

Dream, R.F., Chem, Engg. **106** (January 1999) 90.

Ducla, J.M., Desplanches H. and Chevalier, J.J., *Chem. Eng. Commun.* **21** (1983) 29.

Edney, H.G.S. and Edwards, M.F., *Trans. Inst. Chem. Eng.* **54** (1976) 160.

Edwards, M.F., Godfrey J.C. and Kashim, M.M., *J. Non-Newt. Fluid Mech.* **1** (1976) 309.

Edwards, M.F. and Wilkinson, W.L., *The Chem. Engr.* No. **257** (1972) 310 & 328.

Elson, T.P., Cheesman D.J. and Nienow, A.W., *Chem. Eng. Sci.* **41** (1986) 2555.

Fasano, J.B., Bakker A. and Penney, W.R., *Chem. Eng.* **101** (Aug, 1994) 110.

Galindo, E. and Nienow, A.W., *Chem. Eng. Technol.* **16** (1993) 102.

Gates, L.E., Hicks R.W. and Dickey, D.S., *Chem. Eng.* **83** (Dec 6, 1976) 165.

Giesekus, H., *Rheol. Acta.* **4** (1965) 85.

Gluz, M. and Pavlushenko, I.S., *J. App. Chem. (USSR)*, **39** (1966) 2223.

Godfrey, J.C., in *Mixing in the Process Industries* (edited by Harnby, N. Edwards M.F. and Nienow, A.W., Butterworth-Heinemann, Oxford), 2nd edn. (1992) 185.

Godleski, E.S. and Smith, J.C., *AIChEJ.* **8** (1962) 617.

Grenville, R.K., *et al.* Paper presented at Mixing XV, 15th Biennial North American Mixing Conf., Banff, Alberta, Canada June 18–23, 1995.

Guerin, P., Carreau, P.J., Patterson I. and Paris, J., *Can. J. Chem. Eng.* **62** (1984) 301.

Hagedorn, D. and Salamone, J.J., *Ind. Eng. Chem. Proc. Des. Dev.* **6** (1967) 469.

Hall, K.R. and Godfrey, J.C., *Trans. Inst. Chem. Eng.* **46** (1968) 205.

Harnby, N., Edwards M.F. and Nienow A.W., (editors), *Mixing in the Process Industries.* 2nd edn., Butterworth-Heinemann, Oxford (1992).

Heim, A., *Intl. Chem. Eng.* **20** (1980) 271.
Herbst, H., Schumpe A. and Deckwer, W.-D., *Chem. Eng. Technol.* **15** (1992) 425.
Heywood, N.I., Viney L.J. and Stewart, I.W., *I. Chem. E. Sym. Ser. No. 89 Fluid Mixing II,* (1984) 147.
Hickman, D., *Proc. 6th European Conf. Mixing, Pavia (Italy), BHRA Fluid Eng.,* Cranfield (1988) 369.
Hicks, R.W., Morton J.R. and Fenic, J.G., *Chem. Eng.* **83** (Apr 26, 1976) 102.
Ibrahim, S. and Nienow, A.W. *Trans. Inst. Chem. Eng.* **73A** (1995) 485.
Janssen, L.P.B.M., Twin Screw Extrusion, Elsevier, Amsterdam (1978).
Jeschke, D., *Z. Ver. Deut. Ing.* **69** (1925) 1526.
Johma, A.I. and Edwards, M.F., *Chem. Eng. Sci.* **45** (1990) 1389.
Kai, W. and Shengyao, Y. *Chem. Eng. Sci.* **44** (1989) 33.
Kappel, M., *Int. Chem. Eng.* **19** (1979) 571.
Kawase, Y. and Moo-Young, M., *Chem. Eng. Res. Des.* **66** (1988) 284.
Kelkar, J.V., Mashelkar R.A. and Ulbrecht, J., *Trans. Inst. Chem. Eng.* **50** (1972) 343.
Kelkar, J.V., Mashelkar R.A. and Ulbrecht, J., *Chem. Eng. Sci.* **17** (1973) 3069.
Kuboi, R. and Nienow, A.W., *Proc. Fourth European Conf. Mixing, BHRA Fluid Eng.,* Cranfield (1982) 247.
Kuriyama, M., Arai K. and Saito, S., *J. Chem. Eng. Japan.* **16** (1983) 489.
Lindlay, J.A., *J. Agr. Eng. Res.* **49** (1991) 1.
Machon, V., Vlcek, J. Nienow, A.W. and Solomon, J., *Chem. Eng. J.* **14** (1980) 67.
Manna, L., *Chem. Eng. Jl.* **67** (1997) 167.
Martone, J.A. and Sandall, O.C., *Ind. Eng. Chem. Proc. Des. Dev.* **10** (1971) 86.
Metzner, A.B., Feehs, R.H., Romos, H.L., Otto R.E. and Tuthill, J.P., *AIChEJ.* **7** (1961) 3.
Metzner, A.B. and Otto, R.E., *AIChEJ.* **3** (1957) 3.
Metzner, A.B. and Taylor, J.S., *AIChEJ.* **6** (1960) 109.
Mitsuishi, N. and Hirai, N.J., *J. Chem. Eng. Japan.* **2** (1969) 217.
Nagata, S., *Mixing; Principles and Applications.* Wiley, New York (1975).
Nagata, S., Yanagimoto, M. and Yokoyama, T., *Memoirs of Fac. of Eng.,* Kyoto University (Japan), **18** (1956) 444.
Nagata, S., Nishikawa, M., Tada, H., Hirabayashi H. and Gotoh, S., *J. Chem. Eng. Japan.* **3** (1970) 237.
Nagata, S., Nishikawa, M., Kayama T. and Nakajima, M., *J. Chem. Eng. Japan.* **5** (1972) 187.
Nienow, A.W. and Ulbrecht, J., in *Mixing of Liquids by Mechanical Agitation.* (edited by Ulbrecht J.J. and Patterson, G.K., Gordon & Breach, New York) (1985) Chapter 6.
Nienow, A.W., Wisdom, D.J., Solomon, J., Machon V. and Vlacek, J., *Chem. Eng. Commun.* **19** (1983) 273.
Nishikawa, M., Nakamura, M., Yagi H. and Hashimoto, K., *J. Chem. Eng. Japan.* **14** (1981) 219 & 227.
Norwood K.W. and Metzner, A.B., *AIChEJ.* **6** (1960) 432.
Nouri, J.M. and Hockey, R.M., *J. Chem. Eng. Jpn.* **31** (1998) 848.
Oldshue, J.Y., *Fluid Mixing Technology,* McGraw Hill, New York (1983).
Oliver, D.R., Nienow, A.W., Mitson R.J. and Terry, K., *Chem. Eng. Res. Des.* **62** (1984) 123.
Ottino, J.M., *Sci. Amer.* **260** (1989) 56.
Özcan-Taskin, N.G. and Nienow, A.W., *Trans. Inst. Chem. Eng.* **73C** (1995) 49.
Perez, J.F. and Sandall, O.C., *AlChEJ.* **20** (1974) 770.
Peters, D.C. and Smith, J.M., *Trans. Inst. Chem. Eng.* **45** (1967) 360.
Poggermann, R., Steiff, A. and Weinspach, P.M., *Ger. Chem. Eng.* **3** (1980) 163.
Pollard, J. and Kantyka, T.A., *Trans. Inst. Chem. Eng.* **47** (1969) 21.
Prud'homme, R.K. and Shaqfeh, E., *AlChEJ.* **30** (1984) 485.
Ranade, V.R. and Ulbrecht, J., *AlChEJ.* **24** (1978) 796.
Rauendaal, R.C., *Mixing in Polymer Processing.* Hanser, Munich (1992).

Sandall, O.C. and Patel, K.G., *Ind. Eng. Chem. Proc. Des. Dev.* **9** (1970) 139.

Schenkel, G., *Plastics Extrusion Technology.* Cliffe Books, London (1966).

Skelland, A.H.P., *Non-Newtonian Flow and Heat Transfer.* Wiley, New York (1967).

Skelland, A.H.P., in *Handbook of Fluids in Motion.* (edited by Cheremisinoff N.P. and Gupta, R., Ann Arbor Sci., Ann Arbor) (1983) Chapter 7.

Smith, J.M., in *Mixing of Liquids by Mechanical Agitation.* (edited by Ulbrecht J.J. and Patterson, G.K., Gordon and Breach, New York) (1985) Chapter 5.

Solomon, J., Nienow A.W. and Pace, G.W., *Fluid Mixing I, I. Chem. E. Sym. Ser.* **#64** (1981) Paper No. A1.

Takahashi, K., in *Encyclopedia of Fluid Mechanics.* (edited by Cheremisinoff, N.P. Gulf, Houston) **7** (1988) 869.

Tanguy, P., Lacorix, A., Bertrand, F., Choplin L. and DeLa Fuente, E.B., *AlChEJ.* **38** (1992) 939.

Tatterson, G.B., *Fluid Mixing and Gas Dispersion in Agitated Tanks.* McGraw Hill, New York (1991).

Tatterson, G.B., *Scaleup and Design of Industrial Mixing Processes.* McGraw Hill, New York (1994).

Ulbrecht, J.J. and Carreau, P.J., in *Mixing of Liquids by Mechanical Agitation.* (edited by Ulbrecht, J.J. and Patterson, G.K., Gordon and Breach, New York) (1985) Chapter 4.

van den Bergh, W., *Chem. Eng.* **101** (Dec. 1994) 70.

van der Molen, K. and van Maanen, H.R.E., *Chem. Eng. Sci.* **33** (1978) 1161.

van't Reit, K., *PhD Thesis*, Delft. The Netherlands (1975).

Wichterle, K. and Wein, O., *Int. Chem. Eng.* **21** (1981) 116.

Yagi, H. and Yoshida, F., *Ind. Eng. Chem. Proc. Des. Dev.* **14** (1975) 488.

Yap, C.Y., Patterson W.I. and Carreau, P.J., *AlChEJ.* **25** (1979) 516.

8.9 Nomenclature

		Dimensions in M, N, L, T, θ
A	area for heat transfer (m^2)	$\mathbf{L^2}$
B_i	Bingham number	
C_i	tracer concentration recorded by ith detector (kmol/m^3)	$\mathbf{NL^{-3}}$
C_∞	equilibrium concentration of tracer (kmol/m^3)	$\mathbf{NL^{-3}}$
C_p	heat capacity (J/kg K)	$\mathbf{L^2 T^{-2} \theta^{-1}}$
D	impeller diameter (m)	$\mathbf{L}$
D_c	size of cavity in a shear-thinning fluid, eq. (8.10) (m)	$\mathbf{L}$
D_L	molecular diffusivity (m^2/s)	$\mathbf{L^2 T^{-1}}$
D_T	tank diameter (m)	$\mathbf{L}$
d_c	mean helix diameter (m)	$\mathbf{L}$
d_0	coil tube diameter (m)	$\mathbf{L}$
Fr	Froude number	$-$
Gr	Grashof number	$-$
g	acceleration due to gravity (m/s^2)	$\mathbf{LT^{-2}}$
h	heat transfer coefficient (W/m^{2}K)	$\mathbf{MT^{-3}\theta^{-1}}$
k	thermal conductivity (W/m K)	$\mathbf{MLT^{-3}\theta^{-1}}$
k_s	constant, equation (8.8)	$-$
m	power-law consistency coefficient (Pa·s^n)	$\mathbf{ML^{-1}T^{n-2}}$
N	speed of rotation of agitation (s^{-1})	$\mathbf{T^{-1}}$
N_q	pumping number	$-$
Nu	Nusselt number	$-$

Dimensions in
M, N, L, T, θ

n	power-law index	–
P	power (W)	$\mathbf{ML^2T^{-3}}$
Po	power number	–
Pr	Prandtl number	–
Q_g	volumetric flow rate of gas (m³/s)	$\mathbf{L^3T^{-1}}$
r	radial distance from centre of tank (m)	$\mathbf{L}$
R	thermal resistance due to scale formation (m²K/W)	$\mathbf{M^{-1}\ T^3\theta}$
Re	Reynolds number	–
S_A	scale of agitation	–
t_r	residence time (s)	$\mathbf{T}$
T	temperature (K)	θ
$D_{T_{\text{eff}}}$	effective diameter of vessel $(4V_l/\pi)^{1/3}$ (m)	$\mathbf{L}$
ΔT	temperature difference (K)	θ
t_m	mixing time (s)	$\mathbf{T}$
U	overall heat transfer coefficient (W/m²K)	$\mathbf{MT^{-3}\theta^{-1}}$
V	volume of gas–liquid dispersion in the vessel (m³)	$\mathbf{L^3}$
V_g	gas superficial velocity (m/s)	$\mathbf{LT^{-1}}$
V_l	volume of the liquid batch (m³)	$\mathbf{L^3}$
V_t	terminal rise velocity of a gas bubble (m/s)	$\mathbf{LT^{-1}}$
W_B	width of baffle (m)	$\mathbf{L}$
We	Weber number	–
x_w	wall thickness (m)	$\mathbf{L}$

Greek letters

α	thermal diffusivity (m²/s)	$\mathbf{L^2T^{-1}}$
β	coefficient of thermal expansion (K^{-1})	θ^{-1}
$\dot{\gamma}_{\text{avg}}$	average shear rate, equation (8.8) s^{-1}	$\mathbf{T^{-1}}$
μ	Newtonian viscosity (Pa·s)	$\mathbf{ML^{-1}T^{-1}}$
μ_{eff}	effective shear viscosity evaluated at $\dot{\gamma} = \dot{\gamma}_{\text{avg}}$ (Pa·s)	$\mathbf{ML^{-1}T^{-1}}$
ϕ	gas hold-up	–
θ_m	dimensionless mixing time	
ρ	liquid density (kg/m³)	$\mathbf{ML^{-3}}$
σ	surface tension (N/m)	$\mathbf{MT^{-2}}$
τ_0^B	Bingham yield stress (Pa)	$\mathbf{ML^{-1}T^{-2}}$

Subscripts

b	bulk
cr	critical (to indicate laminar – turbulent transition)
g	with gas present in vessel
i	inside
o	outside
t	fully turbulent
w	wall

Problems

The level of difficulty of problems has been graded: (a): straightforward, (b): somewhat more complex, and (c): most difficult. In any given Chapter the readers are recommended to tackle problems in increasing order of difficulty.

LEVEL

1.1 The following rheological data have been obtained for a liquid at 295.5 K. (a)

Shear rate (s^{-1})	Shear stress (Pa)	Shear rate (s^{-1})	Shear stress (Pa)
2.22	1.32	35.16	20.92
4.43	2.63	44.26	26.33
7.02	4.17	70.15	41.74
8.83	5.25	88.31	52.54
11.12	6.62	111.17	66.15
14	8.33	139.96	83.27
17.62	10.48	176.2	104.84
22.18	13.20	221.82	132.0
27.93	16.62	279.25	166.15

By plotting these data on linear and logarithmic scales, ascertain the type of fluid behaviour, e.g. Newtonian, or shear-thinning, or shear-thickening, etc. Also, if the liquid is taken to have power-law rheology, calculate the consistency and flow-behaviour indices respectively for this liquid.

1.2 The following rheological data have been reported for a 0.6% (by weight) carbopol solution in a 1.5% (by weight) NaOH aqueous solution at 292 K. (a)

$\dot{\gamma}$ (s^{-1})	τ (Pa)	$\dot{\gamma}$ (s^{-1})	τ (Pa)
0.356	1.43	7.12	4.86
0.449	1.54	8.96	5.40
0.564	1.65	11.25	6.03
0.712	1.79	14.17	6.62
0.896	1.97	17.82	7.45
1.13	2.17	22.47	8.40

$\dot\gamma\,(s^{-1})$	$\tau\,(Pa)$	$\dot\gamma\,(s^{-1})$	$\tau\,(Pa)$
1.42	2.44	28.30	9.46
1.78	2.61	35.57	10.41
2.25	2.84	44.89	12.06
2.83	3.27	56.43	13.60
3.56	3.60	71.15	15.14
4.49	3.98	89.55	17.03
5.64	4.38	112.5	19.16

Plot these data in the form of $\tau - \dot\gamma$ and $\mu - \dot\gamma$ on logarithmic coordinates. Evaluate the power-law parameters for this fluid. Does the use of the Ellis fluid (equation 1.15) or of the truncated Carreau fluid (equation 1.14) model offer any improvement over the power-law model in representing these data? What are the mean and maximum % deviations from the data for these three models?

1.3 The following data for shear stress (τ) and first normal stress difference (N_1) have been reported for a 2% (by weight) Separan AP-30 solution in water measured at 289.5 K using a cone and plate rheometer.   (a)

$\dot\gamma\,(s^{-1})$	$\tau\,(Pa)$	$N_1\,(Pa)$	$\dot\gamma\,(s^{-1})$	$\tau\,(Pa)$	$N_1\,(Pa)$
0.004 49	0.26	–	4.49	19.85	57.8
0.005 64	0.33	–	7.12	22.96	72.3
0.007 12	0.42	–	11.25	26.13	90.6
0.008 96	0.53	–	17.83	29.93	125.30
0.0113	0.66	–	28.30	34.44	154.20
0.0142	0.75	–	44.9	40.38	221.70
0.0178	0.96	–	71.2	46.32	318.60
0.0225	1.14	–	112.5	53.44	424.10
0.0356	1.65	–	89.6	49.88	366.30
0.0283	1.39	–	56.4	43.00	269.90
0.0449	1.99	–	35.6	37.5	192.8
0.0564	2.30	–	22.5	32.1	163.8
0.0712	2.85	–	14.2	28	120.5
0.0896	3.33	–	8.96	25.2	96.4
0.113	3.83	4.82	5.64	21.1	84.8
0.178	5.50	7.23	3.56	18.2	62.7
0.283	6.94	11.60	2.25	15.6	43.4
0.449	8.37	19.80	1.42	13.3	38.6
0.712	10.16	22.20	0.896	11	25.1
1.13	12.32	27.95	0.564	9.09	20.3
1.78	14.60	34.70	0.356	7.41	10.6

(i) Plot these shear stress, apparent viscosity and first normal stress difference data against shear rate on log-log scales. Does the

shear stress data extend to the lower Newtonian region? What is
the value of the zero-shear viscosity for this solution?
(ii) Suggest and fit suitable viscosity models covering the entire
range. What is their maximum deviation?
(iii) Calculate and plot the Maxwellian relaxation time (equation 1.25)
as a function of shear rate for this polymer solution.
(iv) Is it possible to characterize the visco-elastic behaviour of this
solution using a single characteristic time (using equation 1.26) in
the higher shear rate region? At what shear rate does it coincide
with the Maxwellian relaxation time calculated in part (iii)?

1.4 The following rheological data for milk chocolate at $313\,\mathrm{K}$ are (a)
available. Determine the Bingham plastic (equation 1.16) and Casson
model (equation 1.18) parameters for this material. What are the mean
and maximum deviations for both these models?

$\dot{\gamma}\,(\mathrm{s}^{-1})$	$\tau\,(\mathrm{Pa})$	$\dot{\gamma}(\mathrm{s}^{-1})$	$\tau\,(\mathrm{Pa})$
0.099	28.6	6.4	123.8
0.14	35.7	7.9	133.3
0.20	42.8	11.5	164.2
0.39	52.4	13.1	178.5
0.79	61.9	15.9	201.1
1.60	71.4	17.9	221.3
2.40	80.9	19.9	236
3.9	100		

It is likely that the model parameters are strongly dependent on the
shear rate range covered by the rheological data. Compare the values
of the model parameters by considering the following shear rate inter-
vals:

a. $\dot{\gamma} \leq 20\,\mathrm{s}^{-1}$
b. $1.6 \leq \dot{\gamma} \leq 20\,\mathrm{s}^{-1}$
c. $\dot{\gamma} \leq 1.6\,\mathrm{s}^{-1}$

1.5 The following shear stress–shear rate data demonstrate the effect of (b)
temperature on the power-law constants for a c ncentrated orange
juice containing 5.7% fruit pulp.

$T = 254.3\,\mathrm{K}$		267.7 K		282.6 K		302.3 K	
$\dot{\gamma}\,(\mathrm{s}^{-1})$	$\tau\,(\mathrm{Pa})$	$\dot{\gamma}\,(\mathrm{s}^{-1})$	$\tau\,(\mathrm{Pa})$	$\dot{\gamma}\,(\mathrm{s}^{-1})$	$\tau\,(\mathrm{Pa})$	$\dot{\gamma}\,(\mathrm{s}^{-1})$	$\tau\,(\mathrm{Pa})$
0.5	14.4	0.6	4.3	1.1	2.6	8	3.6
1	24.3	1	6.5	8	10.3	20	7.6
10	142	10	38.4	15	17	40	13.1
20	240.4	20	65.4	30	29.5	60	17.5
30	327	30	89	60	50.3	120	31.2

$T = 254.3\,K$		267.7 K		282.6 K		302.3 K	
$\dot\gamma\,(s^{-1})$	$\tau\,(Pa)$	$\dot\gamma\,(s^{-1})$	$\tau\,(Pa)$	$\dot\gamma\,(s^{-1})$	$\tau\,(Pa)$	$\dot\gamma\,(s^{-1})$	$\tau\,(Pa)$
40	408	40	111	90	69	240	54.5
50	484	50	132	150	103	480	94.4
60	556	60	152	250	154	800	142
70	635	70	171.3	350	200	1000	170
80	693	80	189.4	450	243	1100	183
150	1120	150	309				
		300	527				

How do the values of the power-law flow behaviour and consistency indices depend upon temperature? Estimate the activation energy of viscous flow (E) from these data by fitting them to the equation $m = m_0 \exp(E/\mathbf{R}T)$ where m_0 and E are constants and $\mathbf{R}$ is the universal gas constant.

1.6 The following shear stress–shear rate data are available for an aqueous (b) carbopol solution at 293 K.

$\dot\gamma\,(s^{-1})$	$\tau\,(Pa)$	$\dot\gamma\,(s^{-1})$	$\tau\,(Pa)$
0.171	53.14	1.382	78.18
0.316	57.86	1.92	84.37
0.421	61.59	2.63	90.23
0.603	66	3.67	98.26
0.812	70	5.07	106.76
1.124	75.47		

By plotting these data on linear and logarithmic scales, ascertain the type of fluid behaviour exhibited by this solution. Suggest a suitable viscosity model and evaluate the parameters for this solution. Does the fluid appear to have a yield stress? Using the vane method (Q.D. Nguyen and D.V. Boger, Annu. Rev. Fluid Mech., 24 (1992) 47), the yield stress was found to be 46.5 Pa. How does this value compare with that obtained by the extrapolation of data to $\dot\gamma = 0$, and that obtained by fitting Bingham, Casson and Herschel–Bulkley fluid models?

1.7 The following rheological data have been reported for a 100 ppm (b) polyacrylamide solution in 96% (by weight) aqueous wheat syrup solution at 294 K.

$\dot\gamma\,(s^{-1})$	$\tau\,(Pa)$	$N_1\,(Pa)$	$\dot\gamma\,(s^{-1})$	$\tau\,(Pa)$	$N_1\,(Pa)$
0.025	0.70	–	0.315	8.79	18.1
0.0315	0.89	–	0.396	11.01	27.7
0.0396	1.12	–	0.500	13.87	42.0

$\dot\gamma\,(\mathrm{s}^{-1})$	$\tau\,(\mathrm{Pa})$	$N_1\,(\mathrm{Pa})$	$\dot\gamma\,(\mathrm{s}^{-1})$	$\tau\,(\mathrm{Pa})$	$N_1\,(\mathrm{Pa})$
0.050	1.42	0.0892	0.628	17.46	65.4
0.0628	1.78	0.158	0.790	21.80	97.5
0.0791	2.25	0.890	0.995	27.16	141.0
0.0995	2.83	1.26	1.25	33.75	196.0
0.125	3.56	2.47	1.58	42.34	283.0
0.158	4.49	3.64	1.99	53.53	440.0
0.199	5.65	6.71	2.50	67.03	661.0
0.250	7.08	11.60	3.15	84.11	863.0

(i) Is this solution shear-thinning?
(ii) Can this solution be treated as a Newtonian fluid? If not, why?
(iii) Is it a visco-elastic fluid? Estimate the value of the Maxwellian relaxation time for this solution.

1.8 The following shear stress–shear rate values have been obtained for aqueous silica (bulk density $= 800\,\mathrm{kg/m^3}$) suspensions to elucidate the effect of concentration on the rheological behaviour of suspensions: (b)

880 kgm^{-3}		905 kgm^{-3}		937 kgm^{-3}		965 kgm^{-3}		995 kgm^{-3}	
$\dot\gamma$ (s^{-1})	τ (Pa)	$\dot\gamma$ (s^{-1})	τ (Pa)	$\dot\gamma$ (s^{-1})	τ (Pa)	$\dot\gamma$ (s^{-1})	τ (Pa)	$\dot\gamma$ (s^{-1})	τ (Pa)
3.9	3.46	1.9	3.79	1.1	5.20	1.9	6.25	0.9	7.55
5.3	3.54	2.6	3.90	2.1	5.40	3.4	6.46	1.8	8.01
5.9	3.61	3.8	4.02	3.3	5.62	4.7	6.67	2.8	8.33
7.0	3.68	4.9	4.12	5.6	5.90	7.2	6.92	4.9	8.75
8.2	3.74	5.9	4.19	7.9	6.12	9.1	7.16	8.0	9.25
9.2	3.80	7.0	4.30	10.8	6.32	10.7	7.27	9.9	9.43
10.3	3.86	8.0	4.38	12.4	6.48	12.0	7.38	11.6	9.64
11.3	3.91	9.0	4.44	13.9	6.58	13.1	7.48	13.0	9.88
12.2	3.96	12.2	4.64	15.2	6.68	14.2	7.55	14.3	9.95
14.9	4.10	13.5	4.70	16.4	6.78	15.1	7.70	15.5	10.10
16.2	4.17	14.9	4.79	17.6	6.88	15.9	7.74	16.6	10.27
19.3	4.30	18.2	5.00	23.1	7.34	17.7	7.98	19.4	10.53
22.5	4.43	21.4	5.15	25.7	7.41	19.4	8.10	22.0	10.80
25.5	4.48	24.7	5.30	28.3	7.59	24.2	8.47	24.5	11.10
28.6	4.70	28.0	5.40	33.2	7.91	27.0	8.62	26.9	11.25
34.6	4.98	34.4	5.82			30.0	8.87	31.8	11.60
40.5	5.15	40.7	5.97			37.1	9.30	36.5	11.91

(i) Plot these data on linear and logarithmic scales and fit the Herschel–Bulkley viscosity model to represent the behaviour of these suspensions.
(ii) How do the model parameters depend upon the concentration?

1.9 The following rheological data have been obtained for a 0.244% Poly-   (b)
isobutylene/92.78% Hyvis Polybutene/6.98% kerosene (by weight)
solution at 293 K.

$\dot\gamma\,(s^{-1})$	$\tau\,(Pa)$	$N_1\,(Pa)$	$\dot\gamma\,(s^{-1})$	$\tau\,(Pa)$	$N_1\,(Pa)$
0.05	0.165	–	0.792	2.55	–
0.0629	0.202	–	0.998	3.18	0.30
0.0792	0.260	–	1.26	3.98	0.57
0.0998	0.330	–	1.58	4.58	1.20
0.126	0.413	–	1.99	6.24	2.21
0.158	0.518	–	2.51	7.72	3.29
0.199	0.663	–	3.15	9.61	5.09
0.251	0.823	–	3.97	12	8.10
0.315	1.03	–	5.00	15	12.10
0.397	1.29	–	6.29	18.6	18.20
0.50	1.61	–	7.92	23.2	28.00
0.629	2.03	–	9.97	29.2	46.90

(i) Does this fluid exhibit Newtonian, or shear-thinning fluid behaviour?
(ii) Is the liquid visco-elastic? Show the variation of the Maxwellian
relaxation time with shear rate.

2.1 The following volumetric flow rate – pressure gradient data have been   (c)
obtained using a capillary viscometer ($D = 10\,mm$ and $L = 0.5\,m$) for
a viscous material. Obtain the true shear stress–shear rate data for this
substance and suggest a suitable viscosity fluid model.

$Q\,(mm^3/s)$	1.3	15	140	1450	14500	1.5×10^5	1.4×10^6
$-\Delta p\,(Pa)$	0.5	1.5	5	15	50	160	500

2.2 A polymer solution was tested in a cone-and-plate viscometer (cone   (b)
angle 0.1 rad and cone radius 25 mm) at various rotational speeds.
Use the following torque – speed data to infer shear stress–shear rate
behaviour and suggest an appropriate fluid model to describe the fluid
behaviour.

$\Omega\,(rad/s)$	10^{-4}	10^{-3}	0.01	0.1	1	10	100
$T\,(Nm)$	0.003	0.033	0.26	1	2.2	3.3	66

2.3 The following data has been obtained with a capillary viscometer for   (c)
an aqueous polymer solution, density $1000\,kg/m^3$ at 293 K.

Capillary data	$\Delta p\,(kPa)$	mass flow rate (kg/s)
$L = 200\,mm$	224.3	1.15×10^{-3}
$D = 2.11\,mm$	431	2.07×10^{-3}
	596.3	2.75×10^{-3}
	720.3	3.88×10^{-3}

Capillary data	Δp (kPa)	mass flow rate (kg/s)
$L = 50$ mm	87.5	1.01×10^{-3}
$D = 2.11$ mm	148.1	1.71×10^{-3}
	361.7	4.29×10^{-3}
	609.8	6.69×10^{-3}
$L = 50.2$ mm	43.40	3.95×10^{-3}
$D = 4.14$ mm	79.22	8.18×10^{-3}
	117.12	1.092×10^{-2}
	160.53	1.48×10^{-2}

3.1 A low molecular weight polymer melt, which can be modelled as a   (a)
power-law fluid with $m = 5$ k Pa·s^n and $n = 0.25$, is pumped through
a 13 mm inside diameter tube over a distance of 10 m under laminar
flow conditions. Another pipe is needed to pump the same material over a distance of 20 m at the same flow rate and with the
same frictional pressure loss. Calculate the required diameter of the
new pipe.

3.2 The flow behaviour of a tomato sauce follows the power-law model,   (b)
with $n = 0.50$ and $m = 12$ Pa·s^n. Calculate the pressure drop per
metre length of pipe if it is pumped at the rate of 1000 kg/h through
a 25 mm diameter pipe. The sauce has a density of 1130 kg/m^3. For a
pump efficiency of 50%, estimate the required power for a 50 m long
pipe.
How will the pressure gradient change if

(a) the flow rate is increased by 50%,
(b) the flow behaviour consistency coefficient increases to 14.75 Pa·s^n
without altering the value of n, due to changes in the composition
of the sauce,
(c) the pipe diameter is doubled,
(d) the pipe diameter is halved.
Is the flow still streamline in this pipe?

3.3 A vertical tube whose lower end is sealed by a movable plate is filled   (b)
with a viscoplastic material having a yield stress of 20 Pa and density
1100 kg/m^3. Estimate the minimum tube diameter for this material to
flow under its own weight when the plate is removed. Does the depth
of the material in the tube have any influence on the initiation of flow?

3.4 A power-law fluid ($m = 5$ Pa·s^n and $n = 0.5$) of density 1200 kg/m^3 (b)
flows down an inclined plane at 30° to the horizontal. Calculate the
volumetric flow rate per unit width if the fluid film is 6 mm thick.
Assume laminar flow conditions.

3.5 A Bingham plastic material is flowing under streamline conditions $\quad$ (c)
in a circular pipe. What are the conditions for one third of the total
flow to be within the central plug region across which the velocity
profile is flat? The shear stress acting within the fluid τ varies with
the velocity gradient dV_x/dr according to the relation:

$$\tau = \tau_0^B + \mu_B \left(-\frac{dV}{dr} \right)$$

where τ_0^B and μ_B are respectively the Bingham yield stress and the
plastic viscosity of the material.

3.6 Tomato purée of density $1100 \, kg/m^3$ is pumped through a 50 mm $\quad$ (c)
diameter pipeline at a flow rate of $1 \, m^3/hr$. It is suggested that, in
order to double production,

(a) a similar line with pump should be put in parallel with the existing
one, or
(b) a larger pump should be used to force the material through the
present line, or
(c) the cross-sectional area of the pipe should be doubled.

The flow behaviour of the tomato pureé can be described by the
Casson equation (1.18), i.e.,

$$(+\tau_{rz})^{1/2} = (+\tau_0^c)^{1/2} + \left(\mu_c \left(-\frac{dV_z}{dr} \right) \right)^{1/2}$$

where τ_0^c, the Casson yield stress, is equal to 20 Pa and μ_c, the Casson
plastic viscosity has a value of 5 Pa·s.
    Evaluate the pressure gradient for the three cases. Also, evaluate
the viscosity of a hypothetical Newtonian fluid for which the pressure gradient would be the same. Assume streamline flow under all
conditions.

3.7 A polymer solution is to be pumped at a rate of 11 kg/min through $\quad$ (b)
a 25 mm inside diameter pipe. The solution behaves as a power-law
fluid with $n = 0.5$ and an apparent viscosity of 63 m Pa·s at a shear
rate of $10 \, s^{-1}$, and a density of $950 \, kg/m^3$.

(a) What is the pressure gradient in the pipe line?
(b) Estimate the shear rate and the apparent viscosity of the solution
at the pipe wall?
(c) If the fluid were Newtonian, with a viscosity equal to the apparent
viscosity at the wall as calculated in (b) above, what would be the
pressure gradient?
(d) Calculate the Reynolds numbers for the polymer solution and for
the hypothetical Newtonian fluid

3.8 A concentrated coal slurry (density 1043 kg/m³) is to be pumped (b) through a 25 mm inside diameter pipe over a distance of 50 m. The flow characteristics of this slurry are not fully known, but the following preliminary information is available on its flow through a smaller tube, 4 mm in diameter and 1 m long. At a flow rate of 0.0018 m³/h, the pressure drop across the tube is 6.9 kPa, and at a flow rate of 0.018 m³/h it is 10.35 kPa. Evaluate the power-law constants from the data for the small diameter tube. Estimate the pressure drop in the 25 mm diameter pipe for a flow rate of 0.45 m³/h.

3.9 A straight vertical tube of diameter D and length L is attached to (c) the bottom of a large cylindrical vessel of diameter $D_T(\gg D)$. Derive an expression for the time required for the liquid height in the large vessel to decrease from its initial value of $H_0(\ll L)$ to $H(\ll L)$ as shown in the following sketch.

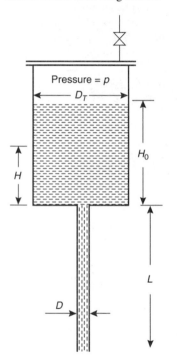

Pressure = p

D_T

H_0

H

L

D

Neglect the entrance and exit effects in the tube as well as the changes in kinetic energy. Assume laminar flow in the tube and (i) power-law behaviour, and (ii) Bingham plastic behaviour.

3.10 Estimate the time needed to empty a cylindrical vessel ($D_T =$ (c) 101 mm), open to the atmosphere, filled with a power-law liquid ($m = 4\,Pa \cdot s^n$ and $n = 0.6$, density $= 1010\,kg/m^3$). A 6 mm ID capillary tube 1.5 m long is fitted to the base of the vessel as shown in the diagram for problem 3.9. The initial height of liquid in the vessel, $H_0 = 230\,\text{mm}$. Estimate the viscosity of a hypothetical Newtonian fluid of same density which would empty in the same time.

3.11 Using the same equipment as in problem 3.10, the following height–time data have been obtained for a coal slurry of density $1135\,\text{kg/m}^3$. Evaluate the power-law model parameters for this slurry (Assume streamline flow under all conditions)

(b)

t (s)	0	380	808
H (m)	0.25	0.20	0.15

3.12 A pharmaceutical formulation having a consistency coefficient of $2.5\,\text{Pa·s}^n$ and a flow behaviour index of 0.65 must be pumped through a stainless steel pipe of 40 mm inside diameter. If the shear rate at the pipe wall must not exceed $140\,\text{s}^{-1}$ or fall below $50\,\text{s}^{-1}$, estimate the minimum and maximum acceptable volumetric flow rates.

(a)

3.13 Two storage tanks, A and B, containing a coal slurry ($m = 2.7\,\text{Pa·s}^n$; $n = 0.5$; density $= 1040\,\text{kg/m}^3$) discharge through pipes each 0.3 m in diameter and 1.5 km long to a junction at D. From D, the slurry flows through a 0.5 m diameter pipe to a third storage tank C, 0.75 km away as shown in the following sketch. The free surface of slurry in tank A is 10 m above that in tank C, and the free surface in tank B is 6 m higher than that in tank A. Calculate the initial rates of discharge of slurry into tank C. Because the pipes are long, the minor losses (in fittings) and the kinetic energy of the liquid can be neglected. Assume streamline flow conditions in all pipes.

(c)

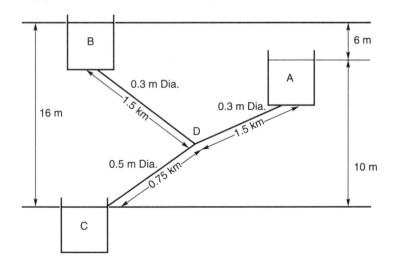

3.14 An engineer having carefully calculated the optimum internal diam-   (c)
eter required of a proposed long pipeline for transporting china clay
slurries (approximating to power-law behaviour) at a given flow rate,
discovers that a full range of pipe sizes is not available at the remote
location. The diameters that are closest are 25% too large and 15% too
small. Not wanting to over design, and fearing to exceed the available
pressure gradient, the engineer decided to use a composite pipeline
made up of the two available sizes in series. If the cost per unit length
of pipe is proportional to its inside diameter, what would be the %
saving as a function of n (0.2 to 1.2) in using the composite line
instead of a uniform line of the larger diameter when the flow in all
pipes is (i) laminar, and (ii) turbulent in which case the friction factor
is given by $f \propto \mathrm{Re}_{MR}^{-1/(3n+1)}$. Assume that in the range of conditions
encountered, the values of the power-law constants (m and n) do not
vary appreciably.

3.15 A power-law liquid is flowing under streamline conditions through   (a)
a horizontal tube of 8 mm diameter. If the mean velocity of flow is
1 m/s and the maximum velocity at the centre is 1.2 m/s, what is the
value of the flow behaviour index?

For a Newtonian, organic liquid of viscosity 0.8 mPa·s, flowing
through the tube at the same mean velocity, the pressure drop is 10 kPa
compared with 100 kPa for the power-law liquid. What is the power-
law consistency coefficient, m, of the non-Newtonian liquid?

3.16 The rheology of a polymer solution can be approximated reasonably   (b)
well by either a power-law or a Casson model over the shear rate
range of $20-100\,\mathrm{s}^{-1}$. If the power law consistency coefficient, m, is
$10\,\mathrm{Pa}\cdot\mathrm{s}^{n}$ and the flow behaviour index, n, is 0.2, what will be the
approximate values of the yield stress and the plastic viscosity in the
Casson model?

Calculate the pressure drop using the power-law model when this
polymer solution is in laminar flow in a pipe 200 m long and 40 mm
inside diameter for a centreline velocity of 1 m/s. What will be the
calculated centreline velocity at this pressure drop if the Casson model
is used?

3.17 Using dimensional analysis, show that the frictional pressure drop   (a)
for the fully developed flow of a Bingham fluid in circular tubes is
given by:

$$f = \phi(\mathrm{Re}, \mathrm{He})$$

where f and Re are the usual fanning friction factor and pipe Reynolds
number ($\rho V D / \mu_B$) respectively; He, the Hedström number, is defined
as $\mathrm{He} = \rho \tau_0^B D^2 / \mu_B^2$.

Use these dimensionless groups to re-arrange equation (3.13) as:

$$f = \frac{16}{Re}\left[1 + \frac{1}{6}\left(\frac{He}{Re}\right) - \frac{1}{3}\frac{He^4}{f^3 Re^7}\right]$$

Give typical plots of f vs Re for a range of values of the Hedstrom number.

A carbopol solution ($\rho = 1000\,kg/m^3$), with Bingham plastic rheology ($\tau_0^B = 8\,Pa$ and plastic viscosity of $50\,m\,Pa\cdot s$), is flowing through a 25 mm diameter pipeline. Estimate the pressure drop per metre of pipe when the mean velocity is 1 m/s. Also, estimate the radius of the unsheared plug in the core region.

During maintenance, it is necessary to use a standby pump and the maximum attainable pressure gradient is 30% less than that necessary to maintain the original flowrate. By what percentage will the flowrate fall? Assume streamline flow conditions in the pipe.

3.18 A viscous plastic fluid of density $1400\,kg/m^3$ and containing 60%   (b)
(by weight) of a pigment is to be pumped in streamline flow through 75 mm, 100 mm and 125 mm diameter pipes within the plant area. The corresponding flow rates of the liquid in these pipes are 6×10^{-3}, 0.011 and $0.017\,m^3/s$, respectively. The following data were obtained on an extrusion rheometer using two different size capillaries:

Gas pressure in reservoir, $-\Delta p$ (kPa)	Time to collect 5 g liquid from the tube, (s)	
	Tube A	Tube B
66.2	913	112
132.4	303	39
323.7	91	11.6
827.4	31	3.9

Tube A: Diameter $= 1.588\,mm$; length $= 153\,mm$
Tube B: Diameter $= 3.175\,mm$; length $= 305\,mm$
Estimate the pressure gradient in each of the pipes to be used to carry this liquid.

3.19 It is necessary to pump a non-Newtonian slurry (density $1000\,kg/m^3$)   (a)
over a distance of 100 m through a 200 mm diameter smooth-walled pipe. What is the maximum possible flow rate at which the flow will be laminar and what then is the pressure drop across the pipe?

The following laboratory results are available for laminar flow of this slurry in small diameter tubes.

$(8V/D), s^{-1}$	35	104	284	677	896	1750	4300	8000	10900	23100	51500
τ_w, Pa	4.7	9.3	17.2	29	37.9	51.4	90.3	153	160	291	429

Can these data be scaled up directly, without assuming a rheological model? Compare the values of the pressure gradient obtained by such a method with that obtained by assuming the power-law model behaviour. Assume streamline flow conditions in all cases.

3.20 A coal-in-oil slurry (density $1640\,kg/m^3$) containing 35% (by weight) of coal (particle size smaller than $50\,\mu m$) is pumped at a rate of $3.5\,m^3/h$ from a storage tank, through a 50 m long 12.5 mm diameter pipe to a boiler where it is burnt to raise steam. The pressure in the gas space in the tank is atmospheric and the fuel slurry must be delivered to the burner at an absolute pressure of 240 kPa.   (b)

(a) Estimate the pumping power required to deliver the slurry to the boiler if the slurry is assumed to be a Newtonian fluid having a viscosity of 200 mPa·s.
(b) Subsequently, preliminary rheological tests suggest the slurry to exhibit Bingham-plastic behaviour with a yield stress of 80 Pa and a plastic viscosity of 200 mPa·s. How will this knowledge affect the predicted value of the pressure drop.

Comment on the difference between the two values of the pump power.

3.21 In a horizontal pipe network, a 150 mm diameter 100 m long pipe   (a) branches out into two pipes, one 100 mm diameter and 20 m long and the other 75 mm diameter and 12 m long; the branches re-join into a 175 mm diameter 50 m long pipe. The volumetric flow rate of a liquid (density $1020\,kg/m^3$) is $3.4\,m^3/min$ and the pressure at the inlet to 150 mm diameter is 265 kPa. Calculate the flow rate in each of the two smaller diameter branches and the pressure at the beginning and end of the 175 mm diameter pipe for:

(a) a liquid exhibiting power-law behaviour, $n = 0.4$; $m = 1.4\,Pa·s^n$.
(b) a liquid exhibiting Bingham plastic behaviour, (yield stress 4.3 Pa and plastic viscosity 43 m Pa·s).

Due to process modifications, the flow rate must be increased by 20%; Calculate the pressure now required at the inlet to the pipe network.

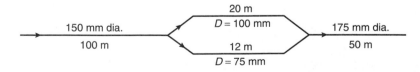

3.22 The fluid whose rheological data are given in Problem 3.19 is to be   (b)
      pumped at $0.34 \, m^3/s$ through a 380 mm diameter pipe over a distance
      of 175 m. What will be the required inlet pressure, and what pump
      power will be needed?

3.23 The same fluid as that used in Problem 3.19 is to be pumped through   (b)
      a 400 mm diameter pipe over a distance of 500 m and the pressure
      drop must not exceed 53 kPa. Estimate the maximum possible flow
      rate achievable in the pipe. (Hint: By analogy with the method used
      for Newtonian fluids, prepare a chart of $\sqrt{Re}$ versus $\sqrt{f \, Re_{MR}^{2/(2-n)}}$ to
      avoid trial and error procedure).

3.24 Determine the diameter of a pipe, 175 m long, required to carry the   (b)
      liquid whose rheology is given in Problem 3.19 at $0.5 \, m^3/s$ with a pres-
      sure drop of 50 kPa. (Hint: Prepare a plot of $\sqrt{Re}$ versus $(f \, Re_{MR}^{5/(4-3n)})$
      to avoid the necessity of using a trial and error procedure; express
      average velocity in terms of the volumetric flow rate).

3.25 The following data have been obtained for the flow of the slurry   (a)
      specified in Problem 3.19 in small tubes of three different diameters:

D (m)	L (m)	$-\Delta p$ (kPa)	V (m/s)
0.016	4.27	136	5.76
		173	6.39
		205	7.14
		236	7.94
		284	9.01
		341	10.1
		395	11
		438	11.8
		500	12.8
		575	14
0.027	4.34	52.2	4.6
		64	5.26
		79.6	6.0
		98.7	6.91
		122	7.91
		152	9.03
		184	10.3
		235	11.7
		281	13.1
		352	15
0.053	2.64	8.41	3.33
		9.6	3.61
		11.4	4.03
		13.6	4.54
		17.6	5.37

Using these data, evaluate A, b, c in Bowen's relation and calculate the frictional pressure drop for a flow rate of 0.34 m³/s in a 380 mm diameter pipe of length of 175 m. How does this value compare with that obtained in Problem 3.22?

3.26 Wilhelm *et al.* (*Ind. Eng. Chem.*, Vol. **3**, p. 622, 1939) presented the (b) following pressure drop–flow rate data for a 54.3% (by weight) rock slurry flowing through tubes of two different diameters.

D (m)	L (m)	$-\Delta p$ (kPa)	V (m/s)
0.019	30.5	338	3.48
		305	3.23
		257	2.97
		165	2.26
		107	1.81
		62	1.38
		48.4	1.20
		43	0.89
		38.6	0.44
		33.4	0.36
0.038	30.5	78.3	2.41
		50.3	1.86
		34.6	1.52
		19.1	1.09
		15.2	0.698
		17.3	0.512
		17.7	0.375

Using Bowen's method, develop a general scale-up procedure for predicting the pressure gradients in turbulent flow of this rock slurry. Estimate the pump power required for a flow rate of 0.45 m³/s in a 400 mm diameter pipe, 500 m long. The pump has an efficiency of 60%. Take the density of slurry 1250 kg/m³.

3.27 What is the maximum film thickness of an emulsion paint that can (a) be applied to an inclined surface (15° from vertical) without the paint running off? The paint has an yield stress of 12 Pa and a density of 1040 kg/m³.

3.28 When a non-Newtonian liquid flows through a 7.5 mm diameter and (a) 300 mm long straight tube at 0.25 m³/h, the pressure drop is 1 kPa.

   (i) Calculate the viscosity of a Newtonian fluid for which the pressure drop would be the same at that flow rate?

(ii) For the same non-Newtonian liquid, flowing at the rate of 0.36 m^3/h through a 200 mm long tube of 7.5 mm in diameter, the pressure drop is 0.8 kPa. If the liquid exhibits power-law behaviour, calculate its flow behaviour index and consistency coefficient.

(iii) What would be the wall shear rates in the tube at flow rates of 0.18 m^3/h and 0.36 m^3/h.

Assume streamline flow.

3.29 Two liquids of equal densities, the one Newtonian and the other   (c)
a power-law liquid, flow at equal volumetric rates down two wide inclined surfaces (30° from horizontal) of the same widths. At a shear rate of 0.01 s^{-1}, the non-Newtonian fluid, with a power-law index of 0.5, has the same apparent viscosity as the Newtonian fluid. What is the ratio of the film thicknesses if the surface velocities of the two liquids are equal?

3.30 For the laminar flow of a time-independent fluid between two parallel   (c)
plates (Figure 3.15), derive a Rabinowitsch–Mooney type relation giving:

$$\left(-\frac{dV_z}{dy} \right)_{wall} = 4 \left(\frac{Q}{Wb^2} \right) + 2 \left(\frac{b}{2} \frac{(-\Delta p)}{L} \right) \frac{d[Q/2Wb^2]}{d[b(-\Delta p/2L)]}$$

where $2b$ is the separation between two plates of width $W (\gg 2b)$.
What is the corresponding shear rate at the wall for a Newtonian fluid?

3.31 A drilling fluid consisting of a china clay suspension of density   (a)
1090 kg/m^3 flows at 0.001 m^3/s through the annular cross-section between two concentric cylinders of radii 50 mm and 25 mm, respectively. Estimate the pressure gradient if the suspension behaves as:

(i) a power-law liquid: $n = 0.3$ and $m = 9.6$ Pa·s^n.
(ii) a Bingham plastic fluid: $\mu_B = 0.212$ Pa·s and $\tau_0^B = 17$ Pa.

Use the rigorous methods described in Section 3.6 as well as the approximate method presented in Section 3.7. Assume streamline flow.

Due to pump malfunctioning, the available pressure gradient is only 75% of that calculated above. What will be the corresponding flow rates on the basis of the power-law and Bingham plastic models?

3.32 A power-law fluid (density 1000 kg/m^3) whose rheological parameters   (a)
are $m = 0.4$ Pa·s^n and $n = 0.68$ is flowing at a mean velocity of 1.2 m/s in ducts of several different cross-sections:

(i) a circular pipe
(ii) a square pipe
(iii) a concentric annulus with outer and inner radii of 26.3 mm and 10.7 mm respectively
(iv) a rectangular pipe with aspect ratio of 0.5.

(v) an elliptic pipe with minor-to-major axis ratio of 0.5.
(vi) an isosceles triangular with apex angle of 40°.

Using the geometric parameter method, estimate the pressure gradient required to sustain the flow in each of these conduits, all of which have the same hydraulic radius as the concentric annulus referred to in (iii). Also, calculate the Reynolds number for each case to test whether the flow is streamline.

How will the pressure gradient in each pipe change if they were all to have same flow area, as opposed to the same hydraulic diameter, as the annulus?

3.33 The relation between cost per unit length C of a pipeline installation and its diameter D is given by

$$C = a + bD$$

where a and b are constants and are independent of pipe size. Annual charges are a fraction β of the capital cost C. Obtain an expression for the optimum pipe diameter on a minimum cost basis for a power-law fluid of density, ρ, flowing at a mass flow rate $\dot{m}$. Assume the flow to be (i) streamline (ii) turbulent with friction factor $f \propto \mathrm{Re}_{MR}^{-1/(3n+1)}$.

Discuss the nature of the dependence of the optimum pipe diameter on the flow behaviour index.

(c)

3.34 For streamline flow conditions, calculate the power needed to pump a power-law fluid through a circular tube when the flow rate is subject to sinusoidal variation of the form:

$$\dot{Q} = Q_m \sin \omega t$$

where Q_m is the maximum flow rate at $t = \pi/2\omega$.
By what factor will the power requirement increase due to the sinusoidal variation in flowrate as compared with flow with uniform velocity?
Repeat this calculation for turbulent flow using the friction factor given by equation (3.38).

(b)

3.35 An aqueous bentonite suspension of density $1300\,\mathrm{kg/m^3}$, used to model an oil drilling mud, is to be pumped through the annular passage between the two concentric cylinders of radii $101.6\,\mathrm{mm}$ and $152.4\,\mathrm{mm}$, respectively. The suspension, which behaves as a Bingham plastic fluid with $\mu_B = 20\,\mathrm{mPa \cdot s}$ and $\tau_0^B = 7.2\,\mathrm{Pa}$, is to be pumped at the rate of $0.13\,\mathrm{m^3/s}$ over a distance of $300\,\mathrm{m}$. Estimate the pressure drop. Over what fraction of the area of the annulus, the material is in plug flow?

Plot the velocity distribution in the annular region and compare it with that for a Newtonian liquid having a viscosity of $20\,\mathrm{mPa \cdot s}$ under otherwise similar conditions.

(a)

3.36 The following data (P. Slatter, PhD thesis, University of Cape Town,   (a)
Cape Town, 1994) have been obtained for the turbulent flow of
a kaolin-in-water suspension (of density $1071 \, kg/m^3$) in pipes of
different sizes:

$D = 207 \, mm$

V (m/s)	2.93	2.67	2.46	2.27	2.04	1.71	1.55	1.35	1.15
τ_w (Pa)	19.2	17.11	13	11.6	7.87	6.14	5.1	4.16	2.89

$D = 140.5 \, mm$

V (m/s)	6.34	5.74	5.39	4.93	4.13	3.49	3	2.56	2	1.4
τ_w (Pa)	71.8	57.3	53.4	43.4	32.4	23.3	17.6	13.37	8.3	4.1

$D = 79 \, mm$

V (m/s)	6.48	5.44	4.5	3.63	2.69	1.92	2.62	2.1	1.71	1.18
τ_w (Pa)	83.5	59.1	41.23	29.5	16.9	12.3	10.7	8.94	7.31	3.93

$D = 21.6 \, mm$

V (m/s)	7.09	6.80	6.48	6.08	5.62	5.1	4.77	4.53	4.1	3.7	2.91	1.47	1.3
τ_w (Pa)	123	114.3	105.6	93.58	82.1	69.3	62.1	55.4	46.2	39	25	7.6	5.9

$D = 13.2 \, mm$

V (m/s)	6.84	6.41	5.82	5.4	4.76	4.1	3.77	3.46	3.18	2.7	2.41	1.91	1.49
τ_w (Pa)	136	120.8	102	89	72	55	47.5	40.6	35	26	21.2	13.74	7.45

$D = 5.6 \, mm$

V (m/s)	4.55	3.89	3.50	3.22	3.08	2.84	2.65	2.08	2.22	1.7	1.51	1.61
τ_w (Pa)	64.6	58.5	48.5	40.5	33.5	27.4	22.9	20.6	18.6	14	13	10.7

(i) Use Bowen's method in conjunction with the second three sets of
data for small diameter tubes to predict the pressure gradient in
the remaining three large diameter pipes and compare them with
the experimental values. Why does the discrepancy increase as
the mean velocity of flow is decreased?

(ii) Slatter also fitted rheological data to the power-law and Bingham-
plastic models and the best values of the parameters were:
$m = 0.56 \, Pa \cdot s^n$ and $n = 0.31$; $\tau_0^B = 2.04 \, Pa$ and $\mu_B = 3.56 \, mPa \cdot s$.
Compare the experimental and the calculated values of pressure
gradient for all sizes of pipes.

3.37 Measurements are made of the yield stress of two carbopol solu-   (b)
tions (density $1000 \, kg/m^3$) and of a 52.9% (by weight) silica-in-water
suspension (density $1491 \, kg/m^3$) by observing their behaviour in an
inclined tray which can be tilted to the horizontal. The values of the
angle of inclination to the horizontal, θ, at which flow commences for
a range of liquid depths, H, are given below. Determine the value of
yield stress for each of these liquids.

0.08% carbopol solution		0.09% carbopol solution		52.9% silica-in-water suspension	
H (mm)	θ (degrees)	H (mm)	θ (degrees)	H (mm)	θ (degrees)
2.0	6.8	6.4	5.5	13.7	3.55
2.6	6.1	7.0	4.6	17.3	2.85
3.2	4.9	12.0	2.8	22.5	2.20
3.9	3.8	12.1	2.6	24.1	1.90
5.2	3.0	15.3	2.1		
8.4	2.0	19.9	1.55		
14.0	1.0	24.1	1.35		
		30.0	1.10		
		32.8	0.95		

3.38 Viscometric measurements suggest that an aqueous carbopol solution (b) behaves as a Bingham plastic fluid with yield stress of 1.96 Pa and plastic viscosity 3.80 Pa·s. The liquid flows down a plate inclined at an angle θ to the horizontal. Derive an expression for the volumetric flow rate per unit width of the plate as a function of the system variables. Then, show that the following experimental results for $\theta = 5°$ are consistent with the theoretical predictions.

Q (mm²/s)	H (mm)	Free surface velocity (mm/s)
4.8	7.19	0.8
10.8	8.62	1.54
18.0	9.70	2.20
26.2	10.51	3.14
35.9	11.50	3.94

4.1 A 19.5% (by volume) kaolin-in-water suspension is flowing under (a) laminar conditions through a horizontal pipe, 42 mm diameter and 200 m long, at a volumetric flow rate of 1.25 m³/h. The suspension behaves as a power-law liquid with $n = 0.16$ and $m = 9.6$ Pa·s^n, and has a density of 1320 kg/m³. Estimate the pressure drop across the pipe. Air at 298 K is now introduced at a upstream point at the rate of 5 m³/h (measured at the pressure at the mid-point of the pipe length). What will be the two-phase pressure drop over the pipe according to:

(i) the simple plug flow model
(ii) the generalised method of Dziubinski and Chhabra, equations (4.19) and (4.24).
(iii) the method of Dziubinski, equation (4.26).

The experimental value of $-\Delta p$ is 105 kPa. Suggest the possible reasons for the discrepancy between the calculated and actual values of $-\Delta p$.
Calculate the average liquid holdup in the pipe.

4.2 A 25% aqueous suspension of kaolin is to be pumped under laminar   (b)
conditions through a 50 mm diameter and 50 m long pipe at the rate of
2 m³/h. The suspension behaves as a power-law fluid with $n = 0.14$,
$m = 28.6\,Pa \cdot s^n$ and has a density of 1400 kg/m³. Calculate the power
needed for this duty when using a pump of 50% efficiency.
     It is proposed to reduce the two-phase pressure drop by 50% by
introducing air into the pipeline at an upstream point. Calculate the
superficial velocity of air required to achieve this if the air at 293 K
enters the pipeline at a pressure of 0.35 MPa. Assume isothermal
expansion of gas. Use all three methods mentioned in problem 4.1 to
obtain the superficial velocity of the air. Using an appropriate model,
determine the maximum reduction in two-phase pressure drop achievable for this slurry. What is the air velocity under these conditions?
What proportion of the volume of the pipe is filled with liquid and
what flow pattern is likely to occur in the pipe under these flow conditions? Neglect the effect of air expansion.

4.3 A 50.4% (by weight) coal-in-water suspension of density 1070 kg/m³   (a)
is to be transported at the flow rate of 5.5 m³/hr through a
pipeline 75 mm diameter and 30 m long. The suspension behaves
as a Bingham-plastic fluid with a yield stress of 51.4 Pa and plastic
viscosity of 48.3 mPa·s. It is decided to introduce air at pressure of
0.35 MPa and at 293 K into the pipeline to lower the pressure drop by
25%, whilst maintaining the same flow rate of suspension. Assuming
that the plug model is applicable, calculate the required superficial
velocity of air.

4.4 It is required to transport gravel particles (8 mm size, density   (a)
2650 kg/m³) as a suspension in a pseudoplastic polymer solution
($m = 0.25\,Pa \cdot s^n$, $n = 0.65$, density 1000 kg/m³) in a 42 mm diameter pipe over a distance of 1 km. The volumetric flow rate of the
mixture is 7.5 m³/h when the volumetric concentration of gravel in the
discharged mixture is 22% (by volume). If the particles are conveyed
in the form of a moving bed in the lower portion of the pipe, estimate
the pressure drop over this pipeline and the pumping power if the
pump efficiency is 45%. What is the rate of conveyance of gravel in
kg/h?

4.5 A 13% (by volume) phosphate slurry ($m = 3.7\,Pa \cdot s^n$, $n = 0.18$,   (b)
density 1230 kg/m³) is to be pumped through a 50 mm diameter horizontal pipe at mean slurry velocities ranging from 0.2 to 2 m/s. It is
proposed to pump this slurry in the form of a two-phase air-slurry
mixture. The following data have been obtained:

Superficial slurry velocity (m/s)	Superficial gas velocity (m/s)	$\dfrac{-\Delta p_{TP}/L}{-\Delta p_L/L}$
0.24	0.2	0.68
	0.5	0.48
	1	0.395
	2	0.40
	3	0.41
	4	0.44
	5	0.52
0.98	0.2	0.92
	0.5	0.90
	1.0	0.96
	1.4	1.08
	2	1.20
	3	1.44
	4	1.62
	5	1.85
1.5	0.5	1.22
	1.0	1.38
	1.5	1.50
	2	1.72
	2.4	1.86
1.95	0.4	1.38
	0.8	1.53
	1.0	1.60
	1.5	1.70

Are these data consistent with the predictions of the simple plug flow model? Compare these values of the two-phase pressure drop with those calculated using equations (4.22), (4.24) and (4.26).

5.1 Calculate the free falling velocity of a plastic sphere ($d = 3.18\,$mm, density $1050\,$kg/m^3) in a polymer solution which conforms to the power-law model with $m = 0.082\,$Pa$\cdot$s^n, $n = 0.88$ and has density of $1000\,$kg/m^3. Also, calculate the viscosity of a hypothetical Newtonian fluid of the same density in which this sphere would have the same falling velocity. To what shear rate does this viscosity correspond for? (a)

5.2 The following data for terminal velocities have been obtained for the settling rate of spherical particles of different densities in a power-law type polymer solution ($m = 0.49\,$Pa$\cdot$s^n, $n = 0.83$, $\rho = 1000\,$kg/m^3). (a)

Sphere diameter (mm)	Sphere density (kg/m^3)	Settling velocity (mm/s)
1.59	4010	6.6
2.00	4010	10.0
3.175	4010	30
2.38	7790	45

Estimate the mean value of the drag correction factor for this power-law fluid. How does this value compare with that listed in Table 5.1?

5.3 For the sedimentation of a sphere in a power-law fluid in the Stokes'   (a)
law regime, what error in sphere diameter will lead to an error of 1% in the terminal falling velocity? Does the permissible error in diameter depend upon the value of the power-law index? If yes, calculate its value over the range $1 \geq n \geq 0.1$. What is the corresponding value for a Newtonian liquid?

5.4 Estimate the terminal falling velocity of a 5 mm steel ball (density   (b)
7790 kg/m^3) in a power-law fluid ($m = 0.3$ Pa·s^n, $n = 0.6$ and density 1010 kg/m^3).

5.5 The rheological behaviour of a china clay suspension of density   (b)
1200 kg/m^3 is well approximated by the Herschel–Bulkley fluid model with consistency coefficient of 11.7 Pa·s^n, flow behaviour index of 0.4 and yield stress of 4.6 Pa. Estimate the terminal falling velocity of a steel ball, 5 mm diameter and density 7800 kg/m^3. What is the smallest steel ball which will just settle under its own weight in this suspension?

5.6 A 7.5 mm diameter PVC ball (density 1400 kg/m^3) is settling in   (b)
a power-law fluid ($m = 3$ Pa·s^n, $n = 0.6$, density 1000 kg/m^3) in a 30 mm diameter cylindrical tube. Estimate the terminal falling velocity of the ball.

5.7 Estimate the hindered settling velocity of a 30% (by volume) defloccu-   (a)
lated suspension of 50 μm (equivalent spherical diameter) china clay particles in a polymer solution following the power-law fluid model with $n = 0.7$ and $m = 3$ Pa·s^n, in a 30 mm diameter tube. The densities of china clay particles and the polymer solution are 2400 and 1000 kg/m^3 respectively.

5.8 In a laboratory size treatment plant, it is required to pump the sewage   (a)
sludge through a bed of porcelain spheres packed in a 50 mm diameter tube. The rheological behaviour of the sludge (density 1008 kg/m^3) can be approximated by a power-law model with $m = 3.8$ Pa·s^n and $n = 0.4$. Calculate the diameter of the spherical packing (voidage 0.4) which will be required to obtain a pressure gradient of 8 MPa/m at a flow rate of 3.6 m^3/h. What will be the flow rates for the same pressure gradient if the nearest available packing sizes are 25% too large and 25% too small? Assume the voidage remains at the same level.

5.9 Estimate the size of the largest steel ball (density 7800 kg/m^3) which   (b)
would remain embedded without settling in a viscoplastic suspension with a density of 1040 kg/m^3 and yield stress of 20 Pa?

5.10 A polymeric melt exhibiting power-law behaviour ($m = 10^4$ Pa·s^n, (b)
n = 0.32, density 960 kg/m^3) is to be filtered by using a sand-pack
composed of 50 μm sand particles; the bed voidage is 37%. The pressure drop across a 100 mm deep bed must be in the range 540 MPa
to 1130 MPa. Estimate the range of volumetric flow rates which can
be processed in a column of 50 mm diameter.

Over the relevant range of shear rates, the flow behaviour of this
melt can also be well approximated by the Bingham plastic model.
What would be the appropriate values of the plastic viscosity and the
yield stress?

5.11 Estimate the minimum fluidising velocity for a bed consisting of (a)
3.57 mm glass spheres (density 2500 kg/m^3) in a 101 mm diameter
column using a power-law polymer solution ($m = 0.35$ Pa·s^n, $n = 0.6$
and density 1000 kg/m^3) if the bed voidage at the incipient fluidised
condition is 37.5%. If the value of the fixed bed voidage is in error
by 10%, what will be the corresponding uncertainty in the value of
the minimum fluidising velocity?

6.1 Calculate the thermal conductivity of 35% (by volume) non- (a)
Newtonian suspensions of alumina (thermal conductivity =
30 W/mK) and thorium oxide (thermal conductivity = 14.2 W/mK)
in water and in carbon tetra chloride at 293 K.

6.2 A power-law non-Newtonian solution of a polymer is to be heated (b)
from 288 K to 303 K in a concentric-tube heat exchanger. The solution will flow at a mass flow rate of 210 kg/h through the inner copper
tube of 31.75 mm inside diameter. Saturated steam at a pressure of
0.46 bar and a temperature of 353 K is to be condensed in the annulus.
If the heater is preceded by a sufficiently long unheated section for
the velocity profile to be fully established prior to entering the heater,
determine the required length of the heat exchanger. Physical properties of the solution at the mean temperature of 295.5 K are:

density = 850 kg/m^3; heat capacity = 2100 J/kg K; thermal conductivity = 0.69 W/mK; flow behaviour index, $n = 0.6$

temperature (K)	288	303	318	333	353	368
consistency coefficient (Pa·s^n)	10	8.1	6.3	4.2	2.3	1.3

Initially assume a constant value of the consistency index and subsequently account for its temperature-dependence using equation (6.36).
Also, ascertain the importance of free convection effects in this case,
using equations (6.37) and (6.38). The coefficient of thermal expansion, β is 3×10^{-4} K^{-1}.

6.3 A coal-in-oil slurry which behaves as a power-law fluid is to be heated   (c)
in a double-pipe heat exchanger with steam condensing on the annulus
side. The inlet and outlet bulk temperatures of the slurry are 291 K
and 308 K respectively. The heating section (inner copper tube of
40 mm inside diameter) is 3 m long and is preceded by a section
sufficiently long for the velocity profile to be fully established. The
flow rate of the slurry is 400 kg/h and its thermo-physical proper-
ties are as follows: density $= 900 \, \text{kg/m}^3$; heat capacity $= 2800 \, \text{J/kg K}$;
thermal conductivity $= 0.75 \, \text{W/mK}$. In the temperature interval $293 \leq$
$T \leq 368 \, \text{K}$, the flow behaviour index is nearly constant and is equal
to 0.52.

temperature (K)	299.5	318	333	353	368
consistency coefficient (Pa·s^n)	8.54	6.3	4.2	2.3	1.3

Calculate the temperature at which the steam condenses on the tube
wall. Neglect the thermal resistance of the inner copper tube wall. Do
not neglect the effects of free convection.

6.4 A power-law solution of a polymer is being heated in a   (c)
1.8 m long tube heater from 291 K to 303 K at the rate
of 125 kg/h. The tube is wrapped with an electrical heating
coil to maintain a constant wall flux of $5 \, \text{kW/m}^2$. Determine
the required diameter of the tube. The physical properties of
the solution are: density $= 1000 \, \text{kg/m}^3$; heat capacity $= 4180 \, \text{J/kg K}$;
thermal conductivity $= 0.56 \, \text{W/mK}$; flow behaviour index $n = 0.33$,
and consistency coefficient, $m = (26 - 0.0756 \, T)\text{Pa} \cdot \text{s}^n$, in the range
$288 \leq T \leq 342 \, \text{K}$.
   Also, evaluate the mean heat transfer coefficient if a mean value of
the consistency coefficient is used. How significant is free convection
in this case? Also, determine the temperature of the tube wall at its
halfway-point and at the exit.

6.5 A power-law solution of a polymer (density $1000 \, \text{kg/m}^3$) is flowing   (b)
through a 3 m long 25 mm inside diameter tube at a mean velocity
of 1 m/s. Saturated steam at a pressure of 0.46 bar and a temperature
of 353 K is to be condensed in the annulus. If the polymer solution
enters the heater at 318 K, at what temperature will it leave? Neglect
the heat loss to the surroundings. The thermo-physical properties of
the solution are: heat capacity $= 4180 \, \text{J/kg K}$; thermal conductivity $=$
$0.59 \, \text{W/mK}$; flow behaviour index, $n = 0.3$.

temperature (K)	303	313	323	333
consistency coefficient (Pa·s^n)	0.45	0.27	0.103	0.081

How many such tubes in parallel would be needed to heat up
17 tonne/h of this solution? What will be the exit fluid temperature

when the flow rate is 20% above, and 20% below, the value considered above?

6.6 A power-law fluid is heated by passing it under conditions of laminar   (c)
flow through a long tube whose wall temperature varies in the direction of flow. For constant thermophysical properties, show that the
Nusselt number in the region of fully-developed (hydrodynamical and
thermal) flow is given by:

$$\mathrm{Nu} = \frac{hd}{k} = \frac{6}{\phi(n)}\left(\frac{n+1}{3n+1}\right)$$

where $\phi(n) = \left(\dfrac{21}{20} - \dfrac{5n}{3n+1} + \dfrac{9n}{5n+1} - \dfrac{4n}{6n+1}\right)$

Assume that the temperature difference between the fluid and the
tube wall is given by a third degree polynomial:

$$\theta = T - T_w = a_0 y + b_0 y^2 + c_0 y^3$$

where y is the radial distance from the pipe wall and a_0, b_0, c_0 are
constant coefficients to be evaluated by applying suitable boundary
conditions.

6.7 A power-law fluid ($\rho = 1040\,\mathrm{kg/m^3}$; $C_p = 2090\,\mathrm{J/kgK}$; $k = 1.21$ (b)
W/mK) is being heated in a 0.0254 m diameter, 1.52 m long heated
tube at the rate of 0.075 kg/s. The tube wall temperature is maintained at 93.3°C by condensing steam on the outside. Estimate the
fluid outlet temperature for the feed temperature of 37.8°C. While the
power-law index is approximately constant at 0.35 in the temperature
interval $37.8 \le T \le 93.3$°C, the consistency coefficient, m of the fluid
varies as

$$m = 1.275 \times 10^4 \exp(-0.01452(T + 273))$$

where m is in Pa·s^n.
Is free convection significant in this example?

7.1 For the laminar boundary layer flows of incompressible Newtonian   (b)
fluids over a wide plate, Schlichting (Boundary Layer Theory, 6th
edn., Mc Graw Hill, New York, 1965) showed that the following two
equations for the velocity distributions give values of the shear stress
and friction factor which are comparable with those obtained using
equation (7.10):

(i) $\dfrac{V}{V_0} = 2\left(\dfrac{y}{\delta}\right) - 2\left(\dfrac{y}{\delta}\right)^3 + \left(\dfrac{y}{\delta}\right)^4$

(ii) $\dfrac{V}{V_0} = \sin\dfrac{\pi}{2}\left(\dfrac{y}{\delta}\right)$

Sketch these velocity profiles and compare them with the predictions of equation (7.10). Do velocity profiles (i) and (ii) above satisfy the required boundary conditions?

Obtain expressions for the local and mean values of the wall shear stress and friction factor (or drag coefficient) for the laminar boundary layer flow of an incompressible power-law fluid over a flat plate? Compare these results with the predictions presented in Table 7.1 for different values of the power-law index.

7.2 A polymer solution (density $1000\,\mathrm{kg/m^3}$) is flowing on both sides   (b)
of a plate 250 mm wide and 500 mm long; the free stream velocity is 1.75 m/s. Over the narrow range of shear rates encountered, the rheology of the polymer solution can be adequately approximated by both the power-law ($m = 0.33\,\mathrm{Pa \cdot s^n}$ and $n = 0.6$) and the Bingham plastic model (yield stress $= 1.75\,\mathrm{Pa}$ and plastic viscosity $= 10\,\mathrm{mPa \cdot s}$). Using each of these models, evaluate and compare the values of the shear stress and the boundary layer thickness 100 mm downstream from the leading edge, and the total frictional force exerted on the two sides of the plate.

At what distance from the leading edge will the boundary layer thickness be half of the value calculated above?

7.3 A dilute polymer solution at 293 K flows over a plane surface   (b)
(250 mm wide × 500 mm long) maintained at 301 K. The thermo-physical properties (density, heat capacity and thermal conductivity) of the polymer solution are close to those of water at the same temperature. The rheological behaviour of this solution can be approximated by a power-law model with $n = 0.43$ and $m = 0.3 - 0.000\,33\,\mathrm{T}$, where m is in $\mathrm{Pa \cdot s^n}$ and T is in K.
Evaluate:

(i) the momentum and thermal boundary layer thicknesses at distances of 50, 100 and 200 mm from the leading edge, when the free stream velocity is 1.6 m/s.
(ii) the rate of heat transfer from one side of the plate.
(iii) the frictional drag experienced by the plate.
(iv) the fluid velocity required to increase the rate of heat transfer by 25% while maintaining the fluid and the surface temperatures at the same values.

7.4 A polymer solution at 298 K flows at 1.1 m/s over a hollow copper   (b)
sphere of 25 mm diameter, maintained at a constant temperature of 318 K (by steam condensing inside the sphere). Estimate the rate of heat loss from the sphere. The thermo-physical properties of the polymer solution are approximately those of water; the power-law constants in the temperature interval $298 \leq T \leq 328\,\mathrm{K}$ are given below: flow behaviour index, $n = 0.40$ and consistency, $m = 30 - 0.05\,T$ ($\mathrm{Pa \cdot s^n}$) where T is in K.

What would be the rate of heat loss from a cylinder 25 mm in diameter and 100 mm long oriented with its axis normal to flow?

Also, estimate the rate of heat loss by free convection from the sphere and the cylinder to a stagnant polymer solution under otherwise identical conditions.

7.5 A 250 mm square plate heated to a uniform temperature of 323 K   (b)
is immersed vertically in a quiescent slurry of 20% (volume) TiO_2 in water at 293 K. Estimate the rate of heat loss by free convection from the plate when the rheology of the slurry can be adequately described by the power-law model: $n = 0.28$; $m = 10 - 0.2(T - 20)$ (Pa·s^n) where T is in °C.

The physical properties of TiO_2 powder are: density = 4260 kg/m^3; heat capacity = 943 J/kg K; thermal conductivity = 5.54 W/mK. The coefficient of thermal expansion of the slurry, $\beta = 2 \times 10^{-4} K^{-1}$.

7.6 A polymer solution (density 1022 kg/m^3) flows over the surface of a   (a)
flat plate at a free stream velocity of 2.25 m/s. Estimate the laminar boundary layer thickness and surface shear stress at a point 300 mm downstream from the leading edge of the plate. Determine the total drag force on the plate from the leading edge to this point. What is the effect of doubling the free stream velocity?

The rheological behaviour of the polymer solution is well approximated by the power-law fluid model with $n = 0.5$ and $m = 1.6$ Pa·s^n.

7.7 A china clay suspension (density 1200 kg/m^3, $n = 0.42$, $m =$ (a)
2.3 Pa·s^n) flows over a plane surface at a mean velocity of 2.75 m/s. The plate is 600 mm wide normal to the direction of flow. What is the mass flow rate within the boundary layer at a distance of 1 m from the leading edge of the plate?

Also, calculate the frictional drag on the plate up to 1 m from the leading edge. What is the limiting distance from its leading edge at which the flow will be laminar within the boundary layer?

8.1 A standard Rushton turbine at a operating speed of 1 Hz is used to   (a)
agitate a power-law liquid in a cylindrical mixing vessel. Using a similar agitator rotating at 0.25 Hz in a geometrically similar vessel, with all the linear dimensions larger by a factor of 2, what will be the ratio of the power inputs per unit volume of fluid in the two cases? Assume the mixing to occur in the laminar region. What will be the ratio of total power inputs in two cases? Plot these ratios as a function of the flow behaviour index, n.

8.2 How would the results in problem 8.1 differ if the main flow in the   (a)
vessel were fully turbulent? Does the flow behaviour index have an influence on these values?

8.3 A fermentation broth (density $890\,kg/m^3$) behaves as a power-law fluid   (a)
with $n = 0.35$ and $m = 7.8\,Pa\cdot s^n$. It must be stirred in a cylindrical
vessel by an agitator 150 mm in diameter of geometrical arrangement
corresponding to configuration A-A in Table 8.1. The rotational speed
of the impeller is to be in the range 0.5 to 5 Hz. Plot both the power
input per unit volume of liquid and the total power input as functions
of the rotational speed in the range of interest.

Due to fluctuations in the process conditions and the composition
of feed, the flow behaviour index remains constant, at approximately
0.35, but the consistency index varies by ±25%. What will be the
corresponding variations in the power input?

8.4 The initial cost of a mixer including its impeller, gear box and motor   (b)
is closely related to the torque, T, requirement rather than its power.
Deduce the relationship between torque and size of the impeller for
the same mixing time as a function of geometrical scale, for turbulent
conditions in the vessel. Does your answer depend upon whether the
fluid is Newtonian or inelastic shear-thinning in behaviour? What will
be the ratio of torques for a scale-up factor of 2?

8.5 A viscous material (power-law rheology) is to be processed in a   (a)
mixing vessel under laminar conditions. A sample of the material
is tested using a laboratory scale mixer. If the mixing time is to be
the same in the small and large-scale equipments, estimate the torque
ratio for a scale-up factor of 5 for the range $1 \geq n \geq 0.2$.

Additionally, deduce the ratio of mixing times if it is desired to
keep the torque/unit volume of material the same on both scales.

8.6 Tests on a small-scale tank 0.3 m in diameter (impeller diameter of   (b)
0.1 m, rotational speed 250 rpm) have shown that the blending of two
miscible liquids (aqueous solutions, density and viscosity approximately the same as for water at 293 K) is completed after 60 s. The
process is to be scaled up to a tank of 2.5 m diameter using the
criterion of constant (impeller) tip speed.

(i) What should be the rotational speed of the larger impeller?
(ii) What power would be required on the large scale?
(iii) What will be the mixing time in the large tank?

For this impeller–tank geometry, the Power number levels off to a
value of 6 for Re > 5×10^4.

8.7 An agitated tank, 3 m in diameter, is filled to a depth of 3 m with   (c)
an aqueous solution whose physical properties correspond to those of
water at 293 K. A 1 m diameter impeller is used to disperse gas and,
for fully turbulent conditions, a power input of $800\,W/m^3$ is required.

(i) What power will be required by the impeller?
(ii) What should be the rotational speed of the impeller?

(iii) If a 1/10 size small pilot scale tank is to be constructed to test the process, what should be the impeller speed to maintain the same level of power consumption, i.e., $800\,W/m^3$?

Assume that at the low gas used, Power number–Reynolds number relationship will not be affected and that under fully turbulent conditions, the Power number is equal to 6.

8.8 A power-law fluid is to be warmed from 293 K to 301 K in a vessel (2 m in diameter) fitted with an anchor agitator of 1.9 m diameter rotating at 60 rpm. The tank, which is filled to a 1.75 m depth, is fitted with a helical copper coil of diameter 1.3 m (25 mm od and 22 mm id copper tubing) giving a total heat transfer area of $3.8\,m^2$. The heating medium, water flowing at a rate of 40 kg/min, enters at 323 K and leaves at 313 K. The thermal conductivity, heat capacity and density of the fluid can be taken as the same as for water. The values of the power-law constants are: $n = 0.45$ and $m = 25 - 0.05\,T\,(Pa\cdot s^n)$ in the range $293\,K \leq T \leq 323\,K$.

(b)

Estimate the overall heat transfer coefficient and the time needed to heat a single batch of liquid.

Index

ABBOTT, M. 200, 203
ABID, M. 352, 389
ACOSTA, A. J. 314, 322
ACRIVOS, A. 297, 310, 319, 321
Activation energy, viscous flow 264
ADDIE, G. R. 199, 203
ADELMAN, M. 319, 322
AGARWAL, P. K. 236, 255
AGARWAL, U. S. 249, 255
Agitated tanks, heat transfer 365
— — scale-up 331
Agitation, scale 342
Air–kaolin suspension mixtures, vertical flow 191
AL-HADITHI, T. S. R. 57, 70
ALLEN, E. 222, 255
ALTENA, E. G. 64, 71
ALVES, G. E. 164, 202
AMATO, W. S. 319, 321
AMIS, E. J. 63, 70
ANDERSSEN, R. S. 65, 66, 70
ANDREWS, G. 262, 286
Annular flow pattern 166
Annulus, flowrate, Bingham plastic fluids 127
— — Herschel–Bulkley fluids 132
— — power-law fluids 125
ANSLEY, R. W. 212, 255
Anti-thixotropy 16
AOYAGI, Y. 133, 135, 159
Apparent viscosity 6
APPUZZO, G. 227, 256
ARAI, K. 369, 390
Archimedes number 252
ARMSTRONG, R. C. 3, 9, 20, 28, 34, 35, 57, 69, 81, 157, 212, 214, 256, 269, 284, 286
ARUMUGAM, M. 341, 389
ASCHOFF, D. 60, 69
ASTARITA, G. 12, 227, 256
ATAPATTU, D. D. 212, 214, 215, 256
Axial flow in an annulus, power-law fluids 124
AYAZI SHAMLOU, P. 369, 389
AZIZ, K. 19, 35, 81, 91, 96, 110, 118, 157, 158, 166, 167, 170, 171, 178, 202, 203

BAGCHI, A. 216, 256
BAGLEY, E. B. 39, 69
Bagley plot 38
BAKKER, A. 337, 341, 342, 383, 389
BANERJEE, T. K. 146, 157
BARDON, J. P. 261, 286
BARNEA, D. 168, 169, 202, 204
BARNES, H. A. 12, 13, 14, 17, 19, 25, 28, 32, 34, 35, 49, 51, 52, 56, 69
BASHIR, Y. 14, 35
BATES, R. L. 334, 389
BAUMGAERTEL, M. 65, 69
BEARD, D. W. 314, 322
BEASLEY, M. E. 146, 157
BEAULNE, M. 214, 215, 256
BECKNER, J. L. 340, 389
Bed expansion characteristics, fluidised beds 252
Beds of particles, equivalent diameter 233
— — — permeability 231
BEGISHEV, V. P. 60, 70
BERG, G. M. 383, 389
BERIS, A. N. 212, 214, 256
BERTRAND, F. 355, 390
BERTRAND, J. 352, 389
BEUKEMA, G. J. 64, 71
Biaxial extension 24
BINDING, D. M. 57, 67, 69
Bingham number 80, 91, 213, 214, 215, 299, 332
— plastic fluids 11, 12, 199
— — — in pipes, heat transfer 272, 273
— — — — laminar flow 78
— — — packed bed flow 237
— — — transitional and turbulent flow in pipes 101
— — model 13, 38, 47
BIRD, R. B. 3, 9, 14, 20, 28, 34, 35, 57, 69, 81, 125, 127, 130, 157, 158, 269, 277, 278, 286
BISHOP, A. A. 178, 181, 203
BITTLESTON, S. H. 127, 133, 158, 159
BLACKERY, J. 214, 256
BLASIUS, H. 303
BOBOK, E. 111, 159

BOBROFF, S. 222, 256
BOERSMA, W. H. 14, 35
BOGER, D. V. 5, 7, 20, 35, 53, 66, 67, 70, 71,
    101, 118, 144, 145, 157, 158, 159, 212,
    257
Boger fluid 5
BOGUE, D. C. 114, 116, 117, 157
BOND, W. N. 228, 256
BONDI, A. 245, 257
Boundary layer, concentration 311
— — flow 289
— — — laminar 293
— — — laminar–turbulent transition 302
— — — turbulent 302
— — — visco-elastic fluids 313
— — heat transfer 303, 304
— — momentum 289
— — over a plate, laminar 290
— — — — — turbulent 302
— — thermal 307
— — thickness, concentration 311
— — — momentum 290, 295, 298
— — — thermal 307
BOURNE, J. R. 353, 358, 389
BOWEN, R. L. 104, 107, 110, 111, 157
BRANDRUP, J. 261, 286
BREA, F. M. 238, 256
BRENNER, H. 232, 257
BRIEND, P. 253, 256
BRIGHAM, O. E. 62, 70
Brinkman number 284
BROADBENT, J. M. 57, 70
BRODKEY, R. S. 39, 114, 157
BROOKES, G. F. 212, 256
Brookfield viscometer 45
BROWN, G. G. 241, 256
BROWN, N. P. 118, 147, 157, 202
BROWN, R. C. 212, 214, 256
BRUGGEMANN, D. A. G. 262, 286
BRUNN, P. O. 247, 257
Bubble flow 165
— rise velocities in polymer solutions 227
— size and hold-up in mixing tanks 363
Bubbles and drops, drag correction factor 225
— — — in power-law fluids 224
Buffer layer, turbulent flow in pipes 112
BUTLER, H. 353, 358, 389

CALDERBANK, P. H. 340, 389
CAMPBELL, G. A. 215, 257
Capillary model, packed bed flow 232
— viscometer 37, 39
— — end correction 38
CARLETON, A. J. 186, 203
Carman–Kozeny equation 250
— — model 232
CARMAN, P. C. 236, 256
CARNALI, J. O. 56, 69

CARREAU, P. J. 9, 10, 20, 34, 35, 157, 285,
    341, 353, 355, 359, 363, 368, 369, 389,
    391
Carreau viscosity model 10
CARTER, G. 154, 158
Casson fluid model 13
CATHEY, C. A. 67, 70
CHAKRAVORTY, S. 67, 71
CHAMBON, F. 60, 70, 71
CHAPMAN, F. S. 355, 389
Characteristic time, fluid 176
CHARLES, M. E. 170, 181, 185, 199, 203
CHARLES, R. A. 199, 203
CHASE, R. C. 114, 157
CHAVAN, V. V. 341, 358, 389
CHAVARIE, C. 253, 256
CHEESMAN, D. J. 215, 257, 349, 389
CHEN, J. J. J. 170, 171, 172, 203, 204
CHENG, D. C.-H. 53, 70, 96, 98, 110, 146,
    158, 186, 203, 356, 389
CHENG, J. 341, 363, 389
CHEVALIER, J. L. 341, 368, 389
CHHABRA, R. P. 9, 20, 34, 35, 52, 70, 88, 89,
    147, 157, 158, 166, 167, 170, 174, 175,
    176, 178, 180, 187, 189, 190, 199, 202,
    203, 206, 207, 209, 210, 212, 214, 215,
    216, 221, 222, 223, 224, 226, 227, 228,
    232, 237, 238, 239, 242, 248, 249, 252,
    253, 254, 255, 256, 257, 285, 310, 313,
    314, 315, 321, 322, 341, 389
CHISHOLM, D. 166, 178, 179, 180, 203
CHO, K. 261, 286
CHO, Y. I. 96, 145, 158, 261, 265, 272, 277,
    278, 285, 286
CHOI, P. K. 63, 71
CHOPLIN, L. 355, 390
CHOU, C. H. 133, 136, 158
CHRISTIANSEN, E. B. 281, 282, 286
CLAPP, R. M. 116, 158
CLARK, N. N. 253, 256
CLIFT, R. 221, 225, 226, 230, 255, 256
COHEN, Y. 240, 241, 249, 256
COLLIAS, D. J. 341, 389
COLLINS, M. 145, 158
COLWELL, R. E. 44, 45, 52, 71
COMITI, J. 242, 257
Complex shear modulus 58
— viscosity 59
Concentration of coarse solids, discharged 200
— — — in-situ 200
Concentric cylinder viscometer 42, 43
— — — wide gap 44
— cylinders, end effects 43
Cone and plate geometry 56, 57
— — — viscometer 47
— — — — slip effects 52
Consistency coefficient 10

Constant wall heat flux condition, analytical
     results 277
— — — — — experimental results 277
Controlled stress rheometer 50
CORPSTEIN, R. R. 334, 389
Correction factor, kinetic energy 83
COUDERC, J.-P. 253, 256
Couette correction, power-law fluids 144
COULSON, J. M. 74, 82, 149, 158, 222, 230,
     241, 244, 255, 256, 278, 286, 303, 331,
     367, 389
COYLE, C. K. 369, 389
CRAIG, S. E. 281, 282, 286
CRAWFORD, T. 167, 204
Creep test 55
Creeping flow past a sphere, criterion 215
— — — — power-law fluid 208, 209
— — — — visco-elastic fluid 215, 216
— — — — viscoplastic fluid 211, 214
Critical Reynolds number 90
— — — Bingham plastic fluids 91, 93
— — — power-law fluids 90
CURTISS, C. F. 57, 69

DAI, G. C. 14, 35, 127, 157
DALLAVALE, J. M. 261, 286
DARBY, R. 101, 110, 157, 158
DAS, M. 146, 157
DAS, S. K. 146, 157
DAVIDSON, J. F. 230, 255, 256
DAVIES, A. R. 65, 66, 70
DAVIES, J. M. 60, 61, 67, 70
DE HAVEN, E. S. 146, 158
DE KEE, D. 9, 20, 34, 35, 157, 226, 227, 228,
     256, 285
DE LA FUENTE, E. B. 355, 390
DEALY, J. M. 65, 71
Deborah number 30, 175, 216, 247, 314, 341,
     350
DECKWER, W.-D. 365, 390
DEGAND, E. 216, 256
DENN, M. M. 30, 35, 68, 71, 314, 322
DESHPANDE, S. D. 178, 181, 203
DESPLANCHES, H. 341, 368, 389
Developing flow, pressure drop 144
Deviatoric stresses 5
DHARAMADHIKARI, R. V. 236, 256
DICKEY, D. S. 342, 390
Die swell 20
DIJKSTRA, F. C. 282, 286
Dilatant behaviour 6, 14
DIMITROV, V. 64, 71
DINH, S. M. 284, 286
Discharge concentration 200
Dispersion coefficient, axial and longitudinal
     243
— in fluidised beds 254
— — packed beds 242

Dispersion of gas in a liquid 360
— — — — — bubble size and hold-up 363
— — — — — cavity formation 361
— — — — — composite impellers 360,
     361, 378
— — — — — helical ribbon impeller 363
— — — — — mass transfer 365
— — — — — mechanism 361
— — — — — pseudoplastic fluids 361
— — — — — visco-elastic fluids 361
DODGE, D. W. 91, 96, 97, 104, 106, 107, 114,
     115, 116, 145, 147, 158
DOLEJS, V. 242, 257
DOMININGHAUS, H. 261, 286
DONATELLI, A. A. 319, 322
DORAISWAMY, D. 340, 389
Drag coefficient 295, 299
— — sphere, creeping flow region 208, 209
— — — high Reynolds numbers 210
— correction factor 208, 209, 214
— — — bubbles and drops 225
— — — rigid spheres 208
— on a sphere 207
— — — — power-law fluids 208, 209
— — — — visco-elastic fluids 215
— — — — viscoplastic fluids 211, 214
— — non-spherical particles 223
— — — — shear-thinning and dilatant fluids
     224
— ratio, modified 187
— — shear-thinning fluids, minimum 181
— — two-phase flow 179
— reduction, gas–liquid flow 181
— — in gas–liquid systems, optimum gas
     flowrate 193
— — — — — practical applications 193
DREAM, R. F. 370, 389
DUAL, J. 64, 70
DUCKHAM, C. B. 154, 158
DUCKWORTH, R. A. 199, 203
DUCLA, J. M. 341, 389
DUKLER, A. E. 168, 169, 170, 181, 204
DULLIEN, F. A. L. 231, 232, 235, 255, 256
DUNCAN, D. 167, 204
DUTTA, A. 249, 255, 262, 266, 286
DZIUBINSKI, M. 187, 188, 190, 194, 203, 232,
     233, 236, 237, 257

Eddy momentum diffusivity 113
— viscosity 112
EDNEY, H. G. S. 368, 389
EDWARDS, M. F. 14, 35, 118, 146, 148, 158,
     238, 256, 326, 331, 335, 340, 341, 360,
     363, 365, 367, 368, 369, 370, 371, 374,
     388, 389, 390
Effect of non-isothermal rheological properties
     281
EGGERS, F. 59, 70

EISSENBERG, F. G. 170, 181, 203
Ellis fluid model 31
— model fluids, laminar flow in pipes 85
— — of fluids 11
Elongational flow 23
— viscosity 24
ELSON, T. P. 215, 257, 349, 389
End effects, concentric cylinders 43
Energy equation, mechanical 147
— — pipe flow, power-law fluids
    265, 266
Entry lengths, power-law fluids 142, 145
Equivalent diameter, bed of particles 233
— — non-circular ducts 133
ERGUN, S. 238, 256
ETCHELLS III, A. W. 340, 389
EVANS, I. D. 12, 35
Extension, uniaxial 25
Extensional effects in packed beds 247
— flow 23
— — tension-thinning 67
— viscosity 68
— — filament stretching method 67
— — measurement methods 66, 67
— — stagnation flows 67
Extra stresses 5
Extruders, mixing 384

Falling ball viscometer 51
False body fluid 16
FAN, L.-S. 254, 257
FANG, T. N. 242, 257
FAROOQI, S. I. 88, 89, 147, 158, 170, 172,
    174, 175, 176, 185, 189, 203
FARRIS, R. G. 60, 71
FASANO, J. B. 342, 389
FEEHS, R. H. 340, 390
FENIC, J. G. 342, 390
FERGUSON, J. 66, 68, 70
FERGUSON, M. E. G. 166, 203
FERREIRA, J. M. 216, 256
FERRY, J. D. 58, 60, 64, 70, 264, 286
First normal stress difference 22, 29, 56, 176
Fixed beds, non-Newtonian flow 206
FLEMMER, R. C. 253, 256
Flow, air–kaolin suspension mixtures 173
— air–polymer solution mixtures 175
— boundary layer 289
— curve 2
— elongational 23
— extensional 23
— field around a sphere, viscoplastic fluids
    212
— in beds of particles 228
— — between plates, laminar 118
— — concentric annulus, laminar 122
— — conduits 73
— — non-circular ducts, laminar 133

Flow in packed beds, Bingham plastic fluids
    237
— — — — power-law fluids 238
— — — — transitional and turbulent 238
— — pipes 73
— — — entrance effects 142
— — — gas–liquid mixtures 162
— — — gas–non-Newtonian liquids 163
— — — pulsating 118
— — — time-dependent fluids 118
— — rough pipes 111
— measurement, non-Newtonian fluids 146
— of Bingham plastic fluids, annulus 127
— — plastics, pipe 80
— patterns, disc turbine impeller 347
— — gas–liquid, bubble flow 165
— — — — plug flow 165
— — — — slug flow 165
— — — — stratified flow 165
— — — — flow 164
— — — — — horizontal 164
— — helical ribbon impeller 353
— — — screw impeller 355
— — in mixing tanks 346
— — mixers, visco-elastic fluids 350
— — — viscoplastic materials 349
— — prediction, shear-thinning fluids 166
— — propeller 347
— — pseudoplastic fluids 348
— — secondary circulation flows 352
— — sigma blade mixer 355
— — vertical upward flow 167
Fluid behaviour, classification 1
— characteristic time 28
— relaxation time 28
Fluidisation 229, 249
— incipient 249
— liquid–solid, bed-expansion behaviour 252
— — minimum fluidising velocity 251
— — power-law fluids 249
— — visco-elastic effects 253
— mass transfer 254
Fluidised beds 229, 249
— — dispersion 254
— — effect of particle shape 253
— — expansion characteristics 252
— — mass transfer 254
— — non-Newtonian flow 206
FONDY, P. L. 334, 389
FORD, E. W. 146, 159
Ford-cup viscometer 51
FORDHAM, E. J. 127, 133, 158
FORREST, G. 282, 284, 286
FOSSA, M. 170
FOSSA. F. 203
Fourier transform mechanical spectroscopy,
    rheometry 60, 61, 62

FREDRICKSON, A. G. 125, 127, 130, 158, 314, 322
Free convection effects 272, 275, 276, 278
FRENCH, R. J. 186, 203
FREUNDLICH, H. 17, 35
Friction factor 96, 101, 133
— — packed beds 236, 238, 247
— — pipe flow 80, 87
— — power-law fluids in pipes 88
— in pipes 110
— losses, minor 140
— — miscellaneous 140
— velocity 113
Frictional drag on a plate 295, 299
Froude number 29, 331, 333, 361
FULLER, G. G. 67, 70

Galileo number 251
GALINDO, E. 363, 389
GANDELSMAN, M. 63, 70
GARCIA, E. J. 110, 158
Gas–liquid flow in pipes 162
— — — vertically upwards 167
— — mixing 359
— — — bubble size and hold-up 363
— — — mass transfer 365
— — — power consumption 362
GATES, L. E. 337, 341, 342, 389
Generalised Newtonian fluids 5
GENTRY, C. C. 319, 322
GERMAN, R. M. 241, 256
GHOSH, T. 203
GHOSH, U. K. 206, 257, 313, 314, 315, 322
GIBSON, J. 167, 204
GIBSON, R. 65, 71
GIESEKUS, H. 350, 389
GINESI, D. 147, 158
GLUZ, M. 368, 389
GODDARD, J. D. 14, 35
GODFREY, J. C. 340, 355, 389
GODLESKI, E. S. 346, 358, 389
GOGOS, G. 157, 159
GOODBREAD, J. 64, 70
GORI, F. 261, 286
GORODETSKY, G. 63, 70, 71
GOTOH, S. 349, 390
GOTTLIEB, M. 63, 70, 71
GOULAS, A. 154, 159
GOVIER, G. W. 19, 35, 81, 91, 96, 110, 118, 157, 158, 166, 171, 178, 202, 203
GRACE, J. 221, 225, 226, 255, 256
Graetz number 267, 268, 269, 271, 272, 277, 281
GRAHAM, D. I. 209, 257
Grashof number 272, 318, 331, 368
GREEN, R. G. 88, 158
GREENKORN, R. A. 231, 232, 257
GREGORY, G. A. 167, 170, 171, 203

GRENVILLE, R. K. 335, 340, 389
GRIGULL, U. 277, 286
GRIMM, R. J. 28, 35
GRISKEY, R. G. 88, 158
GU, D. 209, 257
GÜCÜYENER, H. I. 133, 158
GUERIN, P. 359, 363, 368, 369, 389
GUPTA, O. P. 314, 322
GUPTA, R. K. 5, 26, 35, 67, 68, 71, 262, 286
GUTKIN, A. M. 212, 257

HAGEDORN, D. 370, 389
HALL, K. R. 355, 389
HANBY, R. L. 55, 70
HANKS, R. W. 91, 110, 125, 126, 133, 158
HAPPEL. J. 232, 257
HARIHARAPUTHIRAN, M. 215, 257
HARNBY, N. 326, 331, 335, 341, 360, 363, 365, 374, 388, 389
HARRIS, J. 98, 146, 158, 314, 321
HARRISON, D. 230, 255, 256
HARTNETT, J. P. 52, 70, 96, 110, 135, 136, 145, 158, 159, 261, 265, 272, 277, 278, 285, 286
HASHIMOTO, K. 365, 390
HASSAGER, O. 3, 9, 20, 28, 34, 57, 69, 81, 157, 269, 286
HASSELL, H. L. 245, 257
HAUSLER, K. 64, 70
Heat and mass transfer, free convection 318
— — — to power-law fluids, correlations 314
— — — — — cylinders 315, 316
— — — — — spheres 314
— transfer, Bingham plastic fluids 273
— — coefficients 267, 309
— — — film, class I impellers 368
— — — — II impellers 368
— — — — III impellers 369
— — — in mixing tanks, Newtonian fluids 367
— — — — — non-Newtonian fluids 368
— — helical coils 366
— — in pipes, transitional and turbulent flow 285
— — isothermal tube wall condition 267
— — jacketed tanks 365, 369
— — non-isothermal rheology 281
— — non-Newtonian fluids 260
— — to particles 314
HEDSTRÖM, B. O. A. 80, 158
Hedström number 80, 91, 93, 101
HEIM, A. 369, 389
HERBST, H. 365, 390
HERMANSKY, C. G. 66, 67, 70
HERMES, R. A. 314, 322
Herschel–Bulkley model 13
— — — fluids 132

Herschel–Bulkley model fluids, laminar flow 80
— — modelfluids 214
HETSRONI, G. 166, 178, 203
HEWITT, G. F. 166, 170, 203
HEYWOOD, N. I. 96, 98, 110, 118, 147, 157,
    158, 170, 181, 185, 202, 203, 387, 390
HICKMAN, D. 364, 390
HICKS, R. W. 342, 390
HIGMAN, R. W. 56, 70
Hindered settling velocity 221
HIRABAYASHI, H. 349, 390
HIRAI, E. 272, 286
HIRAI, N. J. 340, 390
HIROSE, T. 225, 257
HIRSCHLAND, H. E. 369
HLAVACEK, B. 253, 256
HOCKEY, R. M. 334, 390
HODGSON, D. F. 63, 70
Hold-up, experimental methods 169, 170
— gas–non-Newtonian liquid flow 168
— in two-phase flow 168
— prediction for gas–liquid flows 170
— predictive methods, horizontal flow 170
— — — vertical flow 176
Hole pressure effect, 57
HOLLAND, F. A. 355, 389
HOLLY, E. E. 60, 70
HONERKAMP, J. 65, 70
Hooke's law 19
Horizontal two-phase flow, hold-up 170
— — — pressure drop 177
HOUGHTON, G. L. 267, 272, 286
HOYT, J. W. 96, 159
HU, R. Y. Z. 52, 70
HUDSON, N. E. 66, 68, 70
HUTTON, J. F. 25, 28, 32, 34, 35, 49, 52, 57,
    69
HWANG, S.-J. 254, 255, 257
Hydraulic mean diameter, non-circular ducts
    133
— transport of solids 197
Hysteresis, time-dependent fluids 16

IBRAHIM, S. 335, 390
IMMERGUT, E. H. 261, 286
Impeller geometries 375
— Scaba 360
— speed, selection 342
— types, double helicone 378
— — Master mix HVS/TS 379
— — sigma blades 380
— — special designs 377
Impellers, types 375, 378, 379
— — Intermig 360
IN, M. 60, 70
Incipient fluidisation 249
Infinite shear viscosity 7
INGARD, K. U. 64, 70

In-line mixers 385
In-situ concentration 200
Instruments, rheological 37
Iron oxide slurry, flow of 101
IRVINE JR., T. F. 97, 98, 135, 158, 261, 265,
    272, 277, 286, 314, 318, 322
Isothermal tube wall condition, analytical
    results 272

Jacketed tanks, Bingham plastic fluids 371
— — heat transfer 369
— — non-Newtonian fluids 370
— — pseudoplastic fluids 371
JACKS, J.-P. 241, 257
JACKSON, R. 56, 70
JACOBSEN, R. T. 46, 70
JADALLAH, M. S. M. 146, 148, 158
JAMES, A. E. 53, 54, 55, 70, 71
JAMES, D. F. 26, 35, 69, 314, 322
JANE, R. J. 356, 389
JANSSEN, L. P. B. M. 385, 390
JENSON, V. G. 273, 286
JESCHKE, D. 367, 390
JOHMA, A. I. 340, 390
JOHNSON, M. M. 90, 159
JONES, D. M. 25, 35, 69
JONES, T. E. R. 60, 61, 67, 70, 209, 257
JONES, W. M. 68, 69, 70
JOSEPH, D. D. 64, 70
JOSHI, S. D. 265, 286
JULISBERGER, F. 17, 35

KAI, W. 369, 390
KALE, D. D. 236, 256
KAMATH, V. M. 65, 70
KAMINSKY, R. D. 190, 203
KANTYKA, T. A. 368, 390
Kaolin suspensions, two-phase flow 173
KAPPEL, M. 355, 390
KARNI, J. 265, 272, 277, 286
KARNI, S. 314, 318, 322
KASEHAGEN, L. J. 68, 71
KASHIM, M. M. 340, 389
KATZ, D. 18, 35
KAWASE, Y. 365, 390
KAYAMA, T. 369
KAYE, A. 56, 57, 70
KEENTOK, M. 53, 70
KELKAR, J. V. 341, 350, 390
KELLER, D. S. 18, 35
KELLER JR., D. V. 18, 35
KELLER, R. J. 118, 159
KELLER, T. A. 146, 159
KEMBLOWSKI, Z. 232, 233, 236, 237, 257
KENCHINGTON, J. M. 200, 201, 203
KEUNINGS, R. 68, 71
KHATIB, Z. 168, 169, 170, 176, 190, 203
KIM, C. B. 319, 322

KIM, I. 261, 286
Kinetic energy correction factor 83
— — of fluid, average 82
KING, I. 170, 203
KISS, A. D. 54, 71
KOSTIC, M. 110, 135, 136, 158, 265, 272, 286
Kozeny–Carman equation 250
— — model 232
KOZICKI, W. 133, 136, 158
KRIEGER, I. M. 45, 46, 70
KUBOI, R. 374, 390
KULICKE, W.-M. 22, 23, 35
KUMAR, S. 245, 246, 254, 257, 313, 314, 322
KUO, Y. 56, 70
KURIYAMA, M. 369, 390
KWACK, E. Y. 136, 158
KWANT, P. B. 282, 286

LABA, D. 34
LACORIX, A. 355, 390
LAIRD, W. M. 127, 128, 158
LAMBERT, D. J. 154
Laminar boundary layer flow, Bingham plastic
        fluids 297
— — — — power-law fluids 293
— — — — velocity distribution 293
— flow 38
— — beds of particles 228
— — between parallel plates 118
— — concentric annuli 122, 134
— — elliptical ducts 134
— — in non-circular ducts 133
— — — pipes 73
— — — — Bingham plastic fluids 78, 81
— — — — heat transfer 264
— — — — power-law fluids 74, 81
— — — rectangular pipes 134
— — — triangular pipes 134
— mixing, shear-thinning fluids 333
— sub-layer 112, 113, 116
Laminar–turbulent transition, boundary layer
        flow 302
— — — in pipes 90
— — — mixing tanks 335
LARSEN, K. M. 125, 126, 158
LARSON, R. G. 28, 34, 35
LAUN, H. M. 38, 70
LAWAL, A. 282, 284, 286
LEE, J. 114, 157
LEE, T.-L. 319, 322
LEE, Y. H. 170, 204
LEIDER, P. J. 28, 35
LEVA, M. 241, 257
LEVAN, J. 14, 35
Leveque approximation 269, 277
LEVEQUE, J. 269, 286
LIEBE, J. O. 170, 204
LIEW, K. S. 319, 322

LIGHTFOOT, E. N. 3, 35
LIM, K. Y. 44, 45, 52, 71
LINDLAY, J. A. 326, 390
LIPTAK, B. G. 147, 158
Liquid–gas flow in pipes 162
Liquid hold-up, average 170, 171
— — non-Newtonian systems, average 172,
        173, 174, 175
— — turbulent flow 176
— — — region, average 176
— — visco-elastic fluids 175
— mixing 324, 327
— — mechanisms 327
— — mixing time 327
— — power consumption 327
— — rate of mixing 327
— — scale-up criteria 327
— — similarity criteria 327
— — visco-elastic fluids 341
Liquid–solid fluidisation, bed-expansion
        behaviour 252
— — — minimum fluidising velocity 251
— — — power-law fluids 249
— — — visco-elastic effects 253
— — mass transfer, fluidised beds 254
— — — — packed beds 245
LLINAS, J. R. 368, 389
Lockhart–Martinelli parameter 171, 173, 174,
        176, 179, 185, 188, 189
— — — modified 174, 187
LOCKHART, R. W. 171, 179, 180, 185, 187,
        188, 203
LOCKYEAR, C. F. 199, 203
LODGE, A. S. 57, 70
LOHNES, R. A. 55, 70
LOULOU, T. 261, 286
LOVEGROVE, P. C. 170, 203
LU, C.-B. 254, 255, 257
LU, W.-J. 254, 255, 257
LYCHE, B. C. 269, 286
LYONS, J. W. 44, 45, 52, 71

MCDONOUGH, R. J. 388
MACHAC, I. 215, 242, 257
MACHON, V. 341, 361, 364, 390
MCKELVEY, J. M. 157, 158
MCKETTA, J. J. 202
MCKINLEY, G. H. 68, 71
MACKLEY, M. R. 65, 70
MCLEAN, A. 241, 257
MACOSKO, C. W. 23, 28, 34, 35, 37, 42, 44,
        48, 49, 67, 68, 69, 70, 71
MACSPORRAN, W. C. 59, 70
MAGNALL, A. N. 146, 158
MAHALINGAM, R. 170, 178, 181, 202, 203,
        278, 286
MALIN, M. R. 91, 158
MALKIN, A. Y. 60, 70

MANDHANE, J. M. 167, 171, 203
MANNA, L. 357, 390
MANSUROV, V. A. 60, 70
MARON, S. 45, 46, 70
MARSH, B. D. 56, 70
MARSHALL, R. J. 248, 257
MARTINELLI, R. C. 171, 179, 180, 185, 187, 188, 203
MARTONE, J. A. 370, 390
MASHELKAR, R. A. 249, 255, 262, 266, 286, 310, 314, 318, 322, 341, 350, 389, 390
Mass transfer coefficients 312
— — gas–liquid systems 365
— — in fluidised beds 254
— — — packed beds 245
— — liquid–solid, fluidised beds 254
— — — — packed beds 245
— — to particles 314
MATTA, J. E. 68, 70
MATTAR, L. 170, 203
MATTHYS, E. G. 69, 70
Maxwell model 26, 27
Maxwellian relaxation time 28
MC *entries are sorted with* MAC
Mechanical energy equation 147
MEHMETEOGLU, T. 133, 158
Memory function 57
MERRILL, R. P. 241, 257
METZNER, A. B. 14, 30, 35, 83, 86, 87, 88, 91, 96, 97, 104, 106, 107, 110, 114, 115, 116, 117, 145, 147, 157, 158, 240, 241, 248, 256, 257, 267, 272, 286, 337, 340, 341, 346, 358, 390
Metzner–Reed Reynolds number 77, 172, 174, 201
— — — — gas–liquid mixtures 183
MEWIS, J. 19, 35
MICHAEL, B. J. 369, 389
MICHELE, H. 248, 257
MILLAN, A. 55, 70
MILLER, C. 133, 158, 236, 257
MILLIKAN, C. B. 96, 158
MILTHORPE, J. F. 53, 70
Minimum fluidising velocity, power-law fluids 251
MISHRA, I. M. 238, 257, 313, 322
MISHRA, P. 238, 257, 313, 322
Mist flow 166
MITSON, R. J. 341, 390
MITSOULIS, E. 214, 215, 256
MITSUISHI, N. 133, 135, 159, 340, 390
Mixers for pastes 380
— in-line 385
— portable 385
— static 385
Mixing, batch 327
— continuous 384

Mixing, effect of visco-elasticity 341
— equipment, baffles 374
— — impellers 375
— — mechanical agitation 374
— — selection 374
— — tanks 374
— — variety 374
— gas–liquid 325, 359
— gas–liquid–solid 326
— immiscible liquids 325
— in continuous systems 384
— — — — extruders 384
— — — — static mixers 385
— — rolling operations 382
— laminar 328
— liquid 324
— liquid–solid 325
— mechanisms, liquid 327
— Newtonian fluids, laminar–turbulent transition 335
— non-Newtonian fluids 324
— rate 356
— solid–solid 326
— static 385
— tanks, average shear rate 337
— — flow patterns 346
— — heat transfer 365
— — scale-up 331
— thixotropic materials 340
— time 356
— — dimensionless 358
— — inelastic fluids 359
— — visco-elastic fluids 359
— turbulent 330
— visco-elastic effects 341
Modified Lockhart-Martinelli parameter 174, 187
MOHAMED, I. O. 146, 159
Momentum balance equation, boundary layer 291, 292
— transfer 2
MOO-YOUNG, M. 225, 257, 340, 365, 389, 390
Mooney–Ewart geometry 44
MOONEY, M. 38, 71
MORGAN, R. G. 149, 159
MORTON, J. R. 342, 383, 389, 390
MOURS, M. 71
MUJUMDAR, A. S. 282, 284, 286
Multiphase flow in pipes 162
— — — — flow patterns 164
MUN, R. 101, 158

NAGATA, S. 340, 349, 353, 369, 388, 390
NAIR, V. R. U. 222, 256
NAKAJIMA, M. 369, 390
NAKAMURA, M. 365, 390
NAKANO, Y. 226, 257

NAKAYAMA, A. 310, 314, 322
NARAIN, A. 64, 70
Natural convection, heat and mass transfer 318
— — — transfer 272, 275, 276, 278
NAVRATIL, L. 111, 159
NELLIST, D. A. 118, 158
NEWITT, D. M. 200, 203
NEWTON, D. A. 228, 256
Newtonian fluid behaviour 10
— — definition 1
NGUYEN, D. A. 67, 68, 71
NGUYEN, Q. D. 16, 35, 53, 71, 212, 257
NICOLAE, G. 242, 257
NIENOW, A. W. 326, 331, 335, 341, 349, 360,
    361, 362, 363, 364, 365, 374, 388, 389,
    390, 391
NISHIKAWA, M. 349, 365, 369, 390
Nominal shear rate 38, 40
— — — at wall 87
Non-circular ducts, equivalent diameter 133
Non-isothermal flow in tubes, power-law fluids
    262
Non-Newtonian flow characteristics 32
— fluid behaviour 5
— — mixing, average shear rate 337
— — — effective shear rate 337
— fluids 37, 74
— — fixed bed flows 228
— — flow metering 146
— — fluidisation 229, 249
— — in a mixer, apparent viscosity 332
— — — pipes, heat transfer 260
— — pumps 149
— — thermo-physical properties 261
— — transient flow 118
Non-spherical particles, drag 223
Normal stress difference coefficients 21
— — — primary 20
— — — secondary 20
— — measurements 56
— stresses 4, 21
NORWOOD, K. W. 346, 358, 390
NOURI, J. M. 334, 390
Nusselt number 277, 281, 308, 331
— — cylinders 315, 316, 319
— — different impeller geometries 370
— — mixing tanks 367, 368
— — pipe flow 267, 269, 271, 272
— — plates 318
— — spheres 314, 319

O'DONOVAN, E. 53, 70
O'NEILL, B. K. 236, 255
OKUBO, T. 64, 71
OLDSHUE, J. Y. 331, 369, 374, 388, 389, 390
OLIVER, D. R. 170, 174, 181, 203, 273, 286,
    341, 390
OOSTERBROEK, M. 64, 71

ORBEY, N. 65, 71
ORR, C. 261, 286
Oscillatory shear, loss modulus 59
— — phase lag 58
— — test 57, 59, 65
Ostwald–de Waele model 9
OTTINO, J. M. 330, 388, 390
OTTO, R. E. 337, 340, 341, 390
OZGEN, C. 133, 159

PACE, G. W. 349, 361, 363, 391
Packed bed flow, anomalous effects 248
— — — capillary model 232
— — — wall effects 240
— beds, anomalous effects 248
— — dispersion 242
— — effect of particle shape 241
— — liquid 61solid mass transfer **I**
— — mass transfer 245
— — non-Newtonian flow 206, 228
— — tortuosity factor 234
— — voidage 231
— — wall effects 240
PADMANABHAN, M. 68, 71
PAL, R. 146, 159
Parallel plate viscometer 48
PARIS, J. 359, 363, 368, 369, 389
PARK, M. G. 91, 158
PARKER, H. W. 245, 257
Particle drag coefficient, effect of particle shape
    223
Particulate suspensions, thermo-physical
    properties 262
— systems, non-Newtonian fluids 206
Paste-like materials, mixing 380
PATEL, K. G. 370, 391
PATTERSON, G. K. 388
PATTERSON, I. 353, 355, 359, 389
PATTERSON, W. I. 341, 391
PAVLUSHENKO, I. S. 368, 389
PAYNE, L. W. 245, 257
PEARSON, J. R. A. 56, 70
Peclet number 245, 254, 255
PEERHOSSAINI, H. 261, 286
PENNEY, W. R. 342, 389
PEREZ, J. F. 365, 390
Permeability of a porous medium 231
PETERS, D. C. 350, 351, 352, 353, 358, 390
PETERSEN, E. E. 297, 310
PETERSEN, F. W. 159
PETRICK, P. 170, 203
PHILLIPS, R. J. 222, 256
PICKETT, J. 253, 256
PIENAAR, V. G. 159
PIGFORD, R. L. 272, 286
PIKE, R. W. 170, 204
Pipe roughness 111
Planar extension 24

Plastic viscosity 11
Plug flow 165
— — model, gas–liquid flow 184
— — region 78
POGGERMANN, R. 367, 368, 371, 390
Poiseuille equation 38
POLLARD, J. 368, 390
Polymer melts, specific heat 261
— — thermal conductivity 261
— solutions, apparent viscosities 8
— — bubble rise velocities 227
— — drag reduction 118
— — heat transfer 271, 272
— — heating and cooling in mixing tanks 371
— — thermal conductivity 261
Porous media, characteristics 231
— — definition and characterisation 230
Portable mixers 383
PORTER, J. E. 261, 272, 286
Power curves, gas–liquid systems 362
— — Newtonian fluids 334, 335
— — non-Newtonian (inelastic) fluids 336, 340
— gas–pseudoplastic liquid mixing 363
— gas–visco-elastic liquid mixing 363
— mixing, inelastic fluids 336
— number 333, 334, 335, 337, 343, 362
— requirement in mixers, high viscosity systems 336
— — — — low viscosity systems 333
— — — — Newtonian fluids 333
— — — — non-Newtonian systems 336
— requirements, typical 342
Power-law consistency coefficient 10
— — — temperature dependence 263, 272, 278
— flow behaviour index 10
— — — — temperature dependence 263
— fluid flow, fixed beds 206
— — — fluidised beds 206
— — — parallel plates 118
— fluids 216, 217
— — bubbles and drop motion 224
— — entry lengths 142, 145
— — flow in an annulus 124
— — in pipes, heat transfer 262, 265
— — laminar flow 74
— — pulsating flow 118
— — transitional and turbulent flow 95, 96, 101
— index, apparent 41
— model 9, 14, 31, 38
— — apparent 87, 235, 238
PRADIPASENA, P. 18, 35
Prandtl mixing length 113
— number 272, 308, 331, 368, 370
Pressure drop across a bed of particles 229

Pressure drop, effect of pipe roughness 111
— — fittings 145
— — horizontal two-phase flow 177
— — in non-circular ducts, laminar flow 133
— — — pipes 73, 101
— — — — Bingham plastic fluids 80, 81
— — — — gas–liquid flows 177
— — — — gas–non-Newtonian systems 181
— — — — power-law fluids 75, 81
— — — — predictive methods, two-phase gas–liquid flow 179
— — liquid–solid flows 200
— — power-law fluid flow in packed beds, effect of particle shape 241
— — — — — — — — streamline flow 232
— — — — — — — — turbulent flow 238
— — — — — — — — — wall effects 240
— — two-phase gas–liquid upward flow 190
— gradient, gas flow 171
— — liquid flow 171
— isotropic 5
— loss, sudden expansion 140
PRILUTSKI, G. 5, 35
Primary normal stress difference 56, 176
PRINCEN, H. M. 54, 71
PRUD'HOMME, R. K. 48, 54, 60, 70, 71, 341, 389, 390
Pseudoplastic behaviour 6
PULLUM, L. 199, 203
Pulsating flow in pipes 118
Pumps, centrifugal 153
— gear 150
— lobe 152
— mono 152
— non-Newtonian fluids 149
— positive displacement 149
— rotary 149
— screw 155

QUADER, A. K. M. A. 98, 147, 159, 285, 286

Rabinowitsch–Mooney equation 86, 133
— — factor 38
RAJAIAH, J. 262, 286
RAMAMURTHY, A. V. 145, 159
RANADE, V. R. 363, 365, 390
RAO, B. K. 136, 158, 188, 204
RAO, M. N. 176, 189, 204
RAO, P. T. 242, 257
Rate of mixing 356
RAUENDAAL, R. C. 385, 390
RAUT, D. V. 176, 189, 204
Recoverable shear 22
REED, J. C. 86, 87, 88, 104, 110, 158
REES, I. J. 68, 70
REINHART, W. H. 64, 70
Relaxation time, fluid 27
— — spectrum 65

Reynolds number 29, 44, 91, 93, 245, 254, 255, 273, 290, 299, 307, 331, 333, 341, 343, 350, 358, 361
— — critical 90
— — — Bingham plastic fluids 91, 93
— — — power-law fluids 90
— — generalised 86, 133
— — laminar–turbulent transition 90
— — Metzner–Reed 77, 87, 96, 98, 103, 172, 174, 201
— — modified 92
— — packed beds 236, 237, 238
— — pipe flow 77
— — sphere 208, 209, 217
RHA, C. 18, 35
Rheogram 2
Rheological instruments 37
— measurements 37
Rheology 37
Rheometer, controlled stress 50
Rheometry 37
— extensional tests 66
— Fourier transform mechanical spectroscopy 60, 61, 62
— high frequency tests 63
— pulse propagation tests 64
— resonance based tests 64
Rheopexy 16, 17, 18
RICCIUS, O. 64, 70
RICHARDSON, J. F. 74, 82, 88, 89, 147, 149, 158, 166, 167, 168, 169, 170, 172, 174, 175, 176, 178, 180, 185, 189, 194, 199, 200, 202, 203, 222, 230, 244, 255, 256, 257, 303, 331, 367, 389
RICHMANN, K.-H. 59, 70
RICHMOND, R. A. 53, 70
RIDES, M. 67, 71
ROCO, M. C. 202
Rod climbing effect 20
RODRIGUE, D. 226, 227, 228, 256
ROHSENOW, W. M. 286
Rolling ball viscometer 51
— processes, as mixing devices 382
ROMOS, H. L. 340, 390
Rotating disc indexer 47
Rotational viscometer, vapour-hood 50
— viscometers 42
— — moisture loss 49
RUCKENSTEIN, E. 262, 286, 314, 322
RYAN, M. E. 5, 35
RYAN, N. W. 90, 159

SABIRI, N. E. 242, 257
SAITO, S. 369, 390
SALAMONE, J. J. 370, 389
SANDALL, O. C. 365, 370, 390
SAYIR, M. 64, 70
Scale of agitation 342

Scale-up method, general 104
SCHALLER, P. 64, 70
SCHECHTER, R. S. 133, 159, 272, 286
SCHENKEL, G. 385, 391
SCHLICHTING, H. 291, 303
Schmidt number 245, 254, 255, 313, 331
SCHOFIELD, C. 356, 389
SCHOWALTER, W. R. 20, 35, 145, 158, 297, 314, 321, 322
SCHRAG, J. L. 60, 71
SCHUMMER, P. 60, 69
SCHUMPE, A. 365, 390
SCHURZ, J. 12, 35
SCRIVEN, L. E. 67, 71
SCRIVENER, O. 96, 159
Second normal stress difference 22, 23, 56
Secondary flows, concentric cylinders 44
SECOR, R. B. 67, 71
SEK, J. 232, 233, 236, 237, 257
SELLIN, R. H. J. 96, 159
SEN, S. 214, 215, 257
SERTH, R. W. 314, 322
Settling velocity, hindered 221
SHAH, M. J. 297, 310
SHAH, R. K. 265, 272, 286
SHANKAR SUBRAMANIAN, R. 215, 257
SHAQFEH, E. 341, 390
SHARMA, M. K. 242, 254, 257
Shear modulus, complex 58
— rate 1, 37
— — concentric cylinders 43
— — cone and plate viscometer 48
— — in mixing tanks 348, 350
— — — — average 337
— — — — effect of rheology 340
— — — — — scale 340
— — — — — — viscoelasticity 341
— — nominal 38, 40
— — parallel plate viscometer 49
— — pore wall 234
— — typical 32
— relaxation modulus 57
— stress 1, 37, 294, 295, 299
— — concentric cylinders 43
— — cone and plate viscometer 48
— — distribution in a pipe 74
— — parallel plate viscometer 49
— — pore wall 234
— viscosity 25, 68
Shear-thickening behaviour 6, 10, 14, 47
Shear-thinning behaviour 6, 10, 47
— fluids, pulsating flow 118
SHEFER, A. 63, 71
SHENGYAO, Y. 369, 390
SHENOY, A. V. 114, 118, 159, 310, 314, 318, 322
Sherwood number 313, 331

Sherwood number, cylinders 315, 316
— — spheres 314
SHIPMAN, R. W. G. 68, 71
SHOOK, C. A. 170, 202, 203, 204
SHU, M. T. 170, 204
Similarity criteria, stirred tanks 331
SINGH, B. 313, 322
SINGH, D. 238, 257
SKELLAND, A. H. P. 81, 90, 116, 128, 147,
    159, 262, 267, 273, 286, 297, 302, 303,
    335, 337, 340, 391
SLATTER, P. T. 92, 93, 111, 146, 159
Slip effects, cone and plate viscometer 52
Slit flow, Bingham plastic fluids 120
— — Ellis model fluids 120
— — power-law fluids 118
Slug flow 165
SMITH, J. C. 346, 358, 389
SMITH, J. M. 340, 350, 351, 352, 353, 358,
    360, 389, 390, 391
SMITH, R. 146, 148, 158
SMITH, T. N. 212, 255
SOARES, A. A. 216, 256
SOEDA, H. 135, 159
SOLOMON, J. 341, 349, 361, 363, 364, 390,
    391
SOUZA MENDES, P. R. 69
SPAANS, R. D. 52, 71
SPEDDING, P. L. 166, 170, 171, 172, 203, 204
Sphere, creeping flow, criterion 215
— — — power-law fluid 208, 209
— — — visco-elastic fluid 215, 216
— — — viscoplastic fluid 211, 214
— drag coefficient, creeping flow region 208,
    209
— — — high Reynolds numbers 210
Sphericity 224, 242
SPIEGELBERG, S. H. 68, 71
SPIERS, R. P. 59, 70
SRIDHAR, T. 5, 26, 35, 67, 68, 71
SRINIVAS, B. K. 242, 249, 252, 253, 256, 257
STANDISH, N. 241, 257
Static equilibrium of particles in viscoplastic
    fluid 212
— mixers 385
STEFFE, J. F. 34, 110, 146, 149, 158, 159
STEG, I. 18, 35
STEIFF, A. 367, 368, 371, 390
STEIN, H. N. 14, 35
STEWART, I. W. 387, 390
STEWART, W. E. 3, 35, 319, 322
Stirred tanks, flow patterns 346
— — heat transfer 365
— — power consumption 332
— — scale-up 331
— — similarity criteria 331
STOIMENOVA, M. 64, 71

Stokes' law 208
— — correction, power-law fluids 208, 209,
    214
Storage modulus 59
Strain-hardening 26, 67
Stratified gas–non-Newtonian liquid flow 165
Streamline flow in packed beds, Bingham
    plastic fluids 237
— — — — — power-law fluids 232
SULLIVAN, J. L. 65, 71
SUNDARARAJAN, T. 209, 223, 257
SWANSON, B. S. 170, 203
SZEMBEK-STOEGER, M. 246, 257
SZILAS, A. P. 111, 159

TADA, H. 349, 390
TADMOR, Z. 157, 159
TAITEL, Y. 168, 169, 170, 181, 202, 204
TAKAHASHI, H. 63, 71
TAKAHASHI, K. 359, 391
TALATHI, M. M. 114, 118, 159
TAM, K. C. 118, 159
TANGUY, P. 355, 390
TANNER, R. I. 18, 20, 28, 34, 35, 56, 70, 209,
    215, 216, 257
TAREEF, B. M. 261, 286
TASSART, M. 253, 256
TATTERSON, G. B. 331, 335, 360, 363, 365,
    374, 388, 391
TAYLOR, J. S. 337, 346, 390
Taylor number 44
TE NIJENHUIS, K. 59, 71
TEHRANI, M. A. 127, 133, 158
Tension-thinning 26, 67
Terminal falling velocity of a sphere 216
— — — — — — effect of particle
    concentration 221
— — — — — — — — — shape 223
— — — — — — — wall effects 219
TERRY, K. 341, 390
Thermal boundary conditions, constant wall
    flux 266, 277
— — isothermal tube wall 266, 267
— — layer 303, 304
— — — power-law fluids 306
— — — temperature distribution 306
— conductivity, structured media 262
— diffusivity 267, 305
Thixotropic materials, mixing 340
Thixotropy 16
THOMAS, A. 67, 70
THOMAS, A. D. 96, 110, 114, 159
THOMAS, D. G. 273, 286
Thoria suspensions, heat transfer 273
TIEN, C. 226, 257, 319, 321
TILTON, L. O. 278, 286
Time-dependent fluid behaviour 15, 19
— — flow, pipes 118

Time-dependent fluids 5
Time-independent fluids 5, 6
— — flow in pipes 104
— — generalised approach 83
TIRTAATMADJA, V. 26, 35, 67, 68, 71
TIU, C. 118, 133, 136, 158, 159, 236, 242, 257
TOMITA, Y. 96, 159
TORRANCE, B. McK. 111, 159
Tortuosity factor, packed beds 234
TOSUN, I. 133, 159
Transient flow in pipes 118
Transitional and turbulent flow, Bingham
    plastic fluids 101
— — — — in pipes, heat transfer 285
— — — — viscoplastic fluids 101
Transport of coarse particles, shear-thinning
    media 199
— — — — viscous media 198
TRIPATHI, A. 209, 223, 257
TROUTON, F. T. 25, 35
Trouton ratio 25, 68
TSAMAPOLOUS, J. 212, 214, 256
Tubular flow, power-law fluids 74
Turbulent boundary layer 302
— core 112, 113
— flow in pipes, buffer layer 112
— — — — friction factor 95, 101
— — — — velocity profiles 111
— — — — viscoplastic fluids 101
— mixing, shear-thinning fluids 334
TURTLE, R. B. 200, 203
TUTHILL, J. P. 340, 390
Two-phase flow, kaolin suspensions 173
— gas–liquid flow, hold-up 168
— — — — pressure drop 177
— liquid–solid flow 197
— pressure drop, estimation methods,
    gas–liquid flow 185
TYABIN, N. V. 212, 257

UHLHERR, P. H. T. 16, 35, 212, 214, 215, 221,
    222, 255, 256, 257
ULBRECHT, J. J. 341, 350, 358, 362, 363, 365,
    388, 389, 390
ULBRICHOVA, I. 215, 257
UNER, D. 133, 159
Uniaxial extension 24, 25
UNNIKRISHNAN, A. 222, 256
UPADHYAY, S. N. 206, 245, 246, 254, 257,
    313, 314, 315, 322
UREY, J. F. 146, 159

VALE, D. G. 57, 70
VALENTIK, L. 212, 257
VALLE, M. A. 170, 181, 203
VAN DEN BERGH, W. 326, 391
VAN DER MOLEN, K. 331, 391
VAN DONSELAAR, R. 59, 71

VAN MAANEN, H. R. E. 331, 391
VAN SITTERT, F. P. 159
VAN WAZER, J. R. 44, 45, 52, 71
VAN'T REIT, K. 361, 391
Vane method, yield stress 53
VAUGHN, R. D. 267, 272, 286
Velocity, average or mean 74
— distribution in boundary layers 293
— — — pipes, Bingham plastic fluids 79
— — — — laminar flow 74
— — — — power-law fluids 75
— — — — shear-thickening fluids 76
— — — — shear-thinning fluids 76
— — parallel plates 119
— no-slip, mixtures 172, 182
— profile, power-law fluids 111
— — turbulent flow in pipes 111
— superficial 167
VENKATARAMAN, S. K. 60, 70
VENU MADHAV, G. 224, 257
Vertical two-phase flow, pressure drop 190
VINEY, L. J. 387, 390
VIRDI, T. S. 69, 70
Visco-elastic behaviour 20, 57, 60, 64
— effects, boundary layer flow 313
— — gas–liquid flow 175
— — in fluidised beds 253
— — — packed beds 246
— — — — excess pressure drop 247
— — mixing 341
— fluid behaviour 19, 20, 26
Viscometer, Brookfield 45
— concentric cylinder 42, 43
— cone and plate 47
— falling ball 51
— parallel plate 48
— rolling ball 51
— rotational 52
Viscoplastic behaviour 6, 11, 12
— flow past a sphere, drag force 214
— — — — — flow field 212
— — — — — sheared cavity 213
— — — — — static equilibrium 212
— — — — — wall effects 213
— fluids, static equilibrium of a particle 212
— — transitional and turbulent flow in pipes
    101
— material, creep test 55
Viscosities, typical values 4
Viscosity 2
— complex 59
— eddy 112
Viscous energy dissipation 283, 284
— flow, activation energy 264
VLACEK, J. 341, 361, 364, 390
Voidage of a bed 231
Voigt model 26, 27

VOLAROVICH, M. P. 212, 257
VORWERK, J. 247, 257
VRATSANOS, M. S. 60, 71

WALKER, C. I. 154, 159
Wall effects in packed bed flows 240
— — on sphere motion 219
— factor, terminal velocity 220
— shear rate 84, 86
— — — apparent 98
— slip 38
— — cone and plate viscometer 52
WALLABAUM, U. 22, 23, 35
WALTERS, K. 12, 20, 23, 25, 26, 28, 32, 34,
    35, 42, 48, 49, 51, 52, 56, 57, 69, 70,
    71, 215, 216, 256, 257, 314, 322
WALTON, I. C. 133, 159
WARD, H. C. 170, 204
WARDHAUGH, L. T. 118, 159
WARDLE, A. P. 186, 203
WATERMAN, H. A. 64, 71
WEBER, M. E. 221, 225, 226, 255, 256
Weber number 331
WEESE, J. 65, 70
WEIN, O. 346, 348, 349, 391
WEINBERGER, C. B. 170, 181, 203, 204
WEINSPACH, P. M. 367, 368, 371, 390
WEISMAN, J. 167, 204
Weissenberg effect 20
— hypothesis 22
WEISSENBERG, K. 21, 35
Weissenberg number 216, 332
Weissenberg–Rabinowitsch equation 38
Weissenberg rheogoniometer 56, 59
WELTMANN, R. N. 146, 159
WEN, C. Y. 245, 254, 257
WEN, Y. F. 65, 71
WHEELER, J. A. 133, 136, 159
WHITE, J. L. 30, 35
WHITLOCK, M. 14, 35
WHITMORE, R. L. 212, 256, 257
WHORLOW, R. W. 38, 42, 56, 64, 69, 71
WICHTERLE, K. 346, 348, 349, 391
WIEST, J. M. 28, 35
WILKINS, B. 170, 204

WILKINSON, W. L. 98, 118, 147, 158, 159,
    238, 256, 282, 284, 285, 286, 367, 368,
    370, 371, 389
WILLIAMS, D. J. A. 54, 55, 64, 65, 70, 71
WILLIAMS, M. C. 52, 71
WILLIAMS, P. R. 25, 35, 54, 55, 64, 65, 69, 70,
    71
WILLIAMS, R. W. 47, 71
WILSON, K. C. 91, 92, 96, 110, 114, 159
WINTER, H. H. 60, 65, 69, 70, 71, 284, 286
WISDOM, D. J. 341, 361, 390
WISSLER, E. H. 133, 136, 159, 272, 286
WOJS, K. 111, 159
WOLLERSHEIM, D. E. 319, 322
WOODCOCK, L. V. 14, 35
WRONSKI, S. 246, 257

XIE, C. 135, 159
XUEREB, C. 352, 389

YAGI, H. 365, 390, 391
YAN, J. 53, 71
YANAGIMOTO, M. 353, 390
YAP, C. Y. 341, 353, 355, 359, 389, 391
YARUSSO, B. J. 14, 35, 127, 157
Yield–pseudoplastic behaviour 11, 12
Yield stress 11, 12, 13, 46, 47, 51, 52, 53, 54,
    55, 78, 130, 199, 211, 212, 297, 332,
    361
— — apparent 45, 46
— — temperature dependence 264
— — vane method 53
YIM, J. 245, 257
YOKOYAMA, T. 353, 390
YOO, S. S. 285, 286
YOSHIDA, F. 365, 391
YOSHIMURA, A. S. 48, 54, 71
YOUNG-HOON, A. 170, 174, 181, 203
Young's modulus 19
YU, A. B. 241, 257

ZAKI, W. N. 222, 257
Zero shear viscosity 7, 11
ZHOU, J. Z. Q. 242, 257
ZWANEVELD, A. 282, 286